The Urban Book Series

The Urban Book Series is a resource for urban studies and geography research worldwide. It provides a unique and innovative resource for the latest developments in the field, nurturing a comprehensive and encompassing publication venue for urban studies, urban geography, planning and regional development.

The series publishes peer-reviewed volumes related to urbanization, sustainability, urban environments, sustainable urbanism, governance, globalization, urban and sustainable development, spatial and area studies, urban management, transport systems, urban infrastructure, urban dynamics, green cities and urban landscapes. It also invites research which documents urbanization processes and urban dynamics on a national, regional and local level, welcoming case studies, as well as comparative and applied research.

The series will appeal to urbanists, geographers, planners, engineers, architects, policy makers, and to all of those interested in a wide-ranging overview of contemporary urban studies and innovations in the field. It accepts monographs, edited volumes and textbooks.

Indexed by Scopus.

Venkatesh Baskaran • S. M. Renuka •
R. Velkennedy
Editors

EcoTech Urbanism

Pioneering Sustainable Technologies for Developing Cities

Springer

Editors
Venkatesh Baskaran
Department of Civil Engineering
Alagappa Chettiar Government College of Engineering and Technology
Karaikudi, Tamil Nadu, India

S. M. Renuka
Department of Civil Engineering
College of Engineering Guindy, Anna University
Chennai, Tamil Nadu, India

R. Velkennedy
Department of Civil Engineering
Thiagarajar College of Engineering
Madurai, Tamil Nadu, India

ISSN 2365-757X ISSN 2365-7588 (electronic)
The Urban Book Series
ISBN 978-3-031-83832-3 ISBN 978-3-031-83833-0 (eBook)
https://doi.org/10.1007/978-3-031-83833-0

This Springer imprint is published by the registered company Springer Nature Switzerland AG
The registered company address is: Gewerbestrasse 11, 6330 Cham, Switzerland

Preface

As global urbanization accelerates, developing cities face environmental degradation, strained infrastructure, and socioeconomic disparities. The imminent threats of climate change, resource depletion, and pollution exacerbate these issues. In response to these challenges, "EcoTech Urbanism: Pioneering Sustainable Technologies for Developing Cities" promotes a comprehensive approach to urban development, combining cutting-edge technology and sustainable practices. EcoTech Urbanism emphasizes balancing human well-being with environmental stewardship by promoting technological solutions in energy, transportation, waste management, and infrastructure to create resilient, adaptable urban environments.

The book focuses on innovative solutions to crucial urban issues such as air and water pollution, demonstrating advanced technologies such as nanoparticle pollution control and sustainable water management systems. By implementing these solutions, cities, particularly those in rapidly developing regions, can reduce their environmental impact while maintaining growth. As climate change exacerbates vulnerabilities such as flooding and droughts, the significance of incorporating advanced modeling techniques and ecological strategies into urban planning grows. This comprehensive approach enables cities to manage water resources more effectively while also building resilience to the effects of climate change.

Sustainable construction and urban design are also essential aspects of the EcoTech Urbanism framework. The book investigates the use of regenerative materials, energy-efficient building practices, and inclusive design principles to reduce the environmental impact of urban spaces while improving residents' quality of life. These approaches ensure that urban development is environmentally responsible and socially equitable, meeting the diverse needs of growing cities. Cities that incorporate sustainability into the construction process can create long-lasting and adaptable infrastructure to future challenges.

In addition to environmental solutions, the book explores the transformative power of smart technologies like the Internet of Things (IoT) and big data analytics. These digital tools can potentially improve urban management systems by allowing for more efficient resource use and better decision-making. The book offers a road map for achieving the United Nations' Sustainable Development Goals (SDGs) in urban settings through practical case studies and technological breakthroughs.

This book challenges readers to rethink the urban future, envisioning sustainable, equitable, and regenerative cities through the lens of EcoTech urbanism. By embracing the ideas and solutions presented, cities can build a more sustainable and prosperous future for all residents.

Karaikudi, India Venkatesh Baskaran
Chennai, India S. M. Renuka
Madurai, India R. Velkennedy

Contents

About the Editors

Venkatesh Baskaran is affiliated with Alagappa Chettiar Government College of Engineering and Technology, Karaikudi. His research has contributed to urban sustainability monitoring and assessment and urban infrastructure planning. He has published his research works in reputed indexed journals and authored some book chapters for popular publishers. He reviewed various journal articles related to urban sustainability studies. He has completed Geographic Information Systems specialization course offered by University of California, Davis. In addition to these specialization courses, he achieved Introduction to Sustainability and Responsive City certifications from the University of Illinois, Urbana, and the Swiss Federal Institute of Technology, Zürich. He is recognized as an Innovative Mentor by Startup TN, a Government of Tamil Nadu unit. He has been involved in various consultancy activities for construction projects implemented under Smart City Mission. Currently, he is actively participating in assessing and monitoring the implementation of different infrastructure-related targets from Sustainable Development Goals at the City and Municipality levels.

S. M. Renuka is working as a Faculty in the Department of Civil Engineering, College of Engineering Guindy, Anna University, Chennai. She completed her doctoral research in Risk Management in Construction Projects. She did her master's in the area of Construction Engineering and management. She is an Anna University 16th Rank Holder in B.E. End Examinations (2005–2009). She is a recipient of Anna Centenary Research Fellowship for the year 2013. She has 10 years of teaching experience and 2 years of research experience. She has published her research works in Q1 and refereed journals. She authored some book chapters in Springer's publication. She reviewed various articles related to smart cities and sustainable materials. She has attended several training programs and certification courses in her area of interest. She is a mentor for student innovative projects. She has guided 25+ UG Projects and 12+ PG students. Currently, four research scholars are working under her guidance. She is contributing her research knowledge as a doctoral committee member and Board of Studies member in various other universities. She is associated with various member societies like the Indian Concrete Institute,

Madras Management Association, and the Institute of Lean Construction. Her current research focus is to address key issues and challenges in Indian construction projects that will help to create sustainable urbanization and facilitate to reach the end users in right time.

R. Velkennedy is Professor of Civil Engineering at Thiagarajar College of Engineering, Madurai (India). He is also serving as the Dean of Extra-Curricular activities. He has 30 years of academic experience. Around eight scholars have completed their Ph.D. degree under his supervision. He successfully completed two sponsored research projects with role of Principal Investigator and Co-Investigator. He has published around 60 papers in reputed journals and presented 20 papers in national and international conferences. His research areas are Transport planning and Policy, Sustainable Pavement Materials, and Urban Infrastructure Planning. He has been actively involved in various construction consultancy activities at southern part of Tamil Nadu, India. He is a life member of Indian Society for Technical Education and Indian Road Congress.

Chapter 1
EcoTech Urbanism for Sustainable Cities

Venkatesh Baskaran, S. M. Renuka, R. Velkennedy, and E. Pounkumar

Abstract Facing increasing urban challenges, especially in developing cities, EcoTech Urbanism introduces a new model that merges ecological practices with advanced technologies to create sustainable, resilient, and livable cities. EcoTech Urbanism is an all-encompassing strategy incorporating state-of-the-art technologies to solve problems often occurring in developing cities. This multidisciplinary approach tackles urbanization's complexities, such as resource management, environmental degradation, and social equity, using innovations like smart grids, IoT, renewable energy, intelligent transportation systems, and public participation platforms. EcoTech Urbanism offers a comprehensive strategy for reducing environmental impact while improving quality of life by emphasizing smart infrastructure, sustainable water and waste management, data-driven governance, and inclusive urban planning. This chapter investigates the critical role of technology in addressing urban development issues, the implementation of smart systems, and the significance of public participation in promoting sustainability. Finally, EcoTech Urbanism promotes ecological harmony within cities and stimulates economic growth and social equity, making it an essential model for future city development.

V. Baskaran (✉)
Department of Civil Engineering, Alagappa Chettiar Government College of Engineering and Technology, Karaikudi, Tamil Nadu, India
e-mail: venkatesh@gcetly.ac.in

S. M. Renuka
Department of Civil Engineering, College of Engineering Guindy, Anna University, Chennai, Tamil Nadu, India

R. Velkennedy
Department of Civil Engineering, Thiagarajar College of Engineering, Madurai, Tamil Nadu, India

E. Pounkumar
Department of Civil Engineering, Thanthai Periyar Government Institute of Technology, Vellore, Tamil Nadu, India

V. Baskaran et al. (eds.), *EcoTech Urbanism*, The Urban Book Series,
https://doi.org/10.1007/978-3-031-83833-0_1

Keywords EcoTech · Urban planning · Smart technologies · IoT · Sustainable cities

1.1 Introduction

Cities worldwide face previously unheard-of difficulties regarding resource management, environmental sustainability, and quality of life in an accelerating era of urbanization (Geissen et al., 2015). When solving these intricate problems, traditional urban planning techniques frequently fall short. Thus, creative solutions seamlessly combining cutting-edge technologies and ecological principles are required. This is where the new urban development paradigm known as "EcoTech Urbanism" comes into play (Deng & Cheshmehzangi, 2018).

EcoTech Urbanism is a multidisciplinary approach that combines ecological sustainability and cutting-edge technology to create resilient, efficient, and livable urban environments. It emphasizes the harmonious coexistence of natural ecosystems and urban infrastructures, utilizing technological advancements to reduce environmental impact while improving urban quality of life (Abid et al., 2022). The importance of EcoTech Urbanism is especially evident in developing cities, which frequently experience rapid population growth, urban sprawl, and environmental degradation. These cities can use EcoTech Urbanism principles to implement sustainable practices that reduce environmental damage, such as deploying renewable energy systems, promoting public transportation, and improving green infrastructure (Young, 2011; Bibri & Krogstie, 2017). This approach reduces severe pollution, deforestation, and biodiversity loss and improves living conditions by providing clean energy, efficient waste management, and green spaces that promote physical and mental health (Staddon et al., 2018; Kumar et al., 2019).

Furthermore, EcoTech Urbanism can stimulate economic growth by establishing new industries and job opportunities, such as in renewable energy, green construction, and smart city technologies. This reduces reliance on nonrenewable resources while promoting economic development (Jonek-Kowalska & Wolniak, 2021). It also promotes social equity by emphasizing inclusive urban planning that considers the needs of all citizens, including marginalized communities. By ensuring equitable access to resources, services, and opportunities, this concept can help bridge socioeconomic divides and promote a more just urban society (Barba-Sánchez et al., 2019). Pollution, waste management, water scarcity, and the deterioration of green places are among the most pressing challenges for developing cities (Koop & van Leeuwen, 2017). Growing vehicle emissions, industrial activity, and the widespread use of low-quality fuels aggravate air pollution. These not only endanger health but also add to global climate change. Another significant issue is trash management, with many developing cities lacking the infrastructure and processes required to handle increasing volumes of solid and hazardous garbage. This results in the spread of illicit dumping sites that pollute land and water resources (Kumar & Agrawal, 2020).

Overall, while developing cities have enormous environmental issues, the smart use of technology provides exciting opportunities to construct more sustainable, livable, and resilient communities. These cities may improve resource management and reduce environmental consequences by implementing creative ideas and smart technologies.

The term "EcoTech" popularity was checked using the Scopus database. Only 19 articles have included this term in the title of an article. While searching for EcoTech in the article's abstract, only 71 documents were found. Even though the number of published papers follows the core meaning of EcoTech, this term is underused. Among the 71 articles, the subject area "Environmental Science" covered 26 papers.

This chapter comprises the importance of emerging technologies for safe and sustainable cities. The other chapters in this book deal with the application and usefulness of specific technologies or relevant initiatives for cities to be resilient against their common challenges.

1.2 The Role of Technology in Urban Development

The urban part of developing nations faces numerous challenges as they expand and urbanize, including rapid population growth, which causes overcrowding, strain on infrastructure, and increased demand for housing, transportation, and services. Urban expansion frequently causes environmental degradation, such as the loss of green spaces, pollution, and increased carbon emissions. Many developing cities face inadequate infrastructure, which causes problems with transportation, water supply, and waste management. Economic inequality is another major issue, with disparities in wealth and access to resources affecting the quality of life for many people. Furthermore, effective urban planning and governance are frequently lacking, resulting in unplanned growth and inefficient use of city resources.

Technology is critical in tackling environmental issues in emerging cities. Innovative clean energy technologies, such as solar and wind power, provide long-term alternatives to fossil fuels while lowering pollution and carbon footprints. Smart grid technologies can optimize electricity delivery, increasing efficiency and lowering waste. In waste management, technologies such as waste-to-energy conversion, recycling, and composting can drastically minimize the amount of garbage in landfills, converting waste into valuable resources.

Smart grids, which improve energy efficiency and reliability, and the Internet of Things (IoT), which optimizes city services by gathering data from multiple sources, are two technologies that are essential in addressing the challenges associated with urban development. Geographic information systems (GIS) use detailed mapping and spatial data analysis to facilitate better urban planning and decision-making. While waste-to-energy technologies help with efficient waste management by converting waste into usable energy like solar panels and wind turbines also reduce carbon emissions.

Fig. 1.1 Significant technologies for sustainable urban development

Green building technologies encourage ecologically responsible and energy-efficient building techniques (Chan et al., 2018). Smart transportation systems optimize traffic flow and public transportation, reducing congestion (Cheng et al., 2020; Zhu et al., 2020). Cutting-edge water management technologies guarantee a reliable supply and effective use of water. Enhanced connectedness through improved broadband and telecommunications supports business, governance, and education. Urban farming technologies, such as vertical farming and hydroponics, provide fresh produce within urban areas, reducing transportation needs and carbon footprints. The abstract of these notable technologies is displayed in Fig. 1.1.

Urban biodiversity infrastructure, urban farming, sustainable urban drainage systems, energy efficient buildings, Geographical Information Systems, electric vehicle charging infrastructure, and advanced building materials were also sustainable technologies for urban development other than the above mentioned.

1.3 Technologies Lined in EcoTech Urbanism

EcoTech Urbanism is a novel approach to urban design that combines ecological principles with cutting-edge technology to produce sustainable, resilient, and habitable communities. As metropolitan areas expand, the necessity for sustainable development becomes more critical.

CleanTech, a fundamental aspect of EcoTech Urbanism, is concerned with innovations contributing to the transition to a sustainable urban future by lowering greenhouse gas emissions, improving energy efficiency, and reducing waste. The following section summarizes the most widely used and active technologies being used globally to address various urban problems in an environmentally friendly way.

1.3.1 Smart Infrastructure

Implementing smart grids, IoT devices, and intelligent transportation systems to optimize resource use, reduce waste, and improve efficiency forms the foundation of smart infrastructure for sustainable cities and EcoTech Urbanism.

1.3.1.1 Smart Grid

Energy management via smart devices is crucial for smart cities, optimizing costs, merging energy sources, and reducing environmental issues (Farmanbar et al., 2019). Smart grids enable the real-time monitoring and management of energy consumption, which improves demand response and load balancing. This efficient energy use minimizes waste and reduces the carbon footprint in urban areas. By utilizing advanced technologies such as sensors, automation, and communication networks, smart grids effectively manage and optimize energy flows (Khalil et al., 2021). They can predict demand, streamline distribution, and integrate renewable sources like solar and wind power. This approach results in more reliable and resilient energy systems, essential for the sustainable development of cities.

Numerous global initiatives, including those in Italian cities and public universities in Brazil, demonstrate the viability and benefits of integrating the smart grid with historic buildings and urban areas to enhance energy performance and efficiency (Martirano et al., 2020; Neto, 2022). A multidisciplinary strategy is needed to establish smart grids in developing country cities, with an emphasis on consumer acceptance, technological improvements, and policy frameworks to enable successful integration and sustainable energy practices (Archana & Singh, 2022).

1.3.1.2 Internet of Things

The Internet of Things (IoT) is vital in developing smart infrastructure for sustainable cities by enabling applications that improve urban living. Integrating IoT technology into smart cities enables real-time data collection, green innovation, and sustainable urban development (Hemanth Kumar et al., 2024). When combined with cloud computing, IoT devices improve many aspects of city life, including traffic management, energy efficiency, public safety, waste management, and parking. These advancements help to make urban environments more efficient and sustainable (Mahobia & Pawar, 2024).

IoT devices embedded in the smart networks collect valuable data to optimize energy consumption, adjust indoor temperatures, and provide real-time traffic updates, improving citizen services and quality of life (Mohan & Mani, 2024). The deployment of an IoT-based waste collection management system in Parakou City, Benin, is a notable example of IoT applications in sustainable cities (Ange Mousse, 2024). This system enhances waste collection efficiency and reduces unnecessary costs by using sensors in waste bins to monitor their filling levels in real-time. Additionally, integrating IoT with image-processing technologies has proven effective for environmental monitoring in various urban contexts, demonstrating IoT's importance in promoting social, economic, and environmental sustainability (Liu & Chen, 2024).

The collaboration between generative AI and IoT transforms urban experiences by emphasizing energy efficiency, behavior prediction, and cybersecurity. IoT is essential for achieving sustainable development goals, supporting smart cities, transportation, disaster management, privacy, security, and emerging applications (Zeng et al., 2024).

1.3.1.3 Intelligent Transportation Systems

Intelligent transportation systems (ITS) integrate advanced technologies to enhance transportation management, public transit, and individual travel by utilizing wireless, electronic, and automated systems (Shaheen & Finson, 2016). These systems employ artificial intelligence techniques, like neural networks, computer vision and natural language processing to optimize safety, behavior, and interactions within transportation networks. ITS applications range from standalone traffic management systems to complex vehicle-to-infrastructure and vehicle-to-vehicle communication networks, often integrating smartphone apps and GPS devices for real-time, informed decision-making (Rani & Sharma, 2023). The network analysis techniques like graph analytics can identify critical nodes, congestion hot spots, and forecast future traffic flows, providing data-driven insights for optimizing transportation systems.

The primary objective of ITS is to improve transportation safety, efficiency, and environmental outcomes, addressing challenges such as energy consumption, road safety, sustainable transport growth, and overall productivity. ITS optimizes traffic flow, reduces emissions, and enhances urban mobility by utilizing technologies like vehicular ad-hoc networks, intelligent traffic lights, and mobility prediction (Guevara & Auat Cheein, 2020). Furthermore, ITS is crucial in developing smart cities, where intelligent technologies and sustainable practices are harmonized to prioritize environmental protection and societal well-being (Guerrero-Ibáñez et al., 2018).

1.3.2 Water Management

Various studies highlight the implementation of smart water management systems that utilize sensors and data analytics to monitor water quality and distribution, leading to improved efficiency and reduced wastage. For instance, integrating GIS facilitates better planning and management of water resources, allowing for targeted interventions in areas facing scarcity (Venkatesh & Velkennedy, 2022). Moreover, innovative treatment technologies, such as decentralized wastewater treatment systems, are being adopted to recycle water, thus alleviating pressure on existing water supplies (Musazura & Odindo, 2021). Additionally, mobile applications are emerging as effective tools for community engagement, enabling residents to report leaks and water quality issues and enhancing local authorities' accountability and responsiveness.

Smart meters and IoT devices allow accurate measurement and real-time monitoring, while predictive analytics and machine learning optimize water distribution and prevent shortages (Andrić et al., 2022). Furthermore, cloud-based platforms centralize data collection, providing a foundation for real-time decision-making. These technologies and public awareness tools promote sustainable water use and community participation.

On the contrary, there are still challenges to overcome, such as the infrastructure development and capacity-building requirements to guarantee these technologies' efficient implementation and upkeep. Other major obstacles include high costs, a need for technical know-how, and privacy concerns regarding data (Roopnarine et al., 2021). Despite these obstacles, developing countries can greatly increase urban water sustainability by strategically applying technology in water management.

1.3.3 Waste Management

Sustainable waste management in urban areas is increasingly facilitated by advanced technologies, including the Internet of Things (IoT), artificial intelligence (AI), blockchain, robotics, and big data analytics. IoT enables the deployment of smart garbage bins equipped with sensors that monitor fill levels and optimize collection routes, enhancing efficiency and reducing operational costs (Sharma et al., 2020). AI further improves waste management by optimizing collection, sorting, and recycling processes, addressing inefficiencies in traditional systems (Bharti et al., 2022; Fang et al., 2023). Blockchain technology provides transparency and traceability, ensuring secure waste tracking from generation to disposal—a feature particularly beneficial for managing electronic waste (França et al., 2020). Robotics and automated systems are increasingly employed in waste sorting facilities, improving precision and speed while minimizing human labor (Macrorie et al., 2021).

Big data analytics is crucial in analyzing waste generation patterns, enabling more effective planning and resource allocation (Khan et al., 2015). Furthermore, renewable energy technologies such as waste-to-energy (WTE) systems convert waste into usable energy, reducing landfill dependency and contributing to a circular economy (Chand Malav et al., 2020). These technologies collectively enhance community engagement and environmental sustainability, aligning with broader smart and intelligent city initiatives and the United Nations' sustainable development goals.

1.3.4 Data-Driven Governance

Data-driven governance employs real-time data and analytics to inform decision-making, optimize resource allocation, and enhance urban management. It is crucial in sustainable city planning, equipping municipalities with tools to efficiently manage urban growth and development (Bokolo, 2023). Cities can transform into smart urban ecosystems that respond dynamically to their inhabitants' needs by utilizing advanced technologies such as the Internet of Things (IoT), artificial intelligence (AI), and big data analytics. According to research, integrating these technologies enables real-time data collection and analysis, which has the potential to significantly improve energy efficiency, pollution mitigation, and urban circulation. This data-driven approach improves operational efficiency and promotes a more sustainable urban environment by minimizing waste and optimizing resource allocation (Matheus et al., 2020).

Furthermore, a structured approach to urban data management can help municipalities develop smart city services by allowing them to use large datasets to make informed decisions. City planners can create tailored solutions to address the unique challenges of each urban area by organizing and analyzing data from various sectors, such as transportation, energy, and public health. Also, predictive models, such as those based on artificial neural networks, can forecast solar radiation, assisting in developing energy-efficient urban infrastructures. These models provide insights into the optimal placement and performance of solar panels, contributing to a city's sustainability goals (Garg & Barach, 2021).

In essence, data-driven governance transforms the way cities operate, making them more resilient, efficient, and capable of adapting to the evolving demands of urban life. As cities continue to grow, integrating these technologies will be crucial in achieving long-term sustainability and enhancing the quality of life for their residents.

1.3.5 *Public Participation Platform*

Public participation platforms empower residents in inclusive, sustainable urban planning and development. Based on citizen science, research shows that by incorporating resident needs into the design process, participatory design can dramatically increase the resilience of urban areas. This partnership will make urban infrastructure more responsive to local demands and more efficient and flexible.

The emergence of platform urbanism is a prime example of how digital tools can improve citizen participation and co-creation in urban planning. By acting as a conduit between citizens, planners, and government agencies, these platforms enable various viewpoints to be heard and taken into account when making important decisions. The conventional top-down approach to urban development is replaced with more democratic, transparent, and creative techniques that support social equity by utilizing technology to enable open discourse. Moreover, public involvement is vital for addressing specific community needs, such as improving infrastructure for active transportation like cycling and walking. By engaging residents directly, cities can prioritize and implement sustainable solutions that support healthier lifestyles and reduce carbon emissions. Community participation ensures that decision-making is enriched with a wide variety of viewpoints, leading to more holistic solutions for urban challenges.

The utilization of digital participation platforms presents an opportunity to collect copious amounts of data via citizen science initiatives, thereby aiding cities in comprehending traffic, energy consumption, and pollution patterns. Cities can use this data to create focused interventions that improve resource efficiency, environmental sustainability, and overall urban resilience. In the end, public participation platforms serve as a bridge to more intelligent, sustainable urban futures while fostering stronger community ties.

1.4 Organization of Chapters

The book contains a total of 19 chapters, including this introductory chapter. This section intends to brief the chapters' organization and their specific relevance to EcoTech Urbanism. Chapter 2 addresses sustainable pollution control technologies for rapidly urbanizing cities, focusing on decentralized, nature-based solutions that integrate with urban green spaces and support sustainable development goals (SDGs). It highlights nature-based treatment as an ecological technology that aligns with the principles of "EcoTech Urbanism." The chapter further explores the core components of "EcoTech Urbanism," which aim to blend technological advancements with environmental sustainability and social inclusiveness.

Chapter 3 demonstrates sustainable nanotechnology solutions for reducing air pollution in urban environments. The use of ZnO nanoparticles (NPs) for air pollution control exemplifies a sustainable, technology-driven solution to urban environmental challenges, which is central to the concept of "EcoTech Urbanism." It showcases the potential for advanced materials like ZnO NPs to mitigate urban air pollution and ties directly to the themes of "EcoTech Urbanism" by presenting them as a technological innovation for enhancing environmental sustainability in cities.

Chapter 4 focuses on urban water management and climate change mitigation in cities, utilizing innovative technologies to enhance resilience, with a particular emphasis on Chennai. It addresses the transformation of Chennai's urban water management through sustainable practices, policy innovations, and green infrastructure, aligning with the principles of "EcoTech Urbanism," which emphasize eco-friendly solutions in urban planning.

Chapter 5 emphasizes integrating surface and groundwater dynamics to enhance urban water management and community well-being. It explores water interactions that promote adaptable and resilient urban landscapes, aligning with "EcoTech Urbanism" goals of fostering cities capable of withstanding environmental challenges and supporting long-term sustainability.

Chapter 6 discusses advanced flood modeling techniques that enhance urban resilience and support effective flood management strategies in response to increasing urban flooding challenges. The chapter covers the coupling of 1D and 2D modeling techniques for urban flood simulation, aligning with the "EcoTech Urbanism" theme, which emphasizes the role of cutting-edge tools and methods in managing environmental risks in cities.

Chapter 7 highlights the role of innovative technologies and holistic approaches in addressing the challenges of SDG 11, aimed at enhancing urban development in Indian cities. It emphasizes the use of geographic information systems (GIS), Internet of Things (IoT), and artificial intelligence (AI) for improving urban planning and infrastructure efficiency, aligning with the book's focus on leveraging technology to create sustainable urban environments.

Chapter 8 showcases strategies for enhancing building energy efficiency and reducing greenhouse gas emissions in urban areas. It highlights the use of building information modeling (BIM) and energy analysis alongside green building certification systems (LEED, IGBC, GRIHA), aligning with "EcoTech Urbanism," which emphasizes technology and data-driven approaches for sustainable urban development. The chapter also discusses government subsidies for green-rated buildings, underscoring the importance of policy measures and financial incentives.

Chapter 9 promotes energy-efficient design for resilient, environmentally responsible infrastructure. It emphasizes the use of eco-friendly construction materials and methods to reduce the ecological footprint of urban infrastructure. The chapter also highlights the role of sustainable building design in addressing climate change challenges, resonating with "EcoTech Urbanism"s' focus on resilient and adaptable urban solutions.

Chapter 10 demonstrates the utilization of industrial by-products for innovative and sustainable urban road infrastructure. It highlights the use of ground granulated blast-furnace slag (GGBS) as a partial cement replacement and HDPE fibers, emphasizing eco-friendly alternatives that contribute to sustainable construction practices. This aligns with "EcoTech Urbanism"s' focus on innovative materials that minimize environmental impact.

Chapter 11 highlights essential practices for resilient infrastructure in developing countries, aiming to overcome resource depletion. The emphasis on sustainable building practices aligns with the core principles of "EcoTech Urbanism," which advocates for environmentally responsible approaches to urban development. The chapter includes using eco-friendly materials and industrial waste, effectively reducing environmental impact.

Chapter 12 focuses on autism-friendly design to enhance urban spatial quality. Ensuring comfort and accessibility for individuals with autism aligns with the principles of EcoTech Urbanism, which advocates for inclusive urban environments that cater to diverse populations. This approach emphasizes universal design criteria that accommodate various user groups, promoting social equity and fostering community cohesion for individuals with autism.

Chapter 13 discusses enhancing indoor environmental quality in urban school classrooms. It underscores the importance of creating energy-efficient, healthy, and welcoming buildings for vulnerable populations like children, fostering more sustainable and resilient communities in alignment with the goals of EcoTech Urbanism. The focus on visual and thermal comfort in classrooms highlights the need for designs that enhance occupant experience, emphasizing that EcoTech Urbanism values not just functionality but also the quality of space and aesthetic appeal.

Chapter 14 emphasizes risk management strategies for resilient highway construction in urban environments. It showcases the use of artificial neural networks (ANN) and regression analysis to identify and evaluate risk factors, demonstrating a data-driven approach to decision-making. This aligns with the EcoTech Urbanism ethos of leveraging technology and data analytics to enhance urban planning.

Chapter 15 examines eco-friendly transportation, highlighting public transit, electric vehicles (EVs), alternative fuels, and innovative mobility solutions. It discusses the importance of building and maintaining resilient infrastructure for public transit and EVs. EcoTech Urbanism emphasizes the need for sustainable infrastructure that supports these eco-friendly transportation options, ultimately reducing the carbon footprint of urban mobility.

Chapter 16 highlights the transformative role of IoT and analytics in enhancing urban sustainability, efficiency, and quality of life. It discusses how smart solutions optimize resource use in water management, traffic control, waste management, and energy systems, contributing to improved environmental health. The potential of these innovations aligns with the overarching goals of EcoTech Urbanism by promoting efficient, innovative, and sustainable urban development practices that enhance the quality of life in urban areas.

Chapter 17 illustrates an artificial neural network (ANN) model to predict the Wastewater Quality Index (WWQI) for protecting public health in urban environments. It emphasizes the necessity of a tailored WWQI for Goa, incorporating advanced predictive models like recurrent neural networks (RNN) and long short-term memory (LSTM) networks to address local wastewater challenges. This approach aligns with EcoTech Urbanism by promoting sustainable urban practices through innovative technologies, ensuring effective wastewater management, and enhancing environmental quality in rapidly urbanizing areas.

Chapter 18 explores the applications of AI and machine learning (ML) in improving rainfall prediction and wind energy forecasting to enhance sustainability. It illustrates advanced machine learning techniques for rainfall classification, such as the XGBoost algorithm, emphasizing their role in improving weather forecasting and disaster preparedness. This aligns with EcoTech Urbanism by demonstrating how data-driven approaches can enhance environmental resilience, support sustainable urban planning, and inform renewable energy decision-making through accurate predictive modeling.

Chapter 19 identifies significant Lean tools in construction, prioritizing root cause analysis to optimize project delivery and management. It highlights the application of the analytic hierarchy process (AHP) in selecting optimal Lean tools, aligning with EcoTech Urbanism by promoting efficient resource management and sustainable practices. By identifying effective methodologies that enhance project workflows and reduce waste, this chapter contributes to the overarching goal of fostering sustainable urban development and efficient construction practices within urban environments.

Overall, the chapters in this book explore innovative technologies and sustainable practices in urban environments, emphasizing EcoTech Urbanism. It addresses key issues like pollution control, water management, energy efficiency, transportation, and infrastructure resilience to promote sustainable urban development and community well-being.

1.5 Conclusion

EcoTech Urbanism has emerged as a critical strategy for addressing rapid urbanization's complex environmental, social, and economic challenges, particularly in developing cities. This approach encourages the development of more efficient, resilient, and inclusive cities by combining smart technologies such as IoT, AI, and intelligent transportation systems with sustainable energy, waste, and water management practices. CleanTech focuses on reducing environmental impact through innovation, while EcoTech integrates sustainable practices across technology and development. EcoTech Urbanism ensures that urban development meets the needs of all citizens by leveraging data-driven governance and public participation platforms, thereby improving social equity and community engagement. Furthermore, by promoting renewable energy and green infrastructure, EcoTech Urbanism helps to mitigate the adverse effects of climate change, reduce pollution, and improve overall quality of life. Overall, the imperatives of EcoTech Urbanism highlight its potential to transform cities into sustainable and thriving ecosystems, laying the robust platform for a more equitable and sustainable future. The upcoming chapters decode the various technological initiatives that drive eco-friendly and resilient cities by exploring innovations.

References

Abid, N., Marchesani, F., Ceci, F., et al. (2022). Cities trajectories in the digital era: Exploring the impact of technological advancement and institutional quality on environmental and social sustainability. *Journal of Cleaner Production, 377*, 134378. https://doi.org/10.1016/j.jclepro.2022.134378

Andrić, I., Vrsalović, A., Perković, T., et al. (2022). IoT approach towards smart water usage. *Journal of Cleaner Production, 367*, 133065. https://doi.org/10.1016/j.jclepro.2022.133065

Ange Mousse, M. (2024). IoT-based waste collection management system for cities: Case study of Parakou city in Benin. *International Journal of Advanced Research, 12*, 654–659. https://doi.org/10.21474/IJAR01/18597

Archana, S. R., & Singh, S. (2022). Development of smart grid for the power sector in India. *Cleaner Energy Systems, 2*, 100011. https://doi.org/10.1016/j.cles.2022.100011

Barba-Sánchez, V., Arias-Antúnez, E., & Orozco-Barbosa, L. (2019). Smart cities as a source for entrepreneurial opportunities: Evidence for Spain. *Technological Forecasting and Social Change, 148*, 119713. https://doi.org/10.1016/j.techfore.2019.119713

Bharti, S., Fatma, S., & Kumar, V. (2022). AI in waste management: The savage of environment. In *Environmental informatics* (pp. 97–123). Springer.

Bibri, S. E., & Krogstie, J. (2017). Smart sustainable cities of the future: An extensive interdisciplinary literature review. *Sustainable Cities and Society, 31*, 183–212. https://doi.org/10.1016/j.scs.2017.02.016

Bokolo, A. J. (2023). Data driven approaches for smart city planning and design: A case scenario on urban data management. *Digital Policy, Regulation and Governance, 25*, 351–367. https://doi.org/10.1108/DPRG-03-2022-0023

Chan, A. P. C., Darko, A., Olanipekun, A. O., & Ameyaw, E. E. (2018). Critical barriers to green building technologies adoption in developing countries: The case of Ghana. *Journal of Cleaner Production, 172*, 1067–1079. https://doi.org/10.1016/j.jclepro.2017.10.235

Chand Malav, L., Yadav, K. K., Gupta, N., et al. (2020). A review on municipal solid waste as a renewable source for waste-to-energy project in India: Current practices, challenges, and future opportunities. *Journal of Cleaner Production, 277*, 123227. https://doi.org/10.1016/j.jclepro.2020.123227

Cheng, Z. (Aaron), Pang, M. -S., & Pavlou, P. A. (2020). Mitigating traffic congestion: The role of intelligent transportation systems. *Information Systems Research, 31*, 653–674. https://doi.org/10.1287/isre.2019.0894.

Deng, W., & Cheshmehzangi, A. (2018). Macro level: Eco-city cases in China. In *Eco-development in China* (pp. 105–155).

Fang, B., Yu, J., Chen, Z., et al. (2023). Artificial intelligence for waste management in smart cities: A review. *Environmental Chemistry Letters, 21*, 1959–1989. https://doi.org/10.1007/s10311-023-01604-3

Farmanbar, M., Parham, K., Arild, Ø., & Rong, C. (2019). A widespread review of smart grids towards smart cities. *Energies, 12*, 4484. https://doi.org/10.3390/en12234484

França, A. S. L., Amato Neto, J., Gonçalves, R. F., & Almeida, C. M. V. B. (2020). Proposing the use of blockchain to improve the solid waste management in small municipalities. *Journal of Cleaner Production, 244*, 118529. https://doi.org/10.1016/j.jclepro.2019.118529

Garg, B., & Barach, R. (2021). Collaborative governance for urban ustainability: Implementing solar cities. *Asia Pacific Journal of Public Administration, 43*, 236–257. https://doi.org/10.1080/23276665.2021.1925132

Geissen, V., Mol, H., Klumpp, E., et al. (2015). Emerging pollutants in the environment: A challenge for water resource management. *International Soil and Water Conservation Research, 3*, 57–65. https://doi.org/10.1016/j.iswcr.2015.03.002

Guerrero-Ibáñez, J., Zeadally, S., & Contreras-Castillo, J. (2018). Sensor technologies for intelligent transportation systems. *Sensors, 18*, 1212. https://doi.org/10.3390/s18041212

Guevara, L., & Auat Cheein, F. (2020). The role of 5G technologies: Challenges in smart cities and intelligent transportation systems. *Sustainability, 12*, 6469. https://doi.org/10.3390/su12166469

Hemanth Kumar, S., Kiran, P., Mahalakshmi, S., et al. (2024). Technology's role in sustainable urban development in smart cities. In *Technology-driven evolution of the corporate social responsibility ecosystem* (pp. 169–208).

Jonek-Kowalska, I., & Wolniak, R. (2021). Economic opportunities for creating smart cities in Poland. Does wealth matter? *Cities, 114*, 103222. https://doi.org/10.1016/j.cities.2021.103222

Khalil, M. I., Jhanjhi, N. Z., Humayun, M., et al. (2021). Hybrid smart grid with sustainable energy efficient resources for smart cities. *Sustainable Energy Technologies and Assessments, 46*, 101211. https://doi.org/10.1016/j.seta.2021.101211

Khan, Z., Anjum, A., Soomro, K., & Tahir, M. A. (2015). Towards cloud based big data analytics for smart future cities. *Journal of Cloud Computing, 4*, 2. https://doi.org/10.1186/s13677-015-0026-8

Koop, S. H. A., & van Leeuwen, C. J. (2017). The challenges of water, waste and climate change in cities. *Environment, Development and Sustainability, 19*, 385–418. https://doi.org/10.1007/s10668-016-9760-4

Kumar, A., & Agrawal, A. (2020). Recent trends in solid waste management status, challenges, and potential for the future Indian cities—A review. *Current Research in Environmental Sustainability, 2*, 100011. https://doi.org/10.1016/j.crsust.2020.100011

Kumar, P., Druckman, A., Gallagher, J., et al. (2019). The nexus between air pollution, green infrastructure and human health. *Environment International, 133*, 105181. https://doi.org/10.1016/j.envint.2019.105181

Liu, W., & Chen, G. (2024). Towards optimal image processing-based internet of things monitoring approaches for sustainable cities. *International Journal of Advanced Computer Science and Applications, 15*, 1134–1142. https://doi.org/10.14569/IJACSA.2024.01505115

Macrorie, R., Marvin, S., & While, A. (2021). Robotics and automation in the city: A research agenda. *Urban Geography, 42*, 197–217. https://doi.org/10.1080/02723638.2019.1698868

Mahobia, S., & Pawar, R. (2024). Comprehensive review of smart cities using IOT in coud. *International Journal of Innovative Research in Computer and Communication Engineering, 12*, 6309–6319. https://doi.org/10.15680/IJIRCCE.2024.1205194

Martirano, L., Rotondo, S., Manganelli, M., & Kermani, M. (2020). A smart microgrid for buildings of the public administration. In *2020 IEEE industry applications society annual meeting* (pp. 1–4). IEEE.

Matheus, R., Janssen, M., & Maheshwari, D. (2020). Data science empowering the public: Data-driven dashboards for transparent and accountable decision-making in smart cities. *Government Information Quarterly, 37*, 101284. https://doi.org/10.1016/j.giq.2018.01.006

Mohan, M., & Mani, H. (2024). The impact of IoT on smart city infrastructure in India. *Recent Research Reviews Journal, 3*, 88–103. https://doi.org/10.36548/rrrj.2024.1.006

Musazura, W., & Odindo, A. O. (2021). Suitability of the decentralised wastewater treatment effluent for agricultural use: Decision support system approach. *Water, 13*, 2454. https://doi.org/10.3390/w13182454

Neto, J. L. M. (2022). An insight into the use of smart grids in public brazilian universities. *Journal of Agricultural Science Research, 2*, 2–7. https://doi.org/10.22533/at.ed.973252201068

Rani, P., & Sharma, R. (2023). Intelligent transportation system for internet of vehicles based vehicular networks for smart cities. *Computers and Electrical Engineering, 105*, 108543. https://doi.org/10.1016/j.compeleceng.2022.108543

Roopnarine, R., Eudoxie, G., Wuddivira, M. N., et al. (2021). Capacity building in participatory approaches for hydro-climatic Disaster Risk Management in the Caribbean. *International Journal of Disaster Risk Reduction, 66*, 102592. https://doi.org/10.1016/j.ijdrr.2021.102592

Shaheen, S. A., & Finson, R. (2016). Intelligent transportation systems. In *Reference module in earth systems and environmental sciences* (pp. 487–496). Elsevier.

Sharma, M., Joshi, S., Kannan, D., et al. (2020). Internet of things (IoT) adoption barriers of smart cities' waste management: An Indian context. *Journal of Cleaner Production, 270*, 122047. https://doi.org/10.1016/j.jclepro.2020.122047

Staddon, C., Ward, S., De Vito, L., et al. (2018). Contributions of green infrastructure to enhancing urban resilience. *Environment Systems and Decisions, 38*, 330–338. https://doi.org/10.1007/s10669-018-9702-9

Venkatesh, B., & Velkennedy, R. (2022). A systematic review on the role of geographical information systems in monitoring and achieving sustainable development goal 6: Clean water and sanitation. *Sustainable Development, 30*, 1–9. https://doi.org/10.1002/sd.2302

Young, R. F. (2011). Planting the living city. *Journal of the American Planning Association, 77*, 368–381. https://doi.org/10.1080/01944363.2011.616996

Zeng, F., Pang, C., & Tang, H. (2024). Sensors on internet of things systems for the sustainable development of smart cities: A systematic literature review. *Sensors, 24*, 2074. https://doi.org/10.3390/s24072074

Zhu, F., Lv, Y., Chen, Y., et al. (2020). Parallel transportation systems: Toward IoT-enabled smart urban traffic control and management. *IEEE Transactions on Intelligent Transportation Systems, 21*, 4063–4071. https://doi.org/10.1109/TITS.2019.2934991

Chapter 2
Exploration of Nature-Based Technologies for Urban Pollution Challenges in Rapidly Growing Cities

Manthiram Karthik Ravichandran and Ravi Sankar V.

Abstract Urbanization in developing countries is occurring at an unprecedented rate, resulting in profound environmental challenges exacerbated by rapid population growth and industrial expansion. This rapid urbanization has led to high-density urban environments facing significant challenges in ensuring adequate and safe water supply, proper sanitation, wastewater disposal, and ambient air quality. The escalating pollution poses significant health risks and threatens the ecological balance, highlighting the pressing necessity to implement pollution reduction initiatives. In addition, the conventional centralized treatment technologies often face constraints such as limited space availability, high capital and operational costs, and substantial energy demands, making them unviable in resource-constrained settings. To address these challenges, there is a critical need for sustainable and decentralized pollution reduction technologies aligned with the sustainable development goal of creating sustainable cities and communities. Nature-based decentralized systems offer several advantages, including lower infrastructure costs, reduced energy consumption, and integration with urban green spaces. These technologies effectively mitigate air and water pollution, and enhance the aesthetic value, exemplify innovative approaches to pollution control. The overarching aim of this book chapter is to provide a comprehensive understanding of sustainable pollution abatement technologies suitable for urban spaces in developing countries, offering insights into their applicability, limitations, and potential for widespread adoption. The chapter explored a diverse array of technologies developed to address pollution in water (water, stormwater, wastewater) and air environments. The factors affecting the system performance, including design and operational conditions, is also dis-

M. K. Ravichandran (✉)
Department of Civil Engineering, Sri Sivasubramaniya Nadar College of Engineering, Kalavakkam, Tamil Nadu, India
e-mail: manthiramkarthikr@ssn.edu.in

Ravi Sankar V.
Department of Civil Engineering, Thiagarajar College of Engineering, Madurai, Tamil Nadu, India
e-mail: environmentengr@tce.edu

V. Baskaran et al. (eds.), *EcoTech Urbanism*, The Urban Book Series,
https://doi.org/10.1007/978-3-031-83833-0_2

cussed. This exploration seeks to pave the way for practical and impactful interventions to address the pressing pollution challenges faced by urban populations in developing countries.

Keywords Sustainable cities · Decentralized approach · Nature-based technologies · Urban environmental management · Urban pollution · Sustainable urban development

2.1 Introduction

Cities in developing countries of the Global South are growing rapidly at an exceptional trend due to the rapid urbanization, population growth, industrial activities, and increased migration from rural to urban areas. This results in improper and unplanned transformation in the urban landscape and infrastructure (Chen et al., 2024). It presents complex environmental and developmental challenges to promote sustainable urban development. Additionally, the inadequate infrastructure and service amenities leads to serious environmental degradation especially water supply, sanitation, and waste management. In most growing cities of India, there is chronic shortage of portable water, inefficient or inadequate infrastructure for sanitation and waste management. Often, untreated or partially treated wastewater and wastes are discharged into the water bodies, causing serious risk to aquatic organisms and human health. It is also associated with the loss of biodiversity and ecosystem (Chen et al., 2024).

Conventional centralized treatment technologies require substantial infrastructure investments and operation and maintenance costs. In most cases, it insists on larger space, higher energy supply, and skilled manpower for regular monitoring. Also, these technologies are significantly affected by climate change and incorporating them in regions of unreliable energy supply turn into more challenging (Zheng & Lam, 2024). In this view, nature-based decentralized treatment techniques (NBT) can be a viable alternative to combat the challenging urban pollution. NBT are green infrastructure that can be easily installed in urban spaces and scaled up for the specific requirements. The European Commission defines NBT as solutions that are inspired and supported by nature, which are cost-effective, simultaneously provide environmental, social, and economic benefits and help build resilience.

Nature-based technologies offer a promising approach to addressing urban pollution by leveraging natural processes and ecosystems to improve environmental quality and resilience. Such interventions help to absorb pollutants, reduce heat island effects, and enhance air and water quality. For instance, urban trees and vegetation can capture airborne particulate matter and pollutants, while green roofs can manage stormwater runoff and lower building temperatures. Additionally, integrating natural elements into urban planning not only mitigates pollution but also provides recreational and aesthetic benefits to residents, fostering a healthier and more sustainable urban environment. Overall, NBT represents a cost-effective and adaptive strategy that harmonizes urban development with ecological health, addressing the challenges of pollution while promoting biodiversity and community well-being.

Nature-based treatment technologies are integrated for the pollution control measures for the development of resilient and sustainable cities. It also aligns with the principle of sustainable urban development. The adoption of NBT can be crucial for achieving the sustainable development goals (SDGs) especially related to clean water and sanitation, climate action, and sustainable cities and communities.

This chapter focuses on exploring the various nature-based treatment techniques for addressing urban pollution in any developing cities. It provides a comprehensive overview of nature-based technologies such as green walls, green roofs, constructed wetlands, bioretention systems, soil biotechnology, and some patented technologies (phytorid, green bridges technology, ecofiltration bank). These techniques can be used to address urban pollution, especially contaminated stormwater or runoff, air pollution, and treatment and reuse of graywater and wastewater. Through detailed review on the working mechanism, applicability, limitation, and potential of the nature-based treatment technologies, the chapter aims to offer insights for the development of sustainable and resilient urban environments. By exploring a range of innovative technologies and their characteristics, this chapter contribute to the advancement of effective and scalable solutions for mitigating pollution and promoting environmental sustainability in rapidly urbanizing regions of the Global South.

2.2 Nature-Based Treatment Systems

The emergence of sustainable development goals and energy neutrality commitment by the nations leads to growing interest in the application of nature-based solutions. In recent days, funding agencies, government authorities, and institutions are considering nature-based solutions as a viable alternative to their projects. The International Union for Conservation of Nature (IUCN) and World Bank introduced the theory of nature-based solutions to highlight the importance of biodiversity conservation for climate change mitigation and adaptation. Any solution can be classified as nature-based solutions if it provides simultaneous benefits to biodiversity and human well-being. Therefore, each solution must either maintain or enhance biodiversity, without which an action cannot be classified as nature-based solutions (WWF, 2021).

From a systems thinking perspective (Fig. 2.1), nature-based solutions for urbanization involve viewing cities as interconnected ecosystems where every component—natural or built—affects the whole. This holistic approach emphasizes integrating nature-based solutions, like permeable pavement, green walls, bioretention system, constructed wetland, and green roofs, to address complex urban challenges like pollution, climate change, greenhouse gas emissions. By

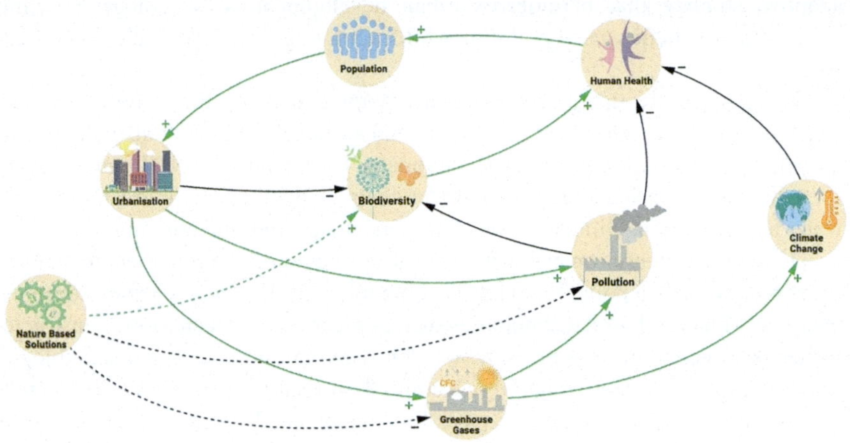

Fig. 2.1 Systems thinking perspective where (+) denotes influence is in the same direction and (−) notates influence is in the opposite direction. (Source: UNEP, 2021)

recognizing the dynamic interplay between human activities and natural systems, systems thinking advocates for designing cities that mimic natural processes.

Emerging technologies grounded in nature-based solutions are revolutionizing how we address environmental challenges by leveraging and enhancing natural processes. For instance, advancements in "living architecture," such as green roofs and vertical gardens, utilize plant systems to improve air quality, manage storm water, and reduce urban heat islands. Various nature-based treatment techniques to combat urban pollution are discussed in the subsequent sections. Additionally, innovative approaches like bioengineering harness natural organisms to remediate contaminated soils and water, exemplified by bioremediation techniques that use microbes to detoxify pollutants. Another cutting-edge development is the use of artificial intelligence and remote sensing to monitor and optimize ecosystem services, such as tracking forest health and biodiversity in real time. These technologies not only capitalize on the inherent efficiencies of natural systems but also offer scalable, adaptable solutions that integrate seamlessly with urban and rural landscapes. As research and development in this field continue to advance, these nature-based technologies promise to enhance our ability to manage environmental impacts while fostering a deeper connection between technology and the natural world.

2.3 Nature-Based Treatment Technologies to Combat Urban Pollution

2.3.1 *Green Walls*

Green walls, also known as vertical gardens, are attached to either interior or exterior walls and consist of planted containers affixed to the structure (Gholami et al., 2023). Unlike green facades, where plants root only at the base and climb upward, green walls allow vegetation to root across the entire vertical structure, creating a more extensive and uniform coverage. These systems are constructed using substrate-filled containers that are suspended on the side of buildings with a vertical support frame, enabling the growth of various plants (Fig. 2.2). Initially developed for aesthetic and insulation purposes, green walls are now being explored for additional applications, including wastewater treatment. In these systems, wastewater is treated as it percolates through planted pots filled with a variety of granular media such as vermiculite, sand, expanded clay, and other lightweight materials, offering a compact and visually appealing alternative. Green walls are classified into two main types based on the arrangement of support materials and plants: green facades and living walls, each offering unique advantages for urban greening and wastewater treatment (Boano et al., 2020; Gholami et al., 2023).

The advantages of green walls compact design, aesthetic appearance, lesser area requirement, etc. It is crucial to consider factors such as their adaptability to local

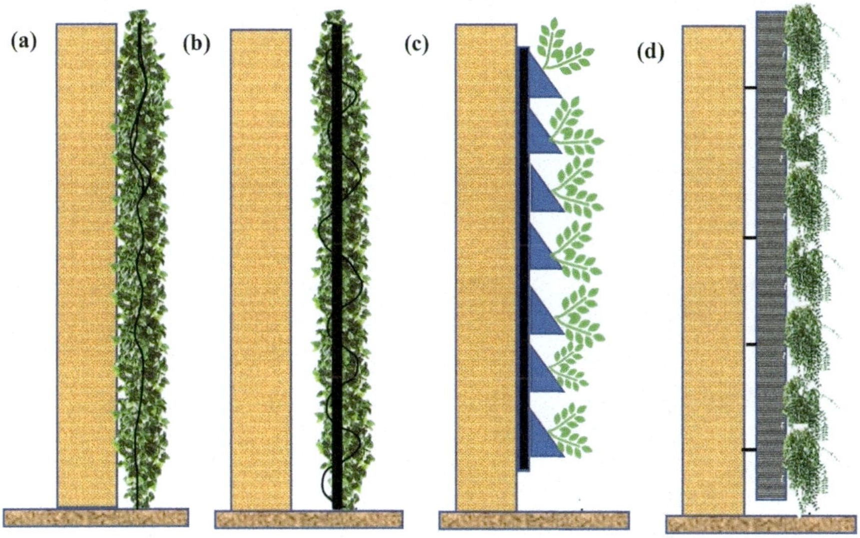

Fig. 2.2 Schematics of green wall systems: (**a**) direct green façade (DGF); (**b**) indirect green façade (IGF); (**c**) modular living wall (MLW); (**d**) continuous living wall (CLW). (Source: Gholami et al., 2023)

climatic conditions, survival rate, aesthetic appeal, root space requirements, and nutrient removal efficiency for the choice of plant species (Boano et al., 2020; Gholami et al., 2023). Additionally, green walls can provide further benefits, such as the potential for agricultural or medicinal uses of the plants, enhancing the overall value of these systems in sustainable urban development.

2.3.2 *Green Roofs*

Green roofs are vegetated surfaces installed on rooftops, typically designed with modular systems that incorporate multiple layers, including an insulation layer, a waterproof membrane, and a vegetation layer embedded in a growing medium. The depth of this medium can vary widely, from 50 mm to over a meter, depending on the roof's load-bearing capacity and the type of plants chosen. Green roofs offer numerous economic and social benefits alongside their environmental advantages. These include stormwater management, reduced energy consumption, improved water and air quality, noise pollution reduction, extended roof life, and the mitigation of the urban heat island effect. Many countries and municipalities have recognized these benefits, leading to the widespread adoption and even mandate of green roofs in urban settings (Vijayaraghavan, 2016) (Fig. 2.3).

Furthermore, green roofs can utilize graywater for irrigation, reducing the need for additional water resources. There is also potential for rainwater harvesting, though the quality and volume of runoff may be affected. To simplify installation

Fig. 2.3 Schematics of green roofs. (Source: Vijayaraghavan, 2016)

and maintenance, modular green roof systems have been developed, offering flexibility for both new and existing buildings. However, the costs and long-term sustainability of these modules remain important considerations for future research and development.

2.3.3 Constructed Wetlands

Constructed wetlands are engineered natural treatment systems that consist of filling materials, plants, and microbes. Various physiochemical and biological processes of natural wetlands take place in a controlled environment. Constructed wetlands are typically properly lined basin structure where the supporting media are filled, planted with macrophytes with inflow and outflow arrangements for the wastewater. There are various types of constructed wetlands based on the flow conditions—free surface wetlands, subsurface wetlands, and floating wetlands.

The free surface wetlands have basins with shallow water depth. Also, the water flows over the bottom supporting medium which holds the roots of the macrophytes (Fig. 2.4a). In predominant cases, the submerged (partly or fully) and floating types of plants are used. In the case of subsurface wetlands, the wastewater flows through the porous spaces of the supporting material filled in the basin and the level of water is maintained below the top surface of the material (Fig. 2.4b–d). The emergent type of plant species such as *Phragmites, Canna, Typha, Scripus*, etc. are widely used. Subsurface wetlands are categorized as horizontal, vertical and hybrid wetlands. In horizontal flow systems (HF), the wastewater enters the inlet zone and flows horizontally through the porous space and exits through the outlet zone (Fig. 2.4b). In vertical flow system (VF), the water is distributed throughout the length of the wetland bed, and it percolated the supporting media and reached the drain layer at the bottom (Fig. 2.4c). Both the horizontal and vertical flow conditions take place in hybrid system where both VF and HF are connected in series (Fig. 2.4d).

The constructed wetlands are facilitated with aerobic zones in the top layer and anoxic/anaerobic zones in the lower layer. Organics are removed through aerobic and anaerobic processes. Nitrogen is removed in constructed wetlands through various biochemical processes, assimilation by macrophytes and microbes, sorption over the filter media, ion exchange, etc. For phosphate removal, processes like sorption and deposition on the substrate, plant uptake, microbial assimilation, and metabolism play key roles. Coliform bacteria are removed through a combination of physical processes like sedimentation and root filtration, chemical processes such as biocide secretion, biological actions like antibiosis, lysis, and bacterial attack, as well as natural die-off. Heavy metals can be removed through several processes: (i) metals can be adsorbed onto soils and sediments, (ii) they can settle out through sedimentation, (iii) cations and anions can exchange or form complexes, (iv) metals can precipitate or coprecipitate as insoluble ions, (v) plants can absorb them, and (vi) microbes can process them through their metabolic activities.

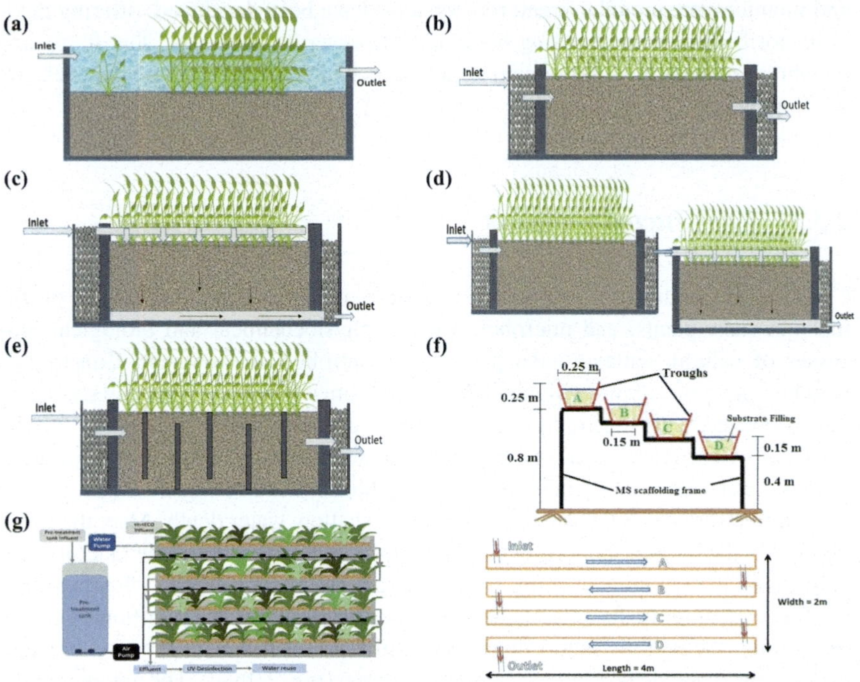

Fig. 2.4 Schematics of different constructed wetland systems: (**a**) free surface wetlands, (**b**) horizontal flow wetlands, (**c**) vertical flow wetlands, (**d**) hybrid wetlands, (**e**) baffled wetlands, (**f**) GROW system, (**g**) vertECO system (Source: Parde et al., 2021 (**a**–**e**), Ramprasad et al., 2017 (**f**), Estelrich et al., 2021 (**g**))

Several studies have documented the efficacy of constructed wetlands in treating a variety of wastewaters. However, the constructed wetland systems require larger area which affects its adoptability in urban areas and also increases the capital costs. To address these issues, several modifications are made in the design aspects and reported in previous studies. Notably, green roof-top water recycling systems (GROW), vertical ecosystem (vert-ECO), baffled constructed wetlands, etc. are effective in treating urban graywater and septic tank effluents.

The GROW system was developed by Water Works UK Ltd., and it has a series of troughs with weirs which aid in achieving horizontal and vertical flows (Fig. 2.4g). The system is planted with various macrophytes in the gravel layers and can be installed in the roofs of the buildings. Ramprasad et al. (2017) noticed the higher efficiency of organics, nutrients, and personal care products (>85%) in the GROW system under Indian climatic conditions. The optimized operating conditions such as hydraulic residence time of 0.8–1.3 days and loading rate of 7.8–15 L/m^2/day. Another constructed wetland system called vertECO (vertical ecosystem) was developed and established in Spain for treating greywater generated from a hotel (Gattringer et al., 2016). This system consists of four cascading stages filled with expanded clay media and a combination of 14 plant species (Fig. 2.4f). The system was fed with pre-aerated wastewater at the first stage and then through the horizontal motion, wastewater flows to the subsequent stages. The removal of COD and suspended solids was observed to be >85% with more than 95% removal of BOD_5 (Estelrich et al., 2021). Ravichandran and Philip (2022) studied the influence of different filling materials of constructed wetland for the removal of organics, nutrients and pharmaceutical compounds (PhCs) in a lab-scale plug flow reactor with three stages. The removal efficiency of the system was observed as 88% (COD), 82% (TN), 78% (TP), and 95% (PhCs). The treated water from all these systems can be reused after proper disinfection to ensure relevant reuse quality. Additionally, these systems can be adopted on the terrace or rooftop of buildings in an urban space to treat the graywater generated in it.

2.3.4 *Bioretention System*

The bioretention system (BRS) is a widely adopted intervention for urban stormwater management. It is also called bioswales or rain garden or biofiltration even though there is a slight difference among them. It consists of retaining structure (concrete, geotextile, kerb stone, etc.), filter materials (gravel, brickbats, sand, etc.), macrophytes, and drainage system (Fig. 2.5). The treated stormwater can drain through overflow weir or any subsurface collection pipe or to underlying soil. BRS is effective in curtailing the various pollutants including suspended solids, heavy

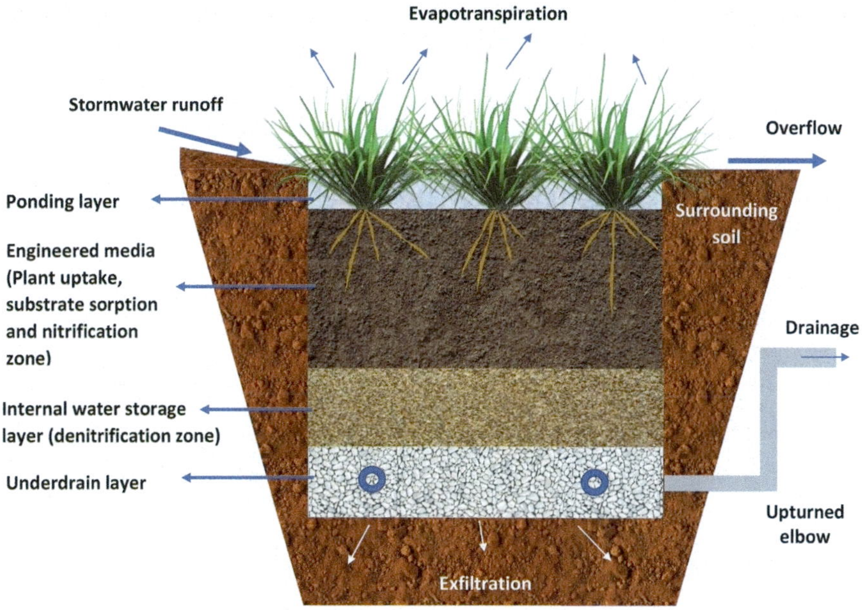

Fig. 2.5 Bioretention systems. (Source: Vijayaraghavan et al., 2021)

metals, nutrients, pathogens, and trace organic contaminants. The efficiency is noticed as TSS (80–97%), TN (58–87%), TP (86–95%), and as COD (80–95%). BRS relies on various processes such as trapping, uptake, and assimilation by biota, and infiltration of contaminants to govern the quantity and quality of stormwater. The pollutant removal mechanism in BRS includes a combination of physical (filtration), chemical (adsorption, hydrolysis, photodegradation, volatilization), and biological processes (biodegradation and phytoremediation) processes (Nazarpour et al., 2023). The performance of BRS in removing these pollutants can vary depending on factors such as design, vegetation, soil media, and climatic conditions.

Bioretention systems has multiple advantages, such as improving the quality of runoff water, managing the runoff volume, can be incorporated in urban space, enhancement of aesthetics and biodiversity, attenuate urban heat island effects (Vijayaraghavan et al., 2021). Thus, they became an integral part of sustainable urban drainage systems (SUDS) promoting blue-green infrastructure.

2.3.5 Phytorid

The National Environmental Engineering Research Institute (NEERI) has developed a nature-based wastewater treatment technology called phytorid. It is a combination of three units where the physiochemical and biological processes take place. The first unit—an anaerobic baffled reactor with gravel and stones of different grades. The second unit is the phytorid reactor containing filter media with plants (Fig. 2.6) (Biswas et al., 2020). The commonly used macrophytes species include *Phragmites, Canna, Scirpus, Phalaris, Typha, Glyceria, etc.* The last unit is the effluent collection cell. The process is cost-effective, eco-friendly, requires minimal or zero energy, and involves no usage of chemicals in the entire process. The typical hydraulic residence time is between 5 and 7 days. The removal efficiency of phytorid system was observed as 80–96% (BOD), 80–94% (COD), 80–92% (TN), 80–92% (TP), and 85–95% (Coliform). The treated effluent can be reused for agricultural activities.

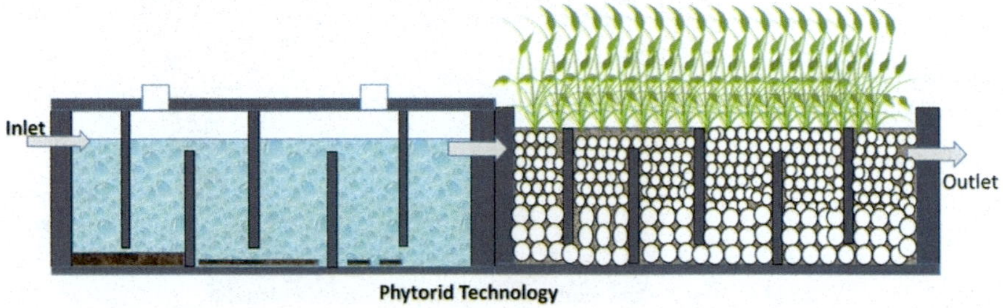

Fig. 2.6 Schematics of phytorid system. (Source: Parde et al., 2021)

2.3.6 *Soil Biotechnology*

Soil biotechnology (SBT) is a natural treatment system where soil microbes help in the removal of pollutants rather than mechanical aerators in conventional activated sludge process. It involves sedimentation, filtration, biochemical reactions occurring during photosynthesis, respiration, nitrogen fixation, uptake and assimilation, ammonification, nitrification, denitrification, etc. SBT can be operated in batch or continuous mode with the hydraulic residence time (HRT) of 0.5–2 h. The treatment capacity of SBT is reported as less than 5 KLD with an area requirement of 0.021 m^2/person/day (Parde et al., 2021). The removal efficiency of COD and BOD is > 90% and nutrients removal (TN, NH_4-N, TKN, nitrate, nitrite, TP) is > 85% with 2–3 log reduction of coliforms (Kamble et al., 2017). The main advantages of SBT includes very low HRT, no chemical addition, natural aeration without any moving parts, odor free and highly oxygenated treated effluent, low energy requirement, and enhanced aesthetics. The treated water can be used for horticulture, flushing, groundwater recharge, etc.

There are various types of soil-based treatment, including constructed soil filter system, soil scape filter, ecofiltration bank (EFB), microbial-earthworm ecofilter, multi-soil layering system, etc. (Kamble et al., 2017). Ecofiltration bank consists of screens, intake structure, soil scape filter with bio-consortium media called ORGANOTREATTM and treated pounds. It has been installed in Ambil stream near the garden of Indradhanushya Centre, Pune, Maharashtra. After the intervention, the rise in DO level was noticed as 8–10 times than the untreated wastewater (Fig. 2.7).

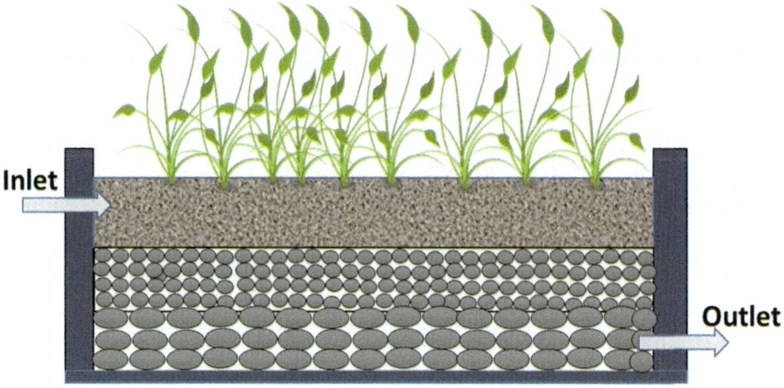

Fig. 2.7 Schematics of Soil Biotechnology. (Source: Parde et al., 2021)

2.3.7 *Other Technologies*

Apart from the above discussed technologies, several other treatment techniques are adopted for in-situ treatment of drains and rivers such as green bridge, microbial dosing, floating islands, NEERI `s RENEU (Restoration of Nallah with ecological units), etc. (Satyendra et al., 2023).

Green bridge technology was developed in 2018 by Shristi Eco Research Institute (SERI), Pune. It consists of biomass, supporting media (gravel and rubble), vegetation, and active microbes where the bridges are introduced as barriers in zigzag manner in the drains. Green bridges are adopted at Ganga River at Allahabad (Rasoolabad stream network) and Ahar River (Udaipur, Rajasthan). Microbial dosing is the addition of beneficial microbes to the contaminated drain to reduce the concentration of organics and nutrients curtailing the formation of hydrogen sulfide and methane. The infusion of bioenergizer into the drain including microbes, buffer solutions, nutrients, and organic acids can foster the biological oxidation of pollutants. In Indore (Eastern part), two natural drains were treated with microbial dosing at eight dosing stations. After 2 months of daily dosing, the BOD removal of 70% was observed.

NEERI has developed a treatment system in 2018 called Restoration of Nallah with ecological units (RENEU) comprising screen and grit trap, sedimentation zone, aeration zone, mobile phytotraps, and florafts. The removal mechanism includes physical and biological processes for the removal of organic and inorganic pollutants and nutrients. It can be readily adopted in narrow drains (6–10 m wide and upto 10MLD flow) but difficult for larger drains. No larger civil infrastructure and electricity is required. The treated effluent after disinfection can be reused for horticulture, floriculture, road washing and flushing. Recently, Thayyil and Philip (2024) proposed an in-situ treatment scheme with anoxic attachment zone, aerobic attachment zone with or without aeration, and hydroponic floating wetlands. Locally available materials can be used in the anoxic and aerobic zones. The overall removal of COD and NH_4-N was found to be 86% and 97%, respectively.

2.4 Factors Affecting Performance and Implementation

The major components of an NBT are substrate materials, plant species, and microbes. The collaborative and synergistic action of these components and the environmental conditions such as temperature, light intensity, etc., significantly influence the pollutant removal by NBT. The performance of NBT depends upon the selection of its components (filter media, macrophytes/vegetation, microbes, etc.), pre- and posttreatment, operating parameters, flow conditions, environmental parameters (temperature, precipitation, etc.), etc. (Fig. 2.8).

Filling materials are an indispensable part of NBTs where predominant physical, chemical, and biological reactions occur. The substrates augment the proliferation

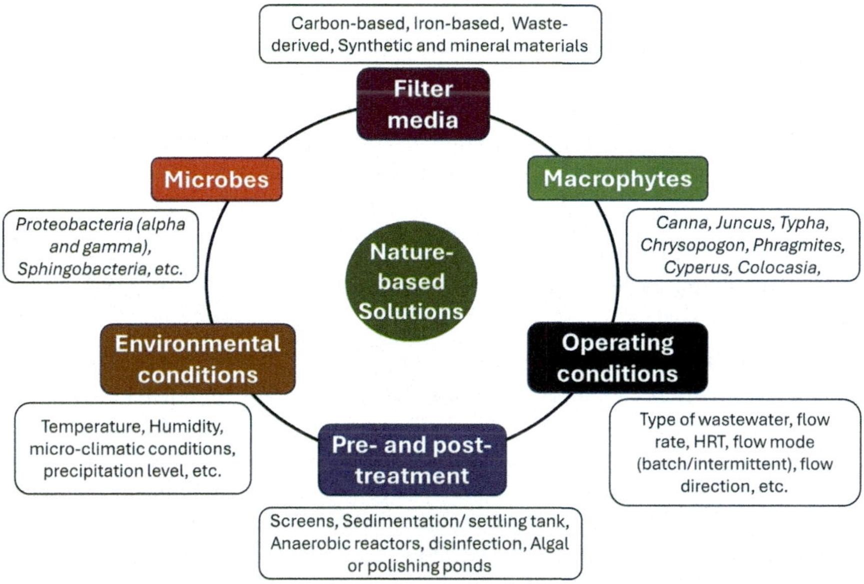

Fig. 2.8 Factors influencing the performance of NBTs

of microbial populations and macrophytes, enhancing the microenvironmental conditions and pollutant removal. The removal processes comprise filtration, sedimentation, sorption, complexation, precipitation by colloidal matters, ion exchange, degradation and transformation by microbes, and bio-immobilization in the filling materials. Various materials including carbon-based materials (biochar derived from different precursor materials (bagasse, bamboo, tyre chips, cork, pet coke, etc.), iron-based materials (volcanic cinder, pyrrhotite, iron slag, pyrite, furnace slag), waste-derived materials (fly ash materials, waste concrete blocks, waste ceramsite, sludge-derived materials, marble waste), and synthetic and mineral materials (zeolite (natural and commercial), LECA, Polonite, activated alumina, silica, montmorillonite) (Ravichandran et al., 2023). The performance of these materials as NBTs substrates largely depends on their characteristics. Substrate properties like particle size distribution, porosity, chemical constituents, surface characteristics, mechanical strength, chemical stability, organic matter contents, hydraulic conductivity, adsorption potential, etc., play important functions in the pollutant attenuation processes.

Macrophytes contribute to the visible structure of constructed wetland systems. The vegetation hosts microbial populations and aids in the attenuation of pollutants. The widely used plant species in NBT include *Phragmites, Canna, Typha, Scripus, Juncus, and Cyperus* (Fig. 2.8). The choice of macrophytes significantly dominates the performance of NBT. The presence of diverse microbial populations, especially *Proteobacteria, Sphingobacteria, etc.* contributes to the removal of organics and nutrients (Ravichandran & Philip, 2022). The incorporation of pretreatment techniques including screens, sedimentation, settling tank, anaerobic reactors, etc., reduces the accumulation of solids in NBTs, reducing clogging and operational issues and enhancing the treatment efficiency. The posttreatment techniques (disinfection, algal ponds, polishing ponds, etc.) help to achieve the quality of treated water for the reuse applications. Additionally, the operating and environmental conditions impact the performance of nature-based treatment systems.

2.5 Barriers to Nature-Based Systems and Way Forward

While nature-based solutions offer substantial benefits for mitigating urban pollution, several barriers can impede their implementation. One significant challenge is the lack of sufficient funding and financial incentives to support the initial investment and ongoing maintenance of green infrastructure projects. Additionally, there can be resistance from stakeholders who prioritize traditional, engineered solutions or who may not fully understand the long-term benefits of NBT. Urban planning and regulatory frameworks may also be inadequately aligned to facilitate the integration of nature-based approaches, often due to outdated policies or competing land-use priorities. Furthermore, there can be challenges related to the scalability and effectiveness of NBT in densely populated urban areas, where space is limited, and pollution levels are high. Addressing these barriers requires a concerted effort from policymakers, urban planners, and community stakeholders to promote awareness, secure funding, and develop supportive frameworks that facilitate the adoption and maintenance of nature-based solutions in urban environments.

To effectively reduce urban pollution through nature-based solutions, cities should adopt a strategic approach that integrates green infrastructure into urban planning and development. This includes investing in projects like green roofs, urban forests, and permeable pavements, which help capture pollutants, manage stormwater, and cool urban areas. Public engagement and education are crucial, as involving residents in the design and maintenance of green spaces can enhance effectiveness and foster a sense of community ownership. Cross-sector collaboration among urban planners, environmental experts, and policymakers ensures that NBT is tailored to specific pollution challenges and integrated into broader city strategies. Supporting policies and funding mechanisms are essential to facilitate the implementation of NBT, while leveraging technology and innovation, such as GIS and remote sensing, can optimize their impact. Additionally, addressing equity by ensuring all communities benefit from these solutions is vital for promoting overall public health and environmental justice. By adopting these measures, cities can create healthier, more sustainable urban environments that effectively combat pollution.

2.6 Conclusion

Nature-based treatment technologies offer effective solutions for pollution control in rapidly urbanizing regions, especially in developing countries. They provide significant advantages such as lower infrastructure costs, reduced energy use, and the potential to integrate with urban green spaces, which helps improve both the environment and urban aesthetics. However, their full potential is hindered by challenges like outdated urban planning and regulatory frameworks that struggle to accommodate NBT. Additionally, implementing these systems in densely populated areas is difficult due to limited space and high pollution levels. Addressing these barriers is crucial for the successful adoption of NBT in urban settings.

To overcome these obstacles, cities must adopt a strategic approach that includes securing funding, fostering public engagement, and updating regulatory frameworks to support the integration of NBT. Collaboration across sectors is essential to tailor these solutions to specific urban challenges. Additionally, leveraging technology and ensuring equitable access to NBT can help maximize their impact. By addressing these challenges, cities can effectively reduce pollution and create healthier, more sustainable urban environments.

References

Biswas, P., Mamatha, S., Varghese, K., Johnson, R., Vijay, R., & Kumar, R. (2020). 3D printing of high surface area ceramic honeycombs substrates and comparative evaluation for treatment of sewage in Phytorid application. *Journal of Water Process Engineering, 37*, 101503.

Boano, F., Caruso, A., Costamagna, E., Ridolfi, L., Fiore, S., Demichelis, F., et al. (2020). A review of nature-based solutions for greywater treatment: Applications, hydraulic design, and environmental benefits. *Science of the Total Environment, 711*, 134731.

Chen, L., Chen, M., Zhang, X., & Xian, Y. (2024). Evaluating inequality divides in urban development intensity between the Global North and South. *Land Use Policy, 145*, 107291.

Estelrich, M., Vosse, J., Comas, J., Atanasova, N., Costa, J. C., Gattringer, H., & Buttiglieri, G. (2021). Feasibility of vertical ecosystem for sustainable water treatment and reuse in touristic resorts. *Journal of Environmental Management, 294*, 112968.

Gattringer, H., Claret, A., Radtke, M., Kisser, J., Zraunig, A., Rodriguez-Roda, I., & Buttiglieri, G. (2016). Novel vertical ecosystem for sustainable water treatment and reuse in tourist resorts. *International Journal of Sustainable Development and Planning, 11*(3), 263–274.

Gholami, M., O'Sullivan, A. D., & Mackey, H. R. (2023). Nutrient treatment of greywater in green wall systems: A critical review of removal mechanisms, performance efficiencies and system design parameters. *Journal of Environmental Management, 345*, 118917.

Kamble, S. J., Chakravarthy, Y., Singh, A., Chubilleau, C., Starkl, M., & Bawa, I. (2017). A soil biotechnology system for wastewater treatment: Technical, hygiene, environmental LCA and economic aspects. *Environmental Science and Pollution Research, 24*, 13315–13334.

Nazarpour, S., Gnecco, I., & Palla, A. (2023). Evaluating the effectiveness of bioretention cells for urban stormwater management: A systematic review. *Water, 15*(5), 913.

Parde, D., Patwa, A., Shukla, A., Vijay, R., Killedar, D. J., & Kumar, R. (2021). A review of constructed wetland on type, treatment and technology of wastewater. *Environmental Technology & Innovation, 21*, 101261.

Ramprasad, C., Smith, C. S., Memon, F. A., & Philip, L. (2017). Removal of chemical and microbial contaminants from greywater using a novel constructed wetland: GROW. *Ecological Engineering, 106*, 55–65.

Ravichandran, M. K., & Philip, L. (2022). Assessment of the contribution of various constructed wetland components for the removal of pharmaceutically active compounds. *Journal of Environmental Chemical Engineering, 10*(3), 107835.

Ravichandran, M. K., Thayyil, M. I., & Philip, L. (2023). Application of engineered natural treatment systems for pollution abatement. In *Technological solutions for water sustainability: Challenges and prospects* (p. 145).

Satyendra, Asoria, S. K., & Vijay, R. (2023). In-situ drain treatment types and technologies for flowing wastewater: A comprehensive review. *Process Safety and Environmental Protection, 170*, 449–463. https://doi.org/10.1016/j.psep.2022.12.035

Thayyil, M. I., & Philip, L. (2024). Sustainable treatment scheme for in-situ remediation of contaminated drains using engineered natural systems. *Chemosphere, 142469*, 142469.

UNEP. (2021). *Nature-based solutions for urban challenges*, March 2021. https://wedocs.unep.org/bitstream/handle/20.500.11822/35864/FB023.pdforg

Vijayaraghavan, K. (2016). Green roofs: A critical review on the role of components, benefits, limitations and trends. *Renewable and Sustainable Energy Reviews, 57*, 740–752.

Vijayaraghavan, K., Biswal, B. K., Adam, M. G., Soh, S. H., Tsen-Tieng, D. L., Davis, A. P., et al. (2021). Bioretention systems for stormwater management: Recent advances and future prospects. *Journal of Environmental Management, 292*, 112766.

WWF. (2021). *Making a case for investing in nature-based solutions: A case study from Tshwane*. WWF South Africa.

Zheng, X., & Lam, K. L. (2024). An overview of environmental co-benefits and trade-offs to reduce greenhouse gas emissions in municipal wastewater management. *Sustainable Production and Consumption, 46*, 1.

Chapter 3
Advancements in Photocatalytic Degradation Using ZnO Nanoparticles for Sustainable Urban Air Pollution Control

Kadarkarai Rajeshkumar, P Andavar, Ponnambalam Nirmal, and S Makesh Kumar

Abstract Air pollution is a major threat to human health and the environment. The main air pollutants that come from different sources in both the indoor and outdoor environment are carbon monoxide (CO), sulfur oxides (SO_X), oxides of nitrogen (NO_X), volatile organic compounds (VOCs), and particulate matter (PM). Nanomaterials are capable of absorbing air pollutants from the source and they help to improve the air pollution level in the atmosphere. Air pollution from indoor and outdoor sources may be effectively managed using photocatalytic oxidation. ZnO nanoparticles were synthesized, characterized, and coated on the inside surface of a quartz tube to remove SO_2 under the UV irradiation process. X-ray diffraction reveals that the average crystalline size of ZnO nanoparticles was found to be 22.63 nm. FTIR analysis for ZnO nanoparticles has shown the spectral peak at 667 cm^{-1} which indicates the ZnO vibrations. The particle size of ZnO nanoparticles was found to be 178.1 nm. FE-SEM analysis of ZnO nanoparticles reveals that the particles are spherical. The EDS analysis of ZnO nanoparticles confirms the presence of Zn and O. TGA/DTA analysis of ZnO nanoparticles shows that total weight loss of 10.95% and it attains stability after 800 °C. The PL spectra of ZnO NPs show an emission peak at 649 nm. The performance of prepared nanomaterial was demonstrated for the removal of SO_2 generated from the source. The removal of SO_2 under UV irradiation for ZnO nanoparticles was 75%. The maximum removal efficiency was achieved after 40 min of irradiation time. There is no decrease in the efficiency of the photocatalyst after three batches and it can be reused and regenerated.

Keywords Air pollution · SO_2 · ZnO · Nanoparticle · Photocatalytic

---

K. Rajeshkumar (✉) · P. Andavar · P. Nirmal · S. Makesh Kumar
Department of Civil Engineering, Government College of Technology, Coimbatore, Tamilnadu, India
e-mail: rajesh.k.civil@gct.ac.in; andavar.p.civil@gct.ac.in; pnirmal@gct.ac.in; makesh.s.civil@gct.ac.in

V. Baskaran et al. (eds.), *EcoTech Urbanism*, The Urban Book Series,
https://doi.org/10.1007/978-3-031-83833-0_3

3.1 Introduction

Fossil fuel combustion emits CO, CO_2, SO_2, NO_X, SO_2, and HCs into the atmospheric air. With urbanization and industrialization growth, air quality levels have decreased in a global level. These air pollutants can cause serious issues to human health and the environment (Manisalidis et al., 2020). The main air pollutants are nitrogen oxides (NO_X), sulfur oxides (SO_X), carbon monoxide (CO), volatile organic compounds (VOCs), ozone (O_3), and particulate matter (PM) (Bikis, 2023).

SO_2 is a nonflammable, nonexplosive, and colorless gas with a strong, odorous smell. Sulfur dioxide is the most significant oxide of sulfur created by the burning of coal, incense sticks, etc., and emitted into the atmosphere. The combustion of sulfur-containing fossil fuel and emissions from thermal power plants, the iron and steel industry, etc., release SO_2 into the atmosphere (Meimand et al., 2020). Human respiratory systems are susceptible to damage due to exposures of SO_2 can harm to the respiratory system of humans and making breathing becomes more difficult. People with asthma, especially children, are more susceptible to the effects of SO_2. Exposure to elevated SO_2 concentrations can lead to respiratory illness, cardiovascular disease, and mortality.

Nanotechnology is the group of emerging technologies that work on a nanometer scale (i.e., between 1 and 100 nm range) to produce materials, devices, and systems with fundamentally new properties and functions for sustainable pollution control (Mohamed, 2017). It offers an excellent opportunity to measure, monitor, manage, and reduce pollutants. It is essential to design and develop nano-enabled filters, membranes, adsorbents, and sensors to demonstrate improved performance for air pollution control (Saleem et al., 2022). It is an advanced treatment technology that controls and remediates air pollutants by utilizing specific properties of nanomaterials such as nano adsorbents, nanocatalysts, nanofilters, and nanosensors. Effectively designed nanomaterials react with a contaminant and degrade into nontoxic products. Numerous manufactured and self-assembled nanoparticles can act as self-cleaning of air pollutants.

Photocatalytic degradation is an advanced oxidation process that requires a photocatalyst and UV or visible light to degrade the air pollutants (Escobedo & de Lasa, 2020). Significant features considered in photo reactors are types of irradiations, the position of light sources, and the photocatalyst coating method. Unlike conventional air treatments, photocatalysis does not demand more energy, disposal of treated materials, or the release of harmful gases. It is one of the promising methods for the control of air pollutant gases (Klinbumrung et al., 2022). Owing to their low cost, chemical stability, and cleanliness, photocatalysts received increased interest in the elimination of gaseous pollutants. The most commonly used photocatalysts were TiO_2, WO_3, and ZnO. Zinc oxide has a wide band gap of 3.37 eV and is expected to have a photocatalytic capacity similar to TiO_2. ZnO can be a suitable alternative for TiO_2 (Rangkooy et al., 2021).

Zinc oxide is a ceramic semiconductor with a wide band gap of 3.37 eV and an excitation energy of 60 MeV. It is a fascinating and one of the significant materials in nanotechnology (Kowsari & Abdpour, 2017). ZnO exhibits impressive properties like immense binding energy, wide band gap, high piezoelectric property, etc. Numerous applications of ZnO include solar cells, UV-light emitting devices, photocatalysis emerging optical devices, gas sensor technology, and pharmaceutical and cosmetic industries.

Owing to its greater surface area, physical and chemical methods were used to synthesize the nanomaterials. The synthesis of nanomaterials has two main approaches. They are top-down approaches and bottom-up approaches (Baig et al., 2021). Various methods are available for synthesizing ZnO nanoparticles, such as microwave-assisted, coprecipitation, chemical vapor deposition, hydrothermal, solvothermal, solgel, and green synthesis (Santika & Rohmawati, 2022). Researchers focus on green synthesis and biosynthesis among the various physical and chemical methods because it is cost-effective and repeatable. Green synthesis is an environment-friendly technique due to the use of the extract of plant parts such as peels, leaves, flowers, roots, stems, seeds, etc. (Klinbumrung et al., 2022; Shah et al., 2023). The current work focuses on the green synthesis of ZnO nanoparticles, its characterization, and coating on the inner surface of the quartz tube for the removal of SO_2 using a photocatalytic reactor.

3.2 Green Synthesis of ZnO Nanoparticles

Pineapple peel extract has been used for the green synthesis of ZnO nanoparticles. The pineapple peels were collected from the juicer shop, thoroughly washed, and rinsed with distilled water. The component of aqueous pineapple peel extract are alkaloids, flavonoids, phenols, etc. Common method was adopted for the preparation of extract. The pineapple peels were washed several times with deionized water, cut into tiny pieces, sundried for 24 h, and kept in the oven at 60 °C overnight (Fig. 3.1a). The peels were ground into a fine powder. A ratio of 1:10 of pineapple peel powder to distilled water was utilized to prepare the extract. The resultant mixture was stirred at 60–70 °C for 15 min. Then, the extract was separated by Whatman filter paper, and the filtered solution was stored at 4 °C for further use. 0.1 M of zinc acetate dihydrate was dissolved in 100 ml distilled water and stirred using a magnetic stirrer for 1 h at 60 °C (Fig. 3.1b). 10 ml of prepared extract was added dropwise under stirring. The pH of the mixture was maintained at 9 using NaOH pellets. The precipitate was separated by centrifuge at 10000 rpm for 15 min. The wet nanoparticles were dried at 80 °C for 12 h and ZnO nanoparticles were obtained (Fig. 3.1c). Pineapple peel is a rich source of phytochemical compounds (alkaloids, flavonoids, and saponin) and bioactive compounds, and these compounds could be used as capping agent in the synthesis of ZnO nanoparticles.

Fig. 3.1 Green synthesis of ZnO nanoparticles. (**a**) Dried Pineapple Peels, (**b**) magnetic stirrer, (**c**) ZnO nanoparticles

3.2.1 Characterization of ZnO Nanoparticles

Characterization studies of ZnO nanoparticles have been carried out using X-ray powder diffraction, Fourier transform infrared spectroscopy, particle size and zeta potential analyzer, field emission scanning electron microscope, energy dispersive X-ray spectroscopy, thermogravimetric analysis, photoluminescence spectroscopy. X-ray diffraction report reveals that the average crystalline size of ZnO nanoparticles was found to be 22.63 nm (Fig. 3.2a). FTIR analysis for ZnO nanoparticles (Fig. 3.2b) has shown that the spectral peak at 667 cm^{-1} indicates the ZnO vibrations. The particle size of ZnO nanoparticles (Fig. 3.2c and d) was found to be 178.1 nm. FE-SEM analysis of ZnO nanoparticles (Fig. 3.2e and f) reveals that the particles are spherical. The EDS analysis of ZnO nanoparticles (Fig. 3.2g) confirms the presence of Zn and O. TGA/DTA analysis of ZnO nanoparticles (Fig. 3.2h) shows a total weight loss of 10.95% (Fig. 3.2i) and it attains stability after 800 °C. The PL spectra of ZnO nanoparticles (Fig. 3.2j) shows an emission peak at 649 nm.

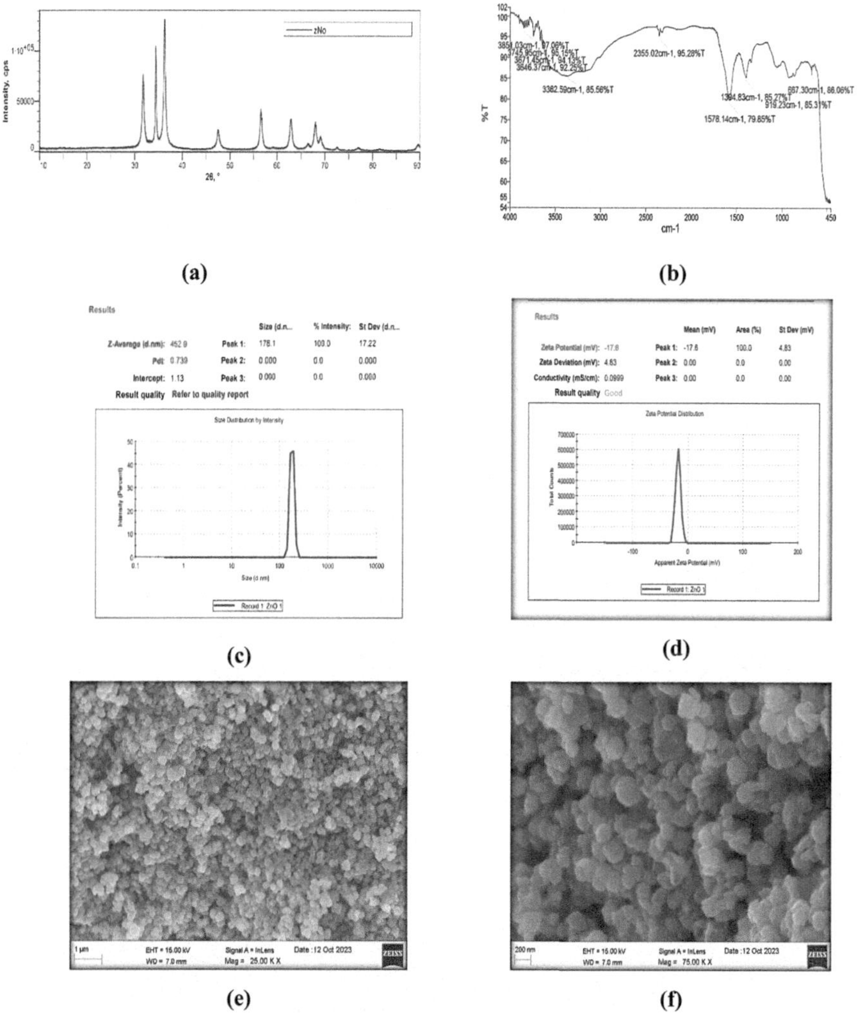

Fig. 3.2 Characterization of ZnO nanoparticles. (**a**) XRD analysis, (**b**) FTIR analysis, (**c**, **d**) Particle Size Distribution, (**e**, **f**) FE-SEM Analysis, (**g**) EDS analysis, (**h**) TGA/DTA analysis, (**i**) Weight loss curve, (**j**) PL Spectra Curve

3.3 Design and Fabrication of Photocatalytic Reactor

The size of the reactor was designed as 30 × 30 × 31 cm. Figure 3.3 shows the schematic diagram of the photocatalytic reactor. A constant distance was kept between the UV lamp and the reactor. A quartz tube with a diameter of 15 mm was coated with nano photocatalytic suspension and kept in the center of the reactor at a

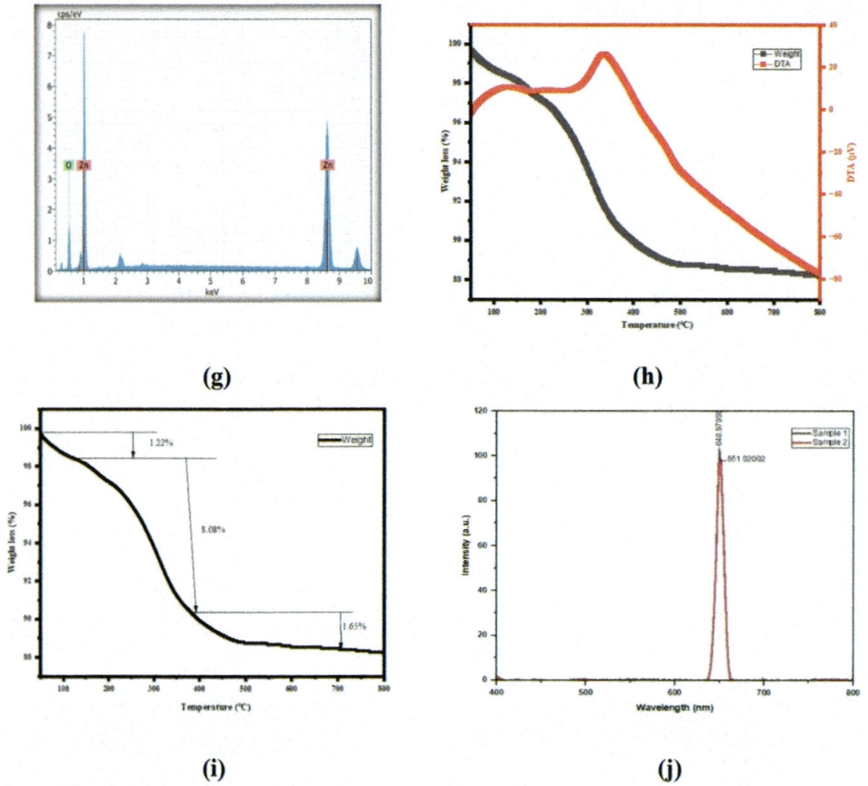

Fig. 3.2 (continued)

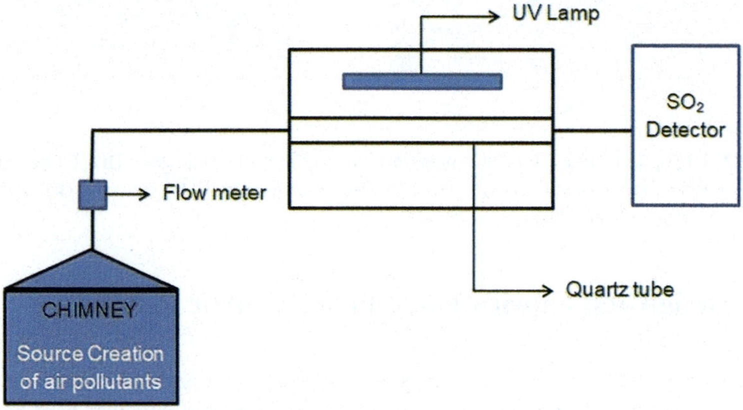

Fig. 3.3 Schematic diagram of photocatalytic reactor

Fig. 3.4 Reactor with a cooling fan

distance of 15.5 cm and 3 cm from the UV lamp (2.5 W Osram, emitting at a wavelength of 200–280 nm). The solution was prepared separately by coating an ethanol suspension of ZnO nanoparticles into a quartz tube. 0.1 g ZnO nanoparticles were suspended in 10 ml of ethanol using a magnetic stirrer at 300 rpm for 1 h. One end of the quartz tube is fully closed and the coating solution was poured and coated all over the tube. To ensure the coating does not affect the flow of inlet gas, in this study we conclude that uniform coating was analyzed by both inlet and outlet flow gas using a flowmeter. To ensure that all ethanol is eliminated from the coated photocatalyst, the quartz tube was heated in the oven for 15 min at 60 °C. Before starting the photocatalytic test, the coated quartz tube was allowed to cool to room temperature. Two cooling fans were installed above the UV lamp to avoid the temperature rising during the operation. Figure 3.4 shows the reactor with a cooling fan.

3.3.1 *Mechanism of Photocatalytic Degradation of SO_2*

Numerous investigations have reported on the mechanism of SO_2 photocatalysis. Under the UV light irradiation process, electron-hole pairs are produced on the catalyst surface, which appears in the form of hydroxyl radicals (•OH) and oxygen radicals (•O_2). Sulfate and sulfite are the primary photocatalytic products of SO_2 photodegradation. Recent studies have demonstrated the crucial role of hydroxyl radical species produced from absorbed H_2O on the photocatalyst surface in the photodegradation of SO_2. The active absorption sites on the surface of hydroxyl groups can absorb pollutant like gases. Additionally, it forms hydrogen bond with H_2O to absorb it, which produces additional hydroxide ions needed for SO_2 photodegradation. The following equations depict a potential chemical route for the elimination of SO_2.

$$\mathbf{SO_2 + \bullet OH \rightarrow HSO_3 \bullet} \tag{3.1}$$

$$\mathbf{HSO_3 \bullet + \bullet OH \rightarrow 2H^+ + SO_4^{2-}} \tag{3.2}$$

$$\mathbf{HSO_3 \bullet + \bullet OH \rightarrow SO_3 + H_2O} \tag{3.3}$$

$$\mathbf{SO_3 + \bullet OH \rightarrow H^+ + \bullet SO_3^-} \tag{3.4}$$

$$\mathbf{\bullet SO_3^- + \bullet OH \rightarrow H^+ + SO_4^{2-}} \tag{3.5}$$

3.3.2 *Working of Photocatalytic Reactor*

The coated quartz tube was carefully placed inside the reactor. The photocatalytic reactor with which the vacuum pump was fixed at the outlet end and the inlet chamber followed by the flowmeter was connected to the fore-end. The flue gas is generated by coal. The flue gas is produced in a controlled environment using a chimney. The main emissions from burning coal are CO_2, particulates, nitrogen oxide (NO_X), and sulfur dioxide (SO_2). Thus, SO_2 was emitted by burning biomass in the form of coal. The chimney is constructed using a galvanized iron sheet and has a dimension of 300 × 450 mm. It has two compartments. The upper part of the component is designed for burning dry fuel, while the lower component is used for collecting ash. Flue gas is transferred to the inlet portion of the reactor and the inlet concentration is noted. The flow rate was maintained at 600 ml/min using a flowmeter. Initial Concentration of SO_2 in the flue gas was determined by using the SO_2 sensor (MQ135 Model gas sensor) and operating voltage is 2.5 V to 5.0 V. Variation of relative humidity between 40% and 70% does not induce any significant variation in SO_2 degradation rate. The humidity ratio was regulated at 40% measured by using humidifier attached with the sensor. The humidity should be maintained for the effective photodegradation of SO_2. Some researchers reported that SO_2 degradation increased with increasing relative humidity. Humidifier is used to maintain the relative humidity throughout the study. The UV lamp and cooling fan were turned on when the adsorption-desorption equilibrium was reached among the components engaged in the photochemical process. The removal ratio of air pollutants was calculated as

$$\eta(\%) = \left(1 - \frac{C}{C_0}\right) \times 100\%$$

Where, C and C_0 are the outlet and inlet concentration of SO_2, respectively.

3.4 Photocatalytic Degradation of SO_2 Using ZnO Nanoparticles

The maximum removal efficiency achieved was 75% with 40 min irradiation time for batch I. The maximum removal efficiency achieved was 73% and 72% for batches II and III with 40 min irradiation time, showing only slight variations from batch I. The high photocatalytic activity of the nanomaterial may be due to the hexagonal structure of ZnO. The details of the removal efficiency of batches I, II, and III are consolidated in Table 3.1 and Fig. 3.5 shows the difference in irradiation time pattern.

3.4.1 Reusability of ZnO Nanoparticles

One of the most important aspects of any kind of catalyst, including photocatalysis, is its reusability. There is no change in terms of the weight, dimension, and thickness of coating were observed. The principal disadvantage of gas-phase photocatalytic processes is the buildup of intermediate products on the surface of the photocatalyst, which diminishes the photocatalytic activity. Three test cycles were conducted to study the reusability of ZnO NPs for the photodegradation of SO_2. There is no significant decrease in the photocatalyst activity after three test cycles. Nonetheless, the intermediate products dissolve and are readily removed from the catalyst surface by the flue gas flow.

3.5 Conclusion

In this study, ZnO NPs were synthesized, characterized, and used for air pollution control. ZnO nanoparticles offer an effective, sustainable, and reusable method for the photocatalytic degradation of SO_2, potentially contributing to air pollution control and environmental protection efforts. ZnO nanoparticles have excellent photocatalytic activity due to their wide band gap and huge surface area, effectively breaking down sulfur dioxide (SO_2) into less hazardous chemicals when exposed to UV light. The average crystalline size of prepared ZnO nanoparticles is 22.63 nm. FTIR analysis for ZnO nanoparticles confirms the ZnO vibrations. The particle size

Table 3.1 Removal efficiency of ZnO nanoparticles in photocatalytic degradation of SO_2

		Batch I				Batch II				Batch III			
S. No	Contact time (min)	Co (mg/m^3)	C (mg/m^3)	C/Co	% removal	Co (mg/m^3)	C (mg/m^3)	C/Co	% removal	Co (mg/m^2)	C (mg/m^3)	C/Co	% removal
1.	5	153.3	126.9	0.827	17	103.9	82.8	0.796	20	118.1	96.9	0.820	18
2.	10	137	108.9	0.794	21	110.6	79.2	0.716	28	124.9	93.3	0.746	25
3.	15	119.9	83.5	0.696	30	127.9	84.3	0.659	34	102.2	70.6	0.690	31
4.	20	108.5	69.1	0.636	36	108.6	67.6	0.622	38	110.6	69.5	0.628	37
5.	25	125.9	76	0.603	40	115.1	65.6	0.569	43	109.4	57	0.521	45
6.	30	112.6	58.3	0.517	48	98.3	51.5	0.523	48	112.8	50.5	0.447	55
7.	35	106.8	42.7	0.399	60	107.9	42.8	0.396	60	103.3	38.2	0.369	63
8.	40	118.1	29	0.245	75	113.7	31.2	0.274	73	95.1	26.9	0.282	72
9.	45	109.7	51.5	0.469	53	98.9	57.8	0.584	42	109.9	64.7	0.588	41
10.	50	95.2	67.7	0.711	29	109.5	77.9	0.711	29	94.7	72.1	0.761	24
11.	55	111.9	91	0.813	18.07	97.4	81	0.813	16.8	102.8	88	0.856	14.4

Where Co and C are the outlet and inlet concentration of SO_2, respectively

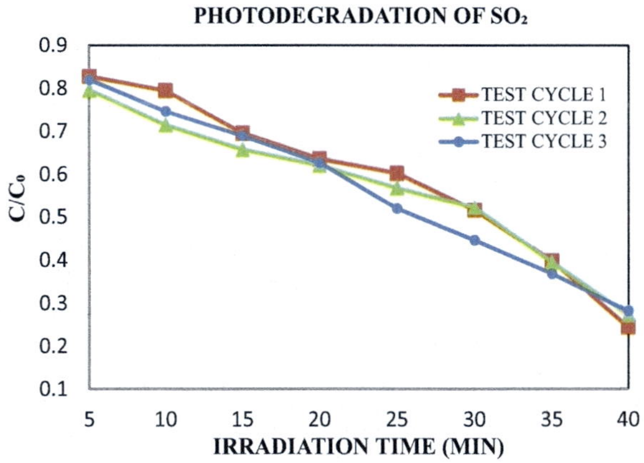

Fig. 3.5 Variation of the irradiation time pattern

of ZnO nanoparticles was found to be 178.1 nm. Zeta potential indicates incipient instability. The TGA-DTA curve of ZnO NPs remains stable after 800 °C with a total weight loss of 10.95%. The PL spectra of ZnO nanoparticles show that there is a structural imperfection or lattice disorder. FE-SEM analysis ZnO particles are spherical and evenly distributed without agglomeration. The EDS analysis of ZnO nanoparticles confirms the presence of Zn and O. During this studies there is no deposition of particulate matter on the quartz tube and the presence of other flue gases does not affect the performance of system. The maximum SO_2 removal efficiency achieved was 75% for ZnO NPs. The reusability of ZnO nanoparticles exhibits excellent stability after three test cycles.

References

Baig, N., Kammakakam, I., & Falath, W. (2021). Nanomaterials: A review of synthesis methods, properties, recent progress, and challenges. *Materials Advances, 2*(6), 1821–1871. https://doi.org/10.1039/D0MA00807A

Bikis, A. (2023). Urban air pollution and greenness in relation to public health. *Journal of Environmental and Public Health, 2023*(1), 8516622. https://doi.org/10.1155/2023/8516622

Escobedo, S., & de Lasa, H. (2020). Photocatalysis for air treatment processes: Current technologies and future applications for the removal of organic pollutants and viruses. *Catalysts, 10*(9), 966.

Klinbumrung, A., Panya, R., Pung-Ngama, A., Nasomjai, P., Saowalakmeka, J., & Sirirak, R. (2022). Green synthesis of ZnO nanoparticles by pineapple peel extract from various alkali sources. *Journal of Asian Ceramic Societies, 10*(4), 755–765. https://doi.org/10.1080/21870764.2022.2127504

Kowsari, E., & Abdpour, S. (2017). In-situ functionalization of mesoporous hexagonal ZnO synthesized in task specific ionic liquid as a photocatalyst for elimination of SO2, NOx, and CO. *Journal of Solid State Chemistry, 256*, 141–150. https://doi.org/10.1016/j.jssc.2017.08.038

Manisalidis, I., Stavropoulou, E., Stavropoulos, A., & Bezirtzoglou, E. (2020). Environmental and health impacts of air pollution: A review. *Frontiers in Public Health, 8*. https://doi.org/10.3389/fpubh.2020.00014

Meimand, M. M., Jafari, A. J., Nasiri, A., & Malakootian, M. (2020). Sulfur dioxide adsorption by iron oxide nanoparticles@clinoptilolite/HCl. *Journal of Air Pollution and Health, 5*(2), 107–120. https://doi.org/10.18502/japh.v5i2.4239

Mohamed, E. F. (2017). Nanotechnology: Future of environmental air pollution control. *Environmental Management and Sustainable Development, 6*(2), 429. https://doi.org/10.5296/emsd.v6i2.12047

Rangkooy, H. A., Jahani, F., Faroji, D. A., & Pour, M. N. (2021). Photocatalytic activity of synthetic ZnO/WO(3) nanocomposites immobilized on a Y-zeolite in removal of gas-phase styrene. *Journal of Environmental Health Science & Engineering, 19*(1), 181–191. https://doi.org/10.1007/s40201-020-00594-6

Saleem, H., Zaidi, S. J., Ismail, A. F., & Goh, P. S. (2022). Advances of nanomaterials for air pollution remediation and their impacts on the environment. *Chemosphere, 287*, 132083. https://doi.org/10.1016/j.chemosphere.2021.132083

Santika, R., & Rohmawati, L. (2022). Analysis of crystalline phase and functional groups of ZnO from pineapple peel extract. *Indonesian Physical Review, 5*(3 SE-Articles), 148–156. https://doi.org/10.29303/ipr.v5i3.160

Shah, T., Surendar, S., & Singh, S. (2023). Green synthesis of zinc oxide nanoparticles using Ananas comosus extract: Preparation, characterization, and antimicrobial efficacy. *Cureus, 15*(10), e47535. https://doi.org/10.7759/cureus.47535

Chapter 4
Urban Water Management and Climate Change Mitigation: A Review for Chennai City

Supriya P.

Abstract Climate change has turned up into a serious problem of current century and will intensify water-related issues in densely populated developed cities. The new pathway from every climate change model is for limiting the global warming to 1.5 °C level dependent on reductions in greenhouse gases (GHG). The GHG are increasing particularly in urban areas like Chennai city located in Southern part of India poses a serious climate change-related threats. This review addresses about the climate change mitigation measures for water conservation and management. The sustainable water conservation practices and management in Chennai city are analyzed to find the suitable mitigation techniques under climate change. The new sustainable development methods can be suggested by considering the following innovative technologies based on renewable energy sources, rainwater harvesting, underground water storages, wastewater reuse, two pipeline system for fresh and treated wastewater usage, climate resilient buildings, etc.

Keywords Water management · Climate change · Urban area · Sustainability · Green infrastructure · Environment

4.1 Introduction

As per the Intergovernmental Panel on Climate Change (IPCC, 2023) report, the hydrological extremes such as floods and drought episodes were increasing worldwide. Water availability is reducing due to sudden downpour of rainfall in a short time period and affects storage in reservoirs. For the safety concern toward impoundment structure, the floodwater is released downstream and remains unutilized. Thereby, it influences the early onset of drought conditions. Highly vulnerable global population approximately 3.6 billion experienced high mortality due to

Supriya P. (✉)
Centre for Water Resources, Department of Civil Engineering, Anna University, Chennai, Tamil Nadu, India

V. Baskaran et al. (eds.), *EcoTech Urbanism*, The Urban Book Series,
https://doi.org/10.1007/978-3-031-83833-0_4

drought, floods, and extreme storm events between 2010 and 2020. In November 2022, the global population touched eight billion. In that total population, 55% of them are living in urban areas and based on population projection, it can rise upto 70% in 2050. Water stress and scarcity are major problems all over the world. 2.4 billion people were living in water-stressed countries, reported as per the year 2020 (UN DESA, 2023).

The minimum water requirement 20 lpcd is mandatory for maintaining personal hygiene and living based on World Health Organization (WHO) report (WHO, 2011). The SDG 6 is clean water and sanitation and it recommends the five strategies for sustainable water management. They are enhancing sector-specific investment, capacity-building, action-based innovation, cross-sectoral networks, and stakeholders cooperation (UN DESA, 2023). As per the Sustainable Development Goal 11 (SDG), sustainable urban expansion depends upon climate resilience, housing with affordable cost, effective transportation, and accessible green spaces to all.

4.2 Study Area

Chennai metropolitan city is the major urban city area all over the world (Demographia, 2020). It is a city in Tamil Nadu, South India near the coast of Bay of Bengal located at 12° 50′ 49″ and 13° 17′ 24″ N latitude and 79° 59′ 53″ and 80° 20′ 12″ E longitude (Fig. 4.1). The length of the coast from north to south is approximately 45 km. The city has a flat geographical terrain with an average elevation of 6.7 m. It is a highly populated region in India with a population density of 26,553 people per km^2 as per the Census 2024. The total current population in Chennai is around 12,053,697 people. In comparison with last year 2023, there is a 2.36% increase rate is reported (Census 2024). There is an exponential growth of population every year is due to industrialization. Chennai city's built-up area is increased from 102 km^2 to 295 km^2 based on the land-use land cover (LULC) analysis of the years 1989 to 2019 (Hemalatha et al., 2024).

The Krishna river, Araniyar, and Kosasthalaiyar are supplying water to the reservoirs. The reservoirs such as Poondi, Cholavaram, Chembarambakkam, and Redhills are the water sources to the city. The total capacity of reservoirs is around 1967 mm^3 is detailed in the Table 4.1. Veeranam and Kandaleru water supply projects is also helping in water supply to the urban area. Water is also distributed from Nemmeli, Perur, and Minjur desalination plants. The city's precipitation is replenishing the groundwater resources situated in Panjetti, Ponneri, and Minjur area. There are 3600 tanks available in and around the Chennai city belongs to Thiruvallur and Kanchipuram districts can be networked and used for city water needs. The treated wastewater from Koyambedu and Kodungaiyur sewage treatment plants is supplied to 691 industries replaced the domestic water supply of 25 MLD. The water sources for the Chennai city is shown in Fig. 4.2.

The Chennai city has a wet and dry trophical climate. The rainfall occurs in the Northeast monsoon (NEM) season during October to December months (Jeganathan

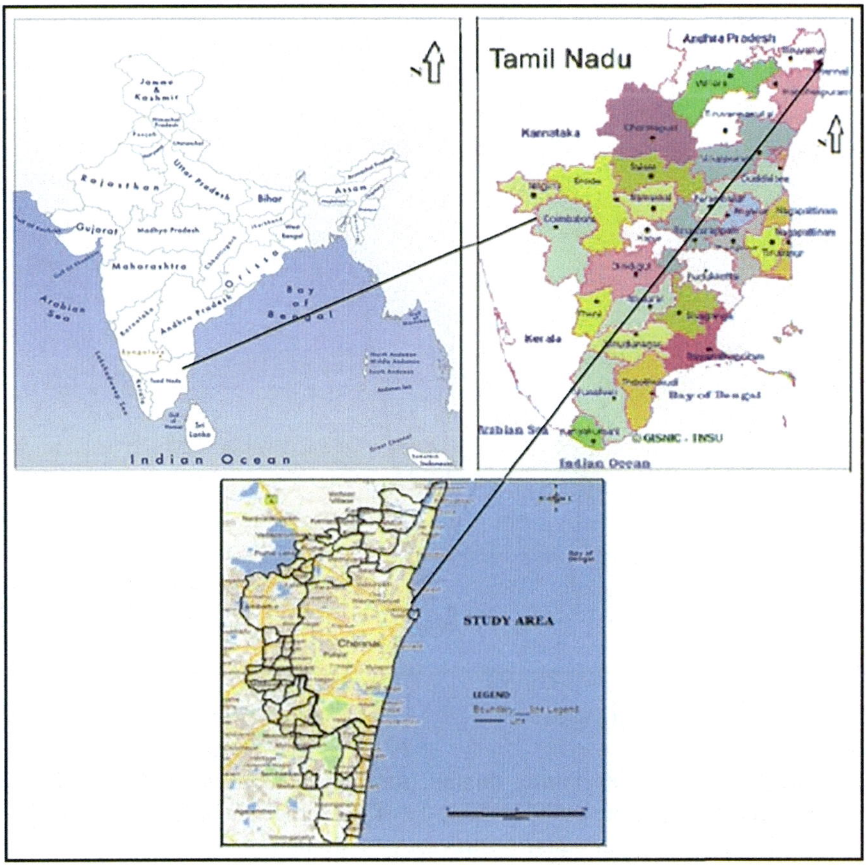

Fig. 4.1 Map showing expanded Chennai city

Table 4.1 Technical features of sources of supply for Chennai city

Reservoir	Total capacity (in mm^3)
Poondi	92
Cholavaram	25
Redhills	93
Chembarambakkam	103
Total	313
Veeranam	41
Kandaleru	1926
Total	1967

Source: WRD

et al., 2021). The annual average rainfall is 1276 mm. The cyclonic precipitation is more common and it causes flooding. The recent flood inundation happened during the NEM season in the year 2023. The weather averages for the meteorological

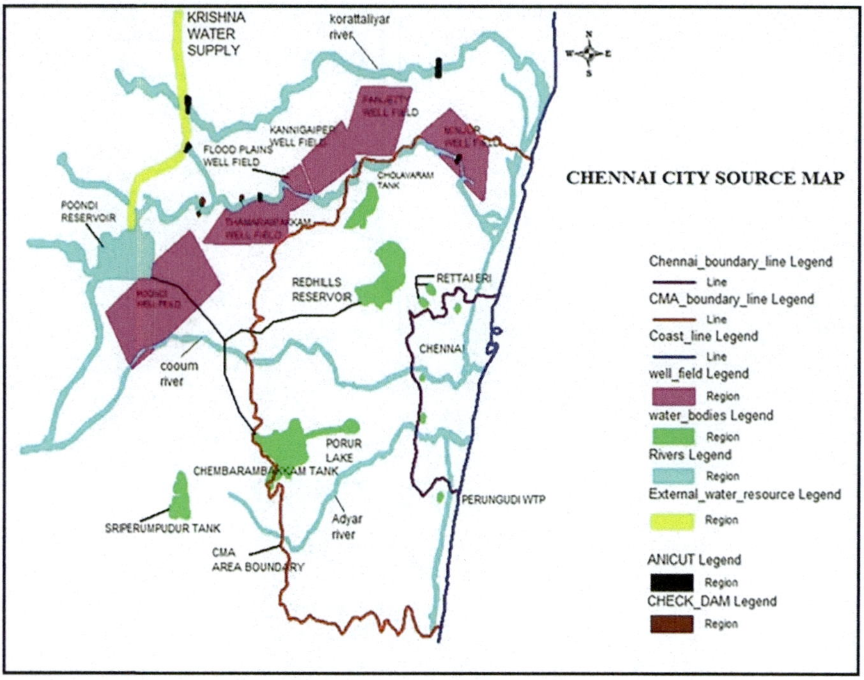

Fig. 4.2 Chennai city water source map

parameters such as temperature, rainfall, sunshine hours, relative humidity, and wind speed is cleary represented in Table 4.2. The average maximum temperature is around 35–40 °C and minimum temperature is around 19–25 °C. The maximum average values of sunshine hours, relative humidity, and wind speed is 9.8 h/day, 80%, and 306.435 km/h, respectively.

4.2.1 Climate Change in Chennai City

The Chennai city's temperature is rapidly increased to 1.2 °C over the 60 years (1951–2010) (Jeganathan & Andimuthu, 2013). Based on the longterm (1813–2009) investigation of rainfall, there is a significant increasing trend noticed in annual and monsoonal timescales (Ramachandran & Anushiya, 2015). The hydrological extremes such as floods and droughts were frequently occurring due to change in climate. The frequent flood events occurred in Chennai city in the years are 2006, 2007, 2008, 2010, 2015, and 2023. The future maximum temperature is simulated by a regional climate model PRECIS reported that 1.0, 2.2, and 3.1 °C for the periods the early 2020s, mid-century 2050s, and end-century 2080s, respectively, based on observed data period 1970–2000 (Bal et al., 2016).

Table 4.2 Weather averages of Chennai, Tamil Nadu

Weather in Chennai, Tamil Nadu	Average maximum temperature in Chennai, Tamil Nadu, India (°C)	Average precipitation/ rainfall (mm)	Average sunlight hours/day	Relative humidity (%)	Average wind speed in Chennai, Tamil Nadu (km/h)
January	29	24	8.6	75	204.29
February	31	7	9.5	73	204.29
March	33	15	9.8	71	204.29
April	35	25	9.2	71	204.29
May	38	52	8.4	65	306.435
June	37	53	6.5	59	306.435
July	35	84	4.8	64	306.435
August	35	124	5.3	68	306.435
September	34	118	6.2	71	204.29
October	32	267	6.5	79	204.29
November	29	309	6.9	80	204.29
December	28	139	7.6	77	306.435

4.3 Sustainable Water Practices in Cities

The urban regions are the known as the forefront of high demand of water, Chennai city requires 1200 MLD to feed the total population. The sustainable practices of water to be implemented in order to reduce the stress on available water resources. The efficient water-use measures and green infrastructure are the most pronounced ways to manage water in cities. The water use optimization through the modern technologies is vital for the urban regions (Bouramdane, 2023)

Automated IoT-based water meters enable real-time monitoring of water usage, making it easy to detect and address any significant fluctuations in consumption (Mehta et al., 2019). The optimum flow water efficient plumbing fixtures can save 28% of water and rooftop rainwater harvesting sytem provide the water supply for longer periods (Nalina et al., 2023). The wastewater treatment plants can treat 60% of Chennai city's sewage if it is increased in its capacity. The two-pipeline system in buildings can save fresh water consumption when replacing the treated wastewater for the non-potable purposes. Approximately 486 MLD can be utilized for non-potable use such as toilet flushing, gardening, vehicle washing, and industries (Paul & Elango, 2018). The prediction-based modeling, leakage detection, and optimum network for distribution with the application of data analytics and artificial intelligence (AI) are useful in precise water management (Kamyab et al., 2023).

Water policies include modifications in building codes to enhance the water-saving technologies, incentives for incorporating water-use efficient equipment, changing the new standards for water use, and encouraging sustainable practices in different sectors (Pluchinotta et al., 2021). Green infrastructure is the innovative method of sustainable water management in cities. Bioswales infiltrates water through the vegetated medium filters pollutants and increases conveyance time of stormwater runoff. It is helpful in increasing recharge of groundwater and safety against stormwater flooding. The natural way of absorbing the rainwater through permeable infrastructure of pebbles, stones, and rocks on the ground is the concept of sponge city. This concept is supportive for the collection and storage of rainwater (Shah et al., 2021). Green roofs are the vegetation-covered roofs on buildings absorb rainwater useful in lowering energy consumption and runoff storage. The abandoned quarries near the city are used as a water-harvesting detention pond in the flooding times, and the saved water is used for the water scarce time. It is a novel method used for the sustainable water management for Chennai (RinishaKartheeshwari et al., 2024).

4.4 Urban Water Systems Under Changing Climate

The current urban water distribution system is facing tremendous pressure due to climate change. Due to climate change, the rainfall pattern, floods, droughts, water quality, and water availability are changing. This impact needs mitigation, adaptation, infrastructure development, and novel planning strategies (Feilberg & Mark, 2016). Irregular intense rainfall cause flooding in stormwater systems. Water scarcity arises by the extreme temperatures and which strain the water supply systems. Saltwater intrusion occurs due to the sea level rise in cities situated in coastal regions is an another threat of groundwater availability (Mitchell, 2007). The climate change resilience needs an integrated approach. Integrated water resources management (IWRM) recommends the coordinated approach between management of stormwater, water supply, ecology, and recycling of wastewater. Based on IWRM principles, water-use optimality, conservation practices, and adaptive structural measures should be combined (Ako et al., 2010).

The effective strategies to be framed by incorporating insights from all the stakeholders, which belong to regional communities, government organizations, and environmental divisions, are based on IWRM principles (Nwokediegwu et al., 2024). The climate change impacts reduce the resilience stormwater management. The penetrable pavements, bioswales, green roof, and green spaces provide additional resilience to stormwater management under climate change. Water-sensitive urban design (WSUD) combines optimized water landscaping, water treatment decentralization, and sustainable drainage methods to improve resilience (Kuller et al., 2017).

Supriya and Saravanan (2016) documented that the demand-supply gap of Chennai city increases at the rate of 35.6% in every 10 years. The demand, supply, and overall population relationship plot of Chennai show an exponential increase in the demand-supply gap in the future, based on baseline data, as clearly depicted in Fig. 4.3. The relationship plot reveals that the current way of water distribution is not be capable to solve the demand-supply gap problem in the future. The increasing trend of demand-supply gap is due to the limited water availability of water in impounding structures with the exponential growth of population. Hence, the study

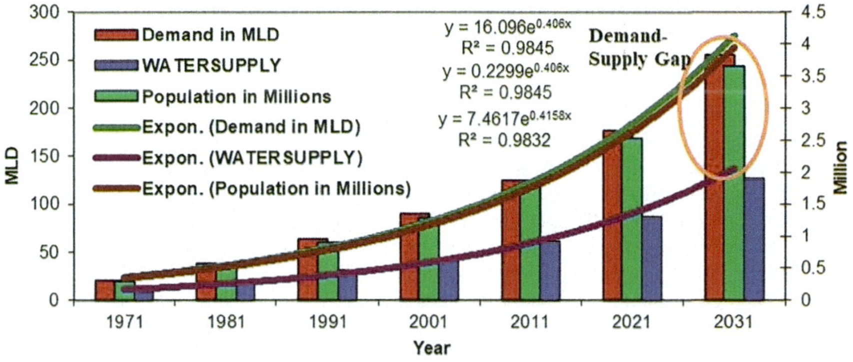

Fig. 4.3 Relation between demand, supply, and overall population plot

suggests a decentralized water distribution system from nearby water bodies to the Peri-urban areas for the sustainable water supply.

4.5 Conclusion

The urban water management in Chennai is critical at this moment and needs holistic interpretation of opportunities and threats that sets its developmental growth. This paper review the sustainable practices, demand-supply dynamics, IWRM-based policy innovations, green infrastructure solutions, and policy suggestions. The regulatory and infrastructure funding are the hurdles to the sustainable water management. The transformation of existing system can be attained with the technologies of advanced treatment, smart grids, nature-based water conservation solutions. Proactive and multidimensional methodology is the need of the hour for the urban water management in order to gain resilience against climate change. The integrated planning and green infrastructure facilities should be coupled with local community involvement, sustainable solutions, and climate change adaptation. The urban water distribution, harvesting systems, and water policies should be upgraded to ensure the sustainable water availability and quality based on future climate change conditions. The adaptable, equitable, efficient, sustainable, and climate change resilient urban water management system can safeguard the environment and human well-being.

References

Ako, A. A., Eyong, G. E. T., & Nkeng, G. E. (2010). Water resources management and integrated water resources management (IWRM) in Cameroon. *Water Resources Management, 24*, 871–888.

Bal, P. K., Ramachandran, A., Geetha, R., Bhaskaran, B., Thirumurugan, P., Indumathi, J., & Jayanthi, N. (2016). Climate change projections for Tamil Nadu, India: Deriving high-resolution climate data by a downscaling approach using PRECIS. *Theoretical and Applied Climatology, 123*, 523–535.

Bouramdane, A. A. (2023). Optimal water management strategies: Paving the way for sustainability in smart cities. *Smart Cities, 6*(5), 2849–2882.

Demographia. (2020). *World urban areas: 16th annual edition* (pp. 1–94). Demographia.

Feilberg, M., & Mark, O. (2016). Integrated Urban Water Management: Improve Efficient Water Management and Climate Change Resilience in Cities. In: Younos, T., Parece, T. (eds) Sustainable Water Management in Urban Environments. *The Handbook of Environmental Chemistry, 47*. Springer, Cham. https://doi.org/10.1007/978-3-319-29337-0_1.

Hemalatha, K., Kamali, R., Pushpalatha, R., Saranya, M., & Vijayalakshmi, P. (2024). Utilizing aerial photography to trace the development of urbanization in the Chennai District. *International Research Journal on Advanced Engineering Hub (IRJAEH), 2*(04), 988–993.

IPCC. (2023). Summary for Policymakers. In Core Writing Team, H. Lee, & J. Romero (Eds.), *Climate Change 2023: Synthesis Report. Contribution of Working Groups I, II and III to the Sixth Assessment Report of the Intergovernmental Panel on Climate Change* (pp. 1–34). IPCC. https://doi.org/10.59327/IPCC/AR6-9789291691647.001

Jeganathan, A., & Andimuthu, R. (2013). Temperature trends of Chennai city, India. *Theoretical and Applied Climatology, 111*(3–4), 417–425.

Jeganathan, A., Andimuthu, R., & Kandasamy, P. (2021). Challenges in Chennai city to cope with changing climate. *European Journal of Climate Change, 03*(01), 33–43.

Kamyab, H., Khademi, T., Chelliapan, S., SaberiKamarposhti, M., Rezania, S., Yusuf, M., et al. (2023). The latest innovative avenues for the utilization of artificial Intelligence and big data analytics in water resource management. *Results in Engineering, 20*, 101566.

Kuller, M., Bach, P. M., Ramirez-Lovering, D., & Deletic, A. (2017). Framing water sensitive urban design as part of the urban form: A critical review of tools for best planning practice. *Environmental Modelling & Software, 96*, 265–282.

Mehta, M., Misra, R., Chaugule, A., & Gupta, J. (2019). Automated water metering system. *International Journal of Advanced Research in Computer Science, 10*(2), 49.

Mitchell, V. G. (2007, November). Impacts of climate change on urban water systems. In *Cutting Edge Science Symposium* (p. 20).

Nalina, M. M., Poornima, B. H., Rashma, S., Deepika, R., & Basavantaraya. (2023, July). Quantification of water efficiencies in residential buildings through a case study. In *International conference on interdisciplinary approaches in civil engineering for sustainable development* (pp. 387–398). Springer Nature Singapore.

Nwokediegwu, Z. Q. S., Ugwuanyi, E. D., Dada, M. A., Majemite, M. T., & Obaigbena, A. (2024). Urban water management: A review of sustainable practices in the USA. *Engineering Science & Technology Journal, 5*(2), 517–530.

Paul, N., & Elango, L. (2018). Predicting future water supply-demand gap with a new reservoir, desalination plant and waste water reuse by water evaluation and planning model for Chennai megacity, India. *Groundwater for Sustainable Development, 7*, 8–19.

Pluchinotta, I., Pagano, A., Vilcan, T., Ahilan, S., Kapetas, L., Maskrey, S., Krivtsov, V., Thorne, C., & O'Donnell, E. (2021). A participatory system dynamics model to investigate sustainable urban water management in Ebbsfleet Garden City. *Sustainable Cities and Society, 67*, 102709.

Ramachandran, A., & Anushiya, J. (2015). Long-term rainfall trends of Indian urban station and its variation in different phases and seasons. *International Journal of Global Warming, 7*(3), 307–321.

RinishaKartheeshwari, M., Sivaraj, K., RaziSadath, P. V., & Elango, L. (2024). Innovative water management using abandoned quarries for urban water supply and flood mitigation. *Environment, Development and Sustainability, 26*, 1–19.

Shah, V. M., Suhane, S. & Gajjar, D. (2021). Green Infrastructure Entities-A Study of Indian and International Cities, PREPRINT (Version 1) available at Research Square. https://doi.org/10.21203/rs.3.rs-1104643/v1.

Supriya, P., & Saravanan, R. (2016). Optimal allocation of surface water resources for peri- urban area of Chennai to minimize the demand-supply gap. *Journal of Applied Hydrology, 29*(1 to 4), 86–93.

UN DESA. (2023). The Sustainable Development Goals Report 2023: New York, USA: UN DESA. https://unstats.un.org/sdgs/report/2023/. (Accessed date 9 July, 2024).

WHO. (2011). *Guidelines for drinking-water quality* (4th ed.). World Health Organization. http://www.who.int/water_sanitation_health/publications/2011/dwq_chapters/en/index.html

Chapter 5
Incorporating the Dynamics of Surface Water and Groundwater Interaction in Urban Ecotechnology

J. Eunice and S. Chandran

Abstract The connection between surface water and groundwater is a crucial part in the water cycle. Urbanization presents challenges to manage water sustainably, especially balancing the interactions between surface water and groundwater. To comprehend the groundwater flow and contaminant movement better, it is vital to examine the different processes of surface and subsurface flow that shape these interactions. Conventional urban growth often disrupts natural water cycles, leading to more floods, water scarcity, and decreased water quality. Consequently, there is a growing interest in incorporating ecotechnological solutions that imitate natural processes for enhancing urban water resilience. This chapter delves into how surface water and groundwater dynamics can be integrated into urban ecotechnology strategies. Successful management of these interactions requires a comprehensive approach that accounts for hydrogeological conditions, land-use patterns, and urban planning principles. Utilizing advanced modeling methods and monitoring systems are crucial for evaluating the effects of ecotechnologies on water quantity and quality. By promoting the synchronization of surface water and groundwater dynamics, urban ecotechnology not only boosts water resource sustainability but also aids in improving ecosystem health and urban livability. Ongoing research and innovation in this realm are vital for devising strong solutions to tackle the intricate challenges posed by urbanization, fostering the environmental stewardship and enhancing community well-being.

Keywords Urbanization · Groundwater dynamics · Surface water and groundwater interactions · Surface water and groundwater dynamics · Ecotechnology · Sustainability

---

J. Eunice (✉) · S. Chandran
Department of Civil Engineering, Thiagarajar College of Engineering, Madurai, Tamil Nadu, India
e-mail: jeeciv@tce.edu

V. Baskaran et al. (eds.), *EcoTech Urbanism*, The Urban Book Series,
https://doi.org/10.1007/978-3-031-83833-0_5

5.1 Introduction

The dynamics of groundwater-surface water interactions in urban streams are crucial for understanding and managing urban ecotechnology. Urbanization and human activities can significantly alter the hydrological regime and channel morphology of urban streams, leading to flashier hydrographs and reduced bedform-driven water flow in the streambed. This can impact the downward transport of oxygen and the associated ecological processes in the hyporheic zone (Musolff et al., 2010). Studies have shown that the magnitude and direction of water fluxes through the streambed in urban streams can vary significantly over both short-term and long-term timescales, due to dynamic changes in hydraulic gradients between the stream, streambed, and shallow aquifer (Musolff et al., 2010; Farokhzad & Ketabchi, 2022). These variable water fluxes can influence redox conditions and water quality in the streambed, with downward flow generally increasing redox potential. However, the complex interactions make it difficult to directly attribute short-term changes in redox to hydraulic conditions (Musolff et al., 2010). Integrating the monitoring of hydraulic gradients, streambed temperature, and water quality parameters is important for understanding the spatial and temporal variability of groundwater-surface water exchange in urban streams. This can help compensate for the negative impacts of degraded urban stream channels (Musolff et al., 2010; Farokhzad & Ketabchi, 2022).

The management of urban water resources has long been a crucial challenge for cities, as they require a reliable supply of clean drinking water along with the need to treat the contaminated groundwater and wastewater and manage stormwater (La Vigna, 2022). The integration of surface water and groundwater dynamics is particularly important for urban ecotechnology, which seeks to leverage natural processes and systems to address environmental challenges in cities (Musolff et al., 2010; Farokhzad & Ketabchi, 2022).

Ecotechnological solutions, such as constructed wetlands, green roofs, and permeable pavements, can help regulate water flows, filter pollutants, and enhance water-related ecosystem services. For example, understanding groundwater-surface water interactions can help optimize the placement and sizing of constructed wetlands, ensuring adequate water supply and effective contaminant removal. However, the successful implementation of these nature-based solutions requires a nuanced understanding of how surface water and groundwater interact in the urban context (Nieuwenhuis et al., 2021). Urban areas often have complex hydrological conditions, with impervious surfaces, underground infrastructure, and altered groundwater flows. Incorporating these dynamics into the design and management of ecotechnological systems is critical to ensure their long-term functionality and resilience (Nieuwenhuis et al., 2021).

5.2 Impacts of Urbanization on Stream Hydrology and Morphology

The continuous expansion of urban areas can support a growing population that disrupts the natural processes like evaporation, infiltration, and runoff, impacting the urban hydrological cycle. Such process can significantly change the stream hydrology, leading to more intense flows, lower baseflow, shifts in sediment movement, and changes in stream structure. Moreover, urban runoff introduces harmful pollutants into surface and groundwater, raising concerns for human health and the environment. It is vital to comprehend and address these effects to properly manage the urban streams. Urbanization has notable effects on streamlogy and morphology, causing a change in water flow patterns and channel characteristics. The key impacts of urbanization on stream hydrology and morphology are labeled in Fig. 5.1.

Urbanization leads to more impervious surfaces like roads, parking lots and rooftops, decreasing infiltration, and increasing surface runoff. This heightened runoff can result in greater streamflow volumes and more frequent peakflows. Urban streams tend to display greater variability in stream stage and discharge compared to rural streams, with more erratic behavior and an increased occurrence of extreme flow events. Research demonstrates that urban channels tend to broaden due to increased peakflows and decreased sediment inputs. Alterations in land cover from urbanization can impact flowpaths and sediment flow to streams, influencing

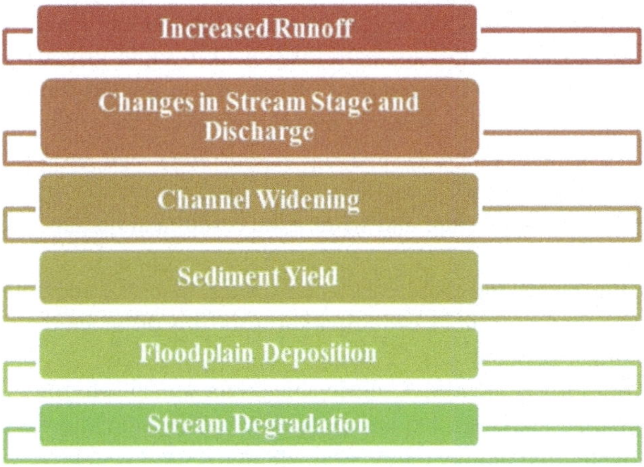

Fig. 5.1 Impacts of urbanization on stream hydrology and morphology

channel structure. Urban development can change sediment yield, with initial growth when bare surfaces are exposed followed by a decrease as urban regions mature. Urbanization can also impact floodplain deposition, evident in studies showing changes in channel shape and floodplain dynamics over time in urbanizing watersheds.

To effectively manage the negative impacts of urbanization on stream hydrology and morphology, it is essential to grasp these effects. Future studies should concentrate on understanding response variations, creating comprehensive water budgets, evaluating the combined effects of land use and climate change and employing interdisciplinary methods to explore urban systems. Growing watershed imperviousness from urbanization has been tied to stream degradation, often known as the urban stream syndrome (O'Driscoll et al., 2010).

Urbanization can result in more impervious surfaces like roads, buildings, and parking lots, decreasing infiltration and increasing surface runoff during storms. This rapid surge in surface runoff can overwhelm natural drainage systems, resulting in flash floods and stream bank erosion. Junior et al. (2010) examined how urbanization affects stream hydrology and morphology, highlighting how changes in land use and surface features because of urban growth can alter natural flow patterns and stream physical characteristics.

Moreover, redirecting urban runoff can modify the upslope area contributing to streams, changing flow patterns and sediment transport within the streams. These hydrological shifts can lead to increased erosion and sediment accumulation in streams, influencing their shape and stability.

By analyzing detailed digital elevation models prior to and after urbanization, researchers were able to pinpoint areas prone to gully erosion and assess erosion potential in urbanized regions. The site encountered with severe gullying due to alterations in drainage patterns linked to rapid urbanization over the past three decades. The increased flow intensity of drainage can result in urban expansion with worsened erosion in an expanding urban settings. The significant influence of urbanization on stream hydrology and morphology underscores the importance of understanding and mitigating these impacts to safeguard urban infrastructure and sustain the ecological balance of streams and watersheds.

To tackle the underlying causes, several challenges must be addressed: managing excess stormwater runoff, increasing riparian space, restoring altered sediment supplies, mitigating legacy impacts on streams from previous land use, and overcoming social and institutional barriers. A sustainable urban stream solution can be realized by addressing the issues depicted in Fig. 5.2, thereby reducing the impact of urbanization on stream morphology (Vietz et al., 2016).

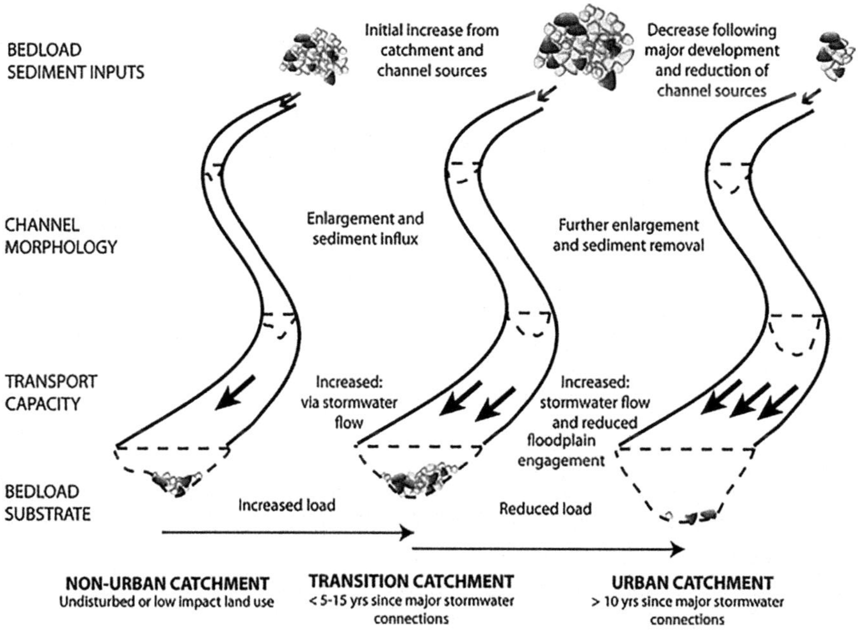

Fig. 5.2 Impacts of urbanization on stream channel. (Source: Vietz et al. (2016))

5.3 Spatial and Temporal Variability of Groundwater-Surface Water Exchange

The dynamics of surface water and groundwater interaction are not static, but rather, are subject to a range of temporal and spatial variabilities. Understanding of the complex interactions between groundwater and surface water systems is essential for the effective management and protection of water resources. Groundwater-surface water exchange is a complex and dynamic process that influences the hydrology, biogeochemistry, and ecology of aquatic environments. Understanding these dynamics is essential for sustainable water management practices and ensuring the long-term health of aquatic ecosystems. Monitoring the spatial and temporal variability of groundwater-surface water exchange is crucial for relating hydrological processes and water resource management. The interaction between surface water and groundwater highlights the influence of factors such as evaporation, human activities, and water-rock interactions on hydrochemical evolution. By integrating field observations, modeling techniques, and interdisciplinary approaches, scientists can continue to unravel the complexities of groundwater-surface water exchange and its significance in the broader context of water resources and environmental conservation. Isotopic analysis, geochemical modeling, and statistical approaches investigate the dynamics of water exchange processes and their implications for water quality and ecosystem health. By incorporating techniques such as hydrochemical diagrams, isotopic analysis, and geostatistical analysis, researchers can effectively monitor and analyze the spatiotemporal variability of groundwater-surface water exchange.

The spatial and temporal variability of groundwater-surface water exchange is a critical component in understanding the hydrological cycle and the transport of nutrients, contaminants, and other solutes between these interconnected systems. The factors that influence the exchange processes include topography, geology, climate, and anthropogenic activities. These factors contribute to the spatiotemporal dynamics of the groundwater-surface water interface. Comprehensive field measurements, numerical modeling, and statistical analyses will be employed to quantify the magnitude, direction, and patterns of groundwater-surface water exchange across different spatial and temporal scales. These findings will have significant implications for water resource management, ecosystem health, and the development of effective strategies for the protection and sustainable utilization of groundwater and surface water resources.

The spatial variability of groundwater-surface water exchange is influenced by a multitude of factors, including the local topography, the geology and hydrogeological properties of the subsurface, the morphology of the surface water bodies, and the presence of any anthropogenic structures or activities that may alter the natural flow paths and exchange dynamics. Surface water-groundwater interactions are dynamic and have varied over time, with many reaches shifting from gaining to losing water throughout the analysis period. This highlights a deficiency in the hydrological models, which cannot replicate these changes. Such flux direction changes can result in errors when predicting low flows (Crosbie et al., 2023).

5.4 Surface Water and Groundwater Integration Through Models

Integrating surface water and groundwater processes through models, researchers, and water resource managers can gain valuable insights into the complex interactions between these systems and develop effective strategies for sustainable water management (Uwamahoro et al., 2024).

The multivariate analysis method, incorporating cluster analysis and stable isotope mass balance approaches, has been employed to quantify the characteristics, processes, and evolution of groundwater-surface water interactions at a large reservoir over different periods. The influence of external factors on the migration and distribution of hydrochemical and isotopic components in groundwater was qualitatively analyzed (Ziwen et al., 2024). The surface and groundwater interactions including their dynamics after stream restoration, multiscale exchange assessed through heat fluxes and their distribution and impacts in urban areas. It also explores the effect of climate change on hydrology beneath tropical glaciers and identifies sources of basin brines affecting drinking water well (Fig. 5.3).

Integrating surface water and groundwater processes through models involves developing numerical simulations that represent the interactions between these two interconnected systems. This integration is crucial for understanding the dynamics of water flow, solute transport, and energy exchange between surface water bodies and groundwater aquifers. Generic integration of surface water and groundwater processes through model is shown in Fig. 5.4.

The first step in integrating surface water and groundwater processes is to conceptualize the hydrological system. This involves identifying the key components, such as rivers, lakes, aquifers, and their interactions, including recharge, discharge, and flow paths between surface water and groundwater. Researchers often use integrated hydrological models that can simulate both surface water and groundwater processes. Models like MODFLOW (groundwater flow model) and SWAT (Soil and Water Assessment Tool) are commonly used for this purpose. These models allow for the representation of complex interactions and feedback mechanisms between surface water and groundwater. Integration is achieved by coupling surface water and groundwater models to simulate the exchange of water, solutes, and energy between the two systems. This coupling involves sharing boundary conditions, fluxes, and hydraulic properties between the surface water and groundwater components of the model. The integrated model is calibrated and validated using observed data from field measurements, remote sensing, and monitoring networks. This step ensures that the model accurately represents the dynamics of surface water-groundwater interactions under different hydrological conditions. Once the

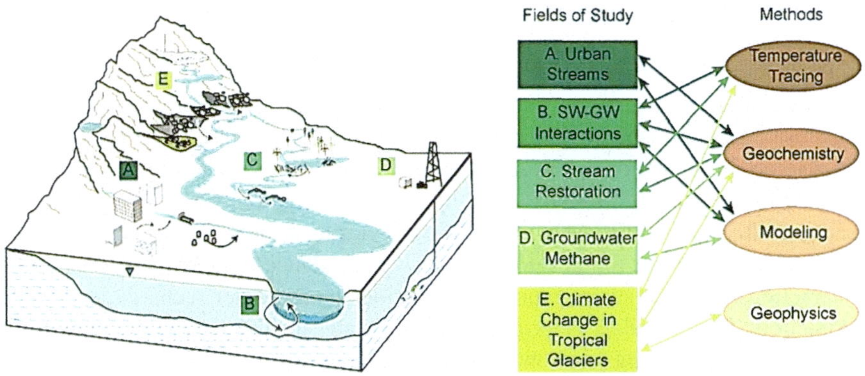

Fig. 5.3 Conceptual model used in the career of Laura Lautz. (Source: Ledford et al. 2022)

Fig. 5.4 Integrating surface water and groundwater processes through models

integrated model is calibrated and validated, researchers can use it to analyze different scenarios and assess the impacts of changes in land use, climate, or water management practices on surface water-groundwater interactions. Scenario analysis helps in understanding the sensitivity of the system to various drivers and stressors. Integrated surface water-groundwater models provide valuable insights for decision-making in water resources management, land-use planning, and environmental protection. These models can help stakeholders evaluate the potential consequences of different management strategies and inform sustainable water resource policies. By integrating surface water and groundwater processes through models, researchers can gain a comprehensive understanding of the complex interactions between these two systems. This integrated approach is essential for addressing water resource challenges, such as water availability, quality, and ecosystem sustainability, in a holistic manner.

The importance of integrating surface water and groundwater through model is to assess the impacts of urbanization on catchment hydrology and morphology. By simulating the complex interactions between surface water and groundwater systems, researchers can better understand the effects of land-use changes and urban development on water resources and ecosystem health. Surface water and groundwater integration through models plays a crucial role in understanding the dynamics of water availability, usage, and sustainability in urban areas (Junior et al., 2010). Palma et al. (2022) discussed the development of a unified hydrogeological conceptual model for the Mexico Basin aquifer has been developed after a century of groundwater exploitation. This model integrates the various aspects of the aquifer's behavior and responses to long-term extraction, providing a comprehensive understanding of its current state and guiding sustainable management practices. This model helps in understanding the interactions between surface water and groundwater systems, including the recharge mechanisms, flow paths, and storage capacities of aquifers.

By integrating surface water and groundwater models, vulnerability assessments can be conducted to evaluate the impact of water extraction, climate change, and land use on water resources (Martinez et al., 2015). The lithology having high conductive aquifers material is more vulnerable and septic system failures have also been reported, leading to increased coliform contamination, which can potentially cause health issues (Mondal et al., 2024). Integrating surface water and groundwater models allows for the development of water balance models that consider the interactions between surface runoff, infiltration, groundwater recharge, and water extraction. These models help in quantifying water availability, identifying water deficits, and optimizing water management strategies.

Surface water and groundwater integration models enable scenario analysis to evaluate the effects of different management strategies on water resources. Models can be used to assess the impact of rainwater harvesting systems (RWHS) on surface water and groundwater resources in urban areas. By simulating the implementation of RWHS in conjunction with existing water systems, these models can evaluate the potential benefits in terms of aquifer recharge, reduced surface water extraction, and improved water sustainability. By simulating scenarios such as increased rainwater harvesting, reduced groundwater pumping, or changes in land use, decision-makers can assess the implications for water availability and quality. Models that integrate surface water and groundwater dynamics provide valuable insights for policy development and water resource management planning. These models can inform decision-making processes by predicting the long-term impacts of urban development, climate change, and water infrastructure projects on water resources (Castelán-Cabañas et al., 2024).

5.5 Implications for Urban Ecotechnology and Water Resource Management

Water management is a crucial aspect of urban planning and urbanization. It significantly affects the environment and the population's standard of living. In developed countries, the importance of addressing the relationship between urbanized areas and water management is now widely recognized (Teichmann et al., 2020). By considering the impact of meteorological conditions on groundwater levels, urban ecotechnology solutions can be designed to optimize the water-use efficiency and reduce reliance on external water sources. Implementing green infrastructure, including rain gardens, green roofs, and permeable pavements, can aid in recharging groundwater and mitigating the effects of droughts on urban ecosystems. These findings can inform water resource management strategies in regions with similar climatic conditions by highlighting the importance of long-term monitoring of groundwater levels. By identifying trends in groundwater levels and understanding the factors influencing fluctuations, water resource managers can develop more effective policies for sustainable groundwater use. Integrating climate change projections into water resource management plans can help anticipate future challenges related to groundwater availability and develop adaptive strategies to ensure water security for urban populations (Dubois & Larocque, 2024).

The concept of ecohydrology explores the intricate connections between water dynamics and living organisms within entire watersheds, aiming to foster sustainable water management. This approach operates on three fundamental principles as shown in Fig. 5.5 (Zalewski & Wagner, 2014).

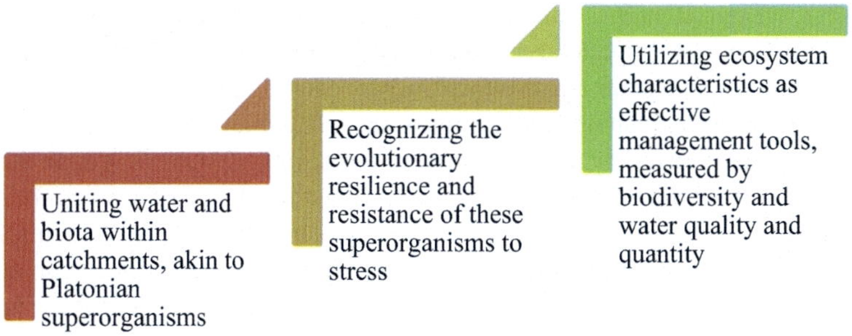

Fig. 5.5 Fundamental principles to achieve the concept of ecohydology

Rainwater harvesting system is one best water resource management option commonly followed over nations. Implementing rainwater harvesting systems can contribute to the sustainability and resilience of urban water systems by reducing reliance on centralized water sources. It provides partial or total water autonomy for users, especially in areas facing water scarcity or quality issues. Rainwater harvesting systems offer environmental benefits such as reducing greenhouse gas emissions, mitigating stormwater runoff, and recharging aquifers. These systems can help in managing urban flooding and promoting a cleaner environmental system. Rainwater harvesting systems promote water awareness among citizens and can contribute to social equity by providing decentralized access to water resources. Understanding the social implications of ecotechnologies like RWHS is crucial for community wellbeing and sustainable development. By diversifying water sources and reducing demand on centralized systems, rainwater harvesting systems enhance water security in urban areas. This is particularly important in regions facing water scarcity and supply challenges. Integrating rainwater harvesting with other water management strategies such as wastewater treatment, infiltration wells, and water culture practices can create a more holistic approach to urban water resource management. This integrated approach is essential for addressing the complex water challenges faced by growing cities. The adoption of rainwater harvesting systems in urban areas can lead to multiple benefits for ecotechnology and water resource management, including sustainability, resilience, environmental protection, social equity, and enhanced water security. By considering the holistic implications of these systems, cities can move toward more sustainable and efficient water management practices (Castelán-Cabañas et al., 2024).

5.6 Conclusion

In urban ecotechnology, it is crucial to consider how surface water and groundwater creates a sustainable and resilient urban environment. Embracing the connection between surface water and groundwater in urban ecotechnology frameworks leads to smarter, more adaptable urban landscapes that can withstand environmental challenges and support sustainable urban development. The intricate relationship between these two enables the development of creative solutions that improve management, prevent flooding, and enhance water quality. Utilizing urban ecotechnological methods, we can effectively utilize these interactions to replenish, decrease surface runoff, and promote healthier urban ecosystems. By integrating these dynamics into urban planning and design, cities can achieve a better ecological balance, increase biodiversity, and secure long-term water resources. While the integration of surface water and groundwater systems has long been recognized as a critical component of sustainable urban development, the complex dynamics governing these interconnected processes have historically posed significant challenges for urban ecotechnology practitioners. However, recent advancements in both theoretical and empirical research have begun to shed new light on the intricate relationships between surface and subsurface hydrological processes, offering valuable insights that can inform the design and implementation of more effective urban ecotechnological solutions.

References

Castelán-Cabañas, R., Padilla-Rivera, A., Muñoz-Villarreal, C., & Güereca-Hernández, L. P. (2024). Advancing urban water autonomy: A social life cycle assessment of rainwater harvesting systems in Mexico City. *Cleaner Environmental Systems, 13*, 100193.

Crosbie, R., Wang, B., Kim, S., Mateo, C., & Vaze, J. (2023). Changes in the surface water–groundwater interactions of the Murray-Darling Basin (Australia) over the past half a century. *Journal of Hydrology, 622*, 129683.

Dubois, E., & Larocque, M. (2024). Contribution of standardized indexes to understand groundwater level fluctuations in response to meteorological conditions in cold and humid climates. *Journal of Hydrology, 634*, 131105.

Farokhzad, P., & Ketabchi, H. (2022). A review of groundwater and surface water interaction using integrated methods. *Water and Energy International, 64*(10), 24–29.

Junior, O. C., Guimaraes, R., Freitas, L., Gomes-Loebmann, D., Gomes, R. A., Martins, E., & Montgomery, D. R. (2010). Urbanization impacts upon catchment hydrology and gully development using mutli-temporal digital elevation data analysis. *Earth Surface Processes and Landforms: The Journal of the British Geomorphological Research Group, 35*(5), 611–617.

La Vigna, F. (2022). Urban groundwater issues and resource management, and their roles in the resilience of cities. *Hydrogeology Journal, 30*(6), 1657–1683.

Ledford, S. H., Briggs, M., Glas, R., & Zimmer, M. A. (2022). Connecting diverse disciplines to improve understanding of surface water-groundwater interactions. *Journal of Hydrology X, 17*, 100141.

Martinez, S., Kralisch, S., Escolero, O., & Perevochtchikova, M. (2015). Vulnerability of Mexico City's water supply sources in the context of climate change. *Journal of Water and Climate Change, 6*(3), 518–533.

Mondal, M., Mukherjee, A., Kumar, P., Ismaeel, N. M., & Das, K. (2024). *A process-based impact of tropical cyclone and hurricane on surface water-groundwater interaction process and contaminant mobilization of coastal aquifers: A semi analytical review* (p. 100318). Progress in Disaster Science.

Musolff, A., Schmidt, C., & Fleckenstein, J. H. (2010, December). Dynamics of groundwater-surface water interactions in urban streams. In *AGU fall meeting abstracts* (Vol. 2010, pp. H33J-08).

Nieuwenhuis, E., Cuppen, E., Langeveld, J., & de Bruijn, H. (2021). Towards the integrated management of urban water systems: Conceptualizing integration and its uncertainties. *Journal of Cleaner Production, 280*, 124977.

O'Driscoll, M., Clinton, S., Jefferson, A., Manda, A., & McMillan, S. (2010). Urbanization effects on watershed hydrology and in-stream processes in the southern United States. *Water, 2*(3), 605–648.

Palma, A., Rivera, A., & Carmona, R. (2022). A unified hydrogeological conceptual model of the Mexico Basin aquifer after a century of groundwater exploitation. *Water, 14*(10), 1584.

Teichmann, M., Helakova, K., & Szeligova, N. (2020). Rainwater management in relation to water bodies and flows in urbanized areas. In *IOP conference series: Earth and environmental science* (Vol. 444, No. 1, p. 012053). IOP Publishing.

Uwamahoro, S., Liu, T., Nzabarinda, V., Frankl, A., Tuyishimire, E., Iradukunda, A., Ingabire, R., & Umugwaneza, A. (2024). *Investigation of groundwater–surface water interaction and land use and land cover change in the catchments, A case of Kivu Lake, DRC-Rwanda* (p. 101236). Groundwater for Sustainable Development.

Vietz, G. J., Rutherfurd, I. D., Fletcher, T. D., & Walsh, C. J. (2016). Thinking outside the channel: Challenges and opportunities for protection and restoration of stream morphology in urbanizing catchments. *Landscape and Urban Planning, 145*, 34–44.

Zalewski, M., & Wagner, I. (2014). Ecohydrology of urban aquatic ecosystems for healthy cities. In *Aquatic habitats in sustainable urban water management* (pp. 95–106). CRC Press.

Ziwen, Z., Zhifang, Z., & Vanapalli, S. K. (2024). A quantitative study of changing groundwater-surface water interactions of a large reservoir due to impoundment based on hydrochemistry and isotope characteristics. *Journal of Hydrology, 634*, 131092.

Chapter 6
Advancements in Urban Flood Modeling: A Comprehensive Review of Models and Capabilities

Vinay Ashok Rangari and Dhirajkumar Sukhwasi Lal

Abstract Major setbacks from urban flooding are experienced by cities worldwide, necessitating the development of sophisticated flood modeling techniques and models that allow for an understanding of the behavior and socioeconomic effects of floods. This chapter gives an overview of the process of urban flood modeling in a broader sense, including the flood deluge and risk mapping in urban expanses. Since the number of urban floods keeps on rising, the research puts emphasis on the need for accurate and detailed information about flood hazards to ensure early warnings and successful implementation of structural and nonstructural measures.

In this review, the authors scrutinized approximately 40 urban stormwater models. Based on their analyses, these models can range from less sophisticated conceptual models to more complex hydraulic models. Once the evaluation of these selected tools is complete, our attention then turns to modeling functionalities in an urban environment. Flood parameters of relevance are deluge extent and depth, which are noted to be measured by GIS-based hydraulic models. Each model's features and capabilities are summarized in a comprehensive table. It also demonstrates the difficulty of urban flood modeling with regard to human intervention, land-use change, changes in climate influenced by urbanization, and rainfall patterns. For its final part, this investigation has looked at the combined use of different modeling techniques needed for accurate results using available dataset.

Keywords Stormwater · Watershed · Urban flood · Modeling

V. A. Rangari (✉) · D. S. Lal
Pimpri Chinchwad College of Engineering, Pune, India
e-mail: vinay.rangari@pccoepune.org; dhirajkumar.lal@pccoepune.org

V. Baskaran et al. (eds.), *EcoTech Urbanism*, The Urban Book Series,
https://doi.org/10.1007/978-3-031-83833-0_6

6.1 Introduction

In the context of the EcoTech Urbanism: Pioneering Sustainable Technologies for Developing Cities, the need for advanced urban flood modeling becomes increasingly evident. Sustainable urban planning is essential as cities all over the world struggle with the twin issues of growing urbanization and intensifying environmental constraints. EcoTech Urbanism prioritizes the incorporation of state-of-the-art technologies to tackle pressing urban concerns such as pollution, resource depletion, and climate change. Advanced models have the potential to greatly improve urban resilience and sustainability by offering thorough insights into flood dynamics and associated repercussions. This review paper, Advancements in Urban Flood Modeling: A Comprehensive Review of Models and Capabilities, aims to bridge the gap between innovative urban technologies and practical applications in flood management, contributing to the book's vision of thriving, sustainable cities.

Riverine floods, which mostly affect rural and coastal communities, were the subject of previous studies on flood disasters. With climate change, there has been a rise in high-intensity rainfall occurring over shorter periods, which has led to more frequent urban flooding (Cavanaugh et al., 2015; Xu & Luo, 2015). Hydrologists from all around the world have been paying close attention to urban flooding lately. It is among the main issues that contemporary towns and cities must deal with. Major Indian cities have been hit by an increase in urban flood events in recent years, and this trend is expected to persist in the coming years (Gupta, 2020; Agilan & Umamahesh, 2016). Because these catastrophes are beyond our control, it is essential to devise plans to lessen their effects and reduce the number of lives and property lost. The challenge of urban floods stems from the unpredictability of flow conditions in urban settings, which is brought about by the quick changes in topography and the lack of large-scale raw datasets (Mondal & Muzumdar, 2017; Bayazit, 2015). As a result, simulating urban floods becomes difficult. In recent years, a large number of sophisticated, commercial numerical models that can map floods have surfaced. These models require large datasets in order to produce accurate findings.

The key focus of urban flood modeling is to understand the characteristics of flooding and its socioeconomic implications in urban environment. Planning floodplain management has become increasingly challenging in urban areas as construction activities in urban centers have nearly peaked (Ramos et al., 2017; Prasad, 2014). Analyzing the suitability of drainage networks in places susceptible to heavy precipitation events and creating hydrologic and hydraulic models that may provide maps of flood hazards are two ways to deal with such problems. Here, we present a brief review of urban flood simulation models is in the context of availability, cost-effectiveness, parameter requirement, and output deliverables. This chapter provides a concise overview of literature concerning the modeling of urban floods and the preparation of flood deluge and flood risk maps in urban environments. It briefly discusses models capable of simulating urban floods, along with a brief review of the functionality, features, and accessibility of several popular 1D and 2D models.

This review will aid new modelers in understanding the modeling requirements and generating meaningful results.

6.2 Chronicles of Model Development

Since 1960, computer models like the Stanford Watershed Model (SWMM) have been used to mimic the behavior of water-carrying networks (Crawford & Linsley, 1966). Early in the 1970s, U.S. government organizations like the U.S. Environment Protection Agency developed the models that could simulate stormwater quality and quantity. Numerous models that simulate the urban watershed have progressed since then. The majority of these one-dimensional models are based on the concepts of momentum, energy, and mass conservation. They consist of continuous simulation types as well as models that are event-based. As long as there is not any overflow from the network inlet or manhole, these urban drainage models can faithfully represent the drainage system. Nevertheless, 1D models find it difficult to replicate actual flood extents when overflows result from insufficient downstream drainage capacity (Chatterjee et al., 2008). To overcome the shortcomings of one-dimensional models, researchers focused on two-dimensional (2D) models and 1D-2D coupled flood simulation models, also known as distributed models. The availability of distributed data, including type of soil, land use, and radar rainfall, has made the creation of simple but physically significant distributed hydrological models easier (Todini, 2007). Technological and computational scientific advances have led to the development of several numerical models for modeling urban floods in recent years (Chatterjee et al., 2008; Todini, 2007; Zoppou, 2001)

Zoppou (2001) reviewed models that simulate stormwater quantity and quality in urban environments, selecting 12 models for detailed analysis based on accessibility, functionality, water quality, and temporal and spatial scales. The review, however, primarily focused on environmental quality aspects and overlooked two-dimensional modeling components. Over time, numerical modeling strategies have significantly progressed from basic 1D models to advanced 1D-2D coupled models, as detailed by Stelling and Verwey (2005). Mitchell et al. (2007) conducted a two-stage review of urban flood models to pinpoint insufficiencies in Integrated Urban Water Management (IUWM) practices, screening 65 models globally and narrowing them down to 7 based on functionality and 1D-2D modeling capabilities. The authors focused on identifying models that included uncertainty analysis and optimization features. Similarly, Elliott and Trowsdale (2007) related models used for low-impact urban development of stormwater to reducing the adverse effects of urbanization on quality of water and hydrology. Further advancements in technology and computational science have led to the advancement of models capable of performing 1D, 2D, or coupled (1D/2D) hydrodynamic simulations for various flooding scenarios. While 1D models simulate water flow in channels, 2D models address surface flooding when water overflows the banks. However, 2D models only consider surface flooding from rainfall, leading to the rise of coupled (1D/2D) simulations that offer a more comprehensive representation of flood extents during extreme events (Mitchell et al., 2007; Dey & Kamioka, 2007).

Pina et al. (2011), in their study, explained the use of free and open-source software in terms of freedom to vary parametrization, user-defined simulations, studying and analyzing output changes, and suggesting changes and improvements for the betterment of the idea. GIS is a powerful and most valuable tool in characterizing the primal base watershed model. Thus, the authors developed an application known as "inp.PINS," which GIS automatically SWMM to enhance its modeling capabilities and spatial extends. Further, Yerramilli (2012), Shrestha et al. (2014), and Nguyen et al. (2015) integrated 1D-2D models with GIS to demonstate the progressive increase in flood flood volume and develop the maps that can directly be used in land planning.

6.3 Review Process and Framework

We have identified around 40 flood simulation models from previously published reviews, journals, conference proceedings, modeling practitioners, and Internet searches. Screening is applied to these models to identify models capable of simulating urban flood scenarios. Table 6.1 presents a brief outline of the screened models, detailing their use, functions, and access. The review helps in selecting a suitable model to analyze the existing stormwater drainage network of the study area and perform 1D-2D flood simulations.

Urban flooding is a problem that affects many cities worldwide and poses serious difficulties for urban planners. This issue can be resolved by managing stormwater drainage properly and choosing the best model from the available choices. Because of the intricacies involved in water quality and urban flow, urban watershed models have changed over time. Numerous models that have been analyzed only consider one aspect of the infrastructure—stormwater, water supply, and wastewater—instead of taking it all into account. Some exceptions, such as SWMM and MIKE URBAN, combine several factors. Urban stormwater models have multiple purposes. Among them are planning tools, design tools, and operational tools (such as Mike Urban, SWMM, and HEC-HMS). Every model that is examined incorporates a rainfall-runoff component and estimates overland flow using basic storage equations. Shallow water wave equations can be solved by models such as MIKE URBAN, HEC-HMS, and SOBEK 1D-2D. In water resources modeling, risk analysis is a more recent development. Commercial models that can model 2D surface flow and create deluge maps include MIKE URBAN, XP SWMM, HEC-HMS, and SOBEK 1D-2D.

Table 6.1 Summary of screened models: use, functionality, and accessibility

Sr. No.	Model & Version	Developer	Model type	Assessibility	Remarks
Urban Models					
1	HEC-1 (V-4.1)	U.S. Army Corps of Engineers	1D	Public domain	With this lumped parameter model of a single storm event, stream routing, losses, rainfall, and unit hydrographs can all be modeled in different ways
2	HEC-HMS (V-2.0 and up)	U S Army Corps of Engineers	1D-2D	Public domain	Designed to take the place of HEC-1, it has more features like single or multiple storm hydrologic modeling and the MOD-Clark quasi-distributed (2D) modeling
3	TR 20	Soil Conservation Services	1D	Public domain	The computer program, which offers a comprehensive graphical user interface to the U.S. Soil Conservation Service, is called Singular Storm Event Lumped Parameter
4	TR 55	U.S. Dept. of Agriculture	1D	Public domain	A computer software that uses the SCS overflow formulas to analyze runoff peak times and overall volume was created for small watersheds. It relies on the lumped parametric method
5	SWMM (V-5.1)	U. S. Environmental Protection Agency	1D	Public domain	Quantity and quality of runoff simulated as a single event or continuously over long time period.
6	XP-SWMM	XP Solutions	1D-2D	Propriatory	Developed to create link-node (1-D) and geographically distributed hydraulic models (2-D), it is an extensive software suite that models storm water, sanitary, and river systems.

(continued)

Table 6.1 (continued)

Sr. No.	Model & Version	Developer	Model type	Assessibility	Remarks
7	SOBEK 1D2D (V-3)	Delft Hydraulics Software	2D	Propriatory	This robust modeling package is useful for predicting floods, controlling irrigation and drainage systems, designing sewer overflows, studying channel morphology, sensing salt intrusion, and examining quality of surface water
8	MIKE-FLOOD	DHI Water and Environment	2D	Propriatory	Connects the two separate software programs, MIKE 21 (2D) and MIKE 11 (1D). The Saint-Venant equations are solved by MIKE 11 using the finite difference scheme
9	MIKE-URBAN	DHI Water and Environment	2D	Propriatory	It is a versatile system created by DHI for stormwater, wastewater, and water distribution network modeling and design. It include a database where data from hydraulic and network models can be stored
10	HSPF (V-12.4)	U.S. Environmental Protection Agency	1D	Public domain	Lumped parameter model to simulate the amount and quality of watersheds outflows over short or long periods of time for both small and big watersheds
11	DR3M-QUAL (V-2.0)	US Geological Survey	1-D	Public domain	In between storms, it enables daily soil moisture accounting and continuous modeling. In the model, the drainage area is represented by a collection of reservoir, channel, and overland flow segments.
12	Penn State	Penn State University	1D	Public domain	This model for combined sewer and drainage study was created in collaboration with the City of Philadelphia
13	STORM (V-0.9.2)	U.S. Army Corps of Engineers	1D	Public domain	Generates hourly runoff depths from hourly rainfall inputs using a simple runoff coefficient and depression storage approach.

(continued)

Table 6.1 (continued)

Sr. No.	Model & Version	Developer	Model type	Assessibility	Remarks
14	Hydro-Planner	CSIRO	1D	Public domain	It makes it possible to evaluate the supply–demand balance by integrating the E2 catchment modeling framework, the REALM supply system simulation program, and the End Use Model software.
			Nonurban models		
15	ILLUDAS	British Road Research Laboratory	1D	Public Domain	Hydrographs from the immediately adjacent paved area and the preceding area are produced using time-area techniques
16	QQS		1D	Propriatory	Using an implicit finite difference approximation, QQS can simulate flow through pipes as well as channels and execute continuous or discrete event simulation at interval of 5 min
17	MUSIC (V-6.0)	eWater	1D	Propriatory	A tool for showcasing how well urban stormwater quality treatment facilities operate
18	Aqua-cycle (V-1.2.1)	Cooperative Research Centre for Catchment Hydrology, Monash University	1D	Public domain	The model is used to assess daily water balance in urban setups, and its outputs include estimates of water demand, evaporation, imported water consumption, stormwater, and wastewater use on a daily, monthly, and annual basis
19	UVQ (V-1.2)	CSIRO Urban Water Program	1D	Public domain	Extension of the water balance model and addition of pollutant balance modeling improved an existing model AQUACYCLE
20	Water-Cress	WaterCress Hydrology	1D	Public domain	It is a comprehensive water cycle model with a continuous time series that mimics the flow through both artificial and natural water systems

(continued)

Table 6.1 (continued)

Sr. No.	Model & Version	Developer	Model type	Assessibility	Remarks
21	WQRRS	U.S. Army Corps of Engineers	1D	Public domain	The reservoir, stream hydraulic, and stream quality modules are its three distinct yet complementary components. The hydraulics modules for reservoirs and streams are standalone programs that can be run, examined, and understood on their own
22	QUAL2E–UNCAS (V-2E)	U.S. Environmental Protection Agency	1D	Public domain	Its intended application is as a planning and regulatory tool for water quality, and it can be run in a steady-state or dynamic manner. The model can be used to determine the amount and quality attributes of nonpoint waste loads as well as look into how waste loads affect the quality of instream water. Steady-state model for typical contaminants in lakes and branching streams with good mixing
23	GSSHA (V-6.1)	U.S. Army Corps of Engineers	2D	Public domain	This 2D hydrologic model is grid-based. Full linkage between the groundwater, vadose zone, waterways, and overland flow are among the features, along with 2D groundwater, 1D stream flow, 1D infiltration, and 2D overland flow. GSSHA can operate in long-term and single-event modes
24	PRMS (V-2.1)	U.S. Geological Survey	1D	Public domain	PRMS was first created as a single FORTRAN 77 application for the management of natural resources and water. It assesses how changes in watershed conditions affect the hydrologic response

6.4 Overview Modeling Capability in an Urban Environment

The primary objective of flood modeling is to comprehend the flood characteristics in urban areas and their socioeconomic implications. In particular, the extent and depth of flood deluge are crucial parameters, especially for mapping flood hazards. With the integration of GIS-supported hydraulic models, water surface elevation profiles can be translated into deluge maps, aiding in identifying areas at risk during rainfall events and enabling early warning systems and risk reduction measures. Several models, including SWMM, PCSWMM, Mike Urban, Mike Flood, SOBEK 1D-2D, HEC-RAS, and TELEMAC 2D, incorporate topographical data for flood deluge mapping. Past studies have demonstrated the effectiveness of these models in accurately modeling urban floods and producing reliable results. However, most of these models, except for SWMM and HEC-RAS, are commercial software, relying on precise raw data availability. Additionally, apart from MIKE URBAN and SWMM, these models typically focus on only one aspect of the analysis. Some models are useful for planning or designing, but only a select few—like MIKE-URBAN, SWMM, and HEC-RAS—are useful for operations. These models usually include rainfall-runoff modules and use modest storage equations for approximating surface flows, with some capable of solving shallow water wave equations and modeling two-dimensional surface flows to produce flood deluge maps. For a comprehensive overview of these models' features and capabilities, refer Table 6.2.

6.5 Concluding Remarks

This chapter briefly discusses an overview of literature related to modeling of urban floods and preparation of flood deluge and flood risk maps in an urban environment. It is seen that urban flooding is one of the major challenges faced in modern towns and cities lately and has gained considerable attention from hydrologists over the world in the past few years. Flooding leaves a deep and long-lasting impact on the economic and social health of people. The majority of the global population is concentrated in the urban areas and flooding of such a highly-populated area will produce huge damage to the life of people and property. Therefore, understanding the behavior of floods in an urban environment and their consequences is of utmost importance in recent times.

Floods are not new to humankind but flooding in urban setup has been amplified significantly in the recent past. Understanding the flood behavior and its consequences will help in planning accurate remedial measures. Initial flood studies were focused on investigating changes in flood trends and damage assessment (Dutta & Herath, 2004; Douglas et al., 2008; Ramachandraiah, 2011). The technological advancement in modeling methodologies has made available tools to link 1D channel flows with 2D surface flows. The literature presented a wide range of flood models and modeling techniques used to regenerate critical flood events in the past.

Table 6.2 Description of models with features and capabilities

Program name	Capabilities	Reference
SWMM	It is a dynamic simulation model for rainfall, runoff, and routing. It functions on a group of sub-catchment sites that are subject to precipitation, produce runoff, and accumulate pollutants. The routing system monitors the amount and quality of runoff produced, as well as the rate of flow, depth of flow, and water quality. It is made up of nodes, conduits, pumps, regulators, storages and treatment units. SWMM-5 is the GUI (Graphical User Interface) of SWMM and is publicly available	Rossman (2010)
SWMM-XP	It is XP Software's commercial rendition of SWMM. There is a GIS interface in this version of SWMM	XP Software
PC-SWMM	PCSWMM, an automated decision-support system for SWMM that was first created in 1984 and has been maintained and supported by CHI Water ever since. The Graphical User Interface-based EPA SWMM implementation for Windows, which is compatible with all SWMM versions	CHI Water
SOBEK 1D-2D	Previously known as Delft-FLS, it is an collective software package that integrates 1D and 2D components for overland flow. The flow parameters for rural, urban, and river morphology are dynamically linked to create complicated models. Utilizing a finite difference approach, SOBEK 1-D satisfies the Saint Venant equations. Whereas, SOBEK 2-D (Overland Flow) solves the shallow water equations using a rectangular grid and the same finite difference scheme as SOBEK 1-D	Vanderkimpen et al. (2009)
MIKE URBAN	It offers an adaptable approach for designing and simulating water distribution networks as well as stormwater and wastewater collecting systems. It includes quality of water, water hammer parameters to real-time simulation. It also includes GIS-based model development and management capabilities. It offers integrated dynamic result visualization and thematic mapping that can be completed	Graham and Butts (2005); DHI

(continued)

Table 6.2 (continued)

Program name	Capabilities	Reference
MIKE FLOOD	Creates a dynamic link between MIKE 21 (2D) and MIKE 11 (1D), two separate software products. MIKE 11 formulates 1D Saint-Venant calculations using finite difference procedure. The "classic" MIKE 21 program simulate the shallow-water calculations with a finite difference scheme on a rectangular grid. It can handle wind friction, eddy viscosity, Coriolis forces, floods, and drying. It can also handle surface roughness that varies spatially.	Vanderkimpen et al. (2009); DHI
HEC-RAS	Addresses issues related to open-channel flow and is typically employed for calculating water surface profiles, stage, and velocity. Determines steady flow stage profiles based on channel shape, energy-loss model parameters, and steady flow rate. Determines unsteady flow by using the energy-loss model parameters, channel geometry, and upstream hydrograph. Has the ability to calculate hydraulics in one and two dimensions for an entire network of created and natural channels. The most recent version 5.0.6 is freely accessible	HEC-RAS Technical reference manual, 2000 (2016)

However, the 2D modeling was limited to riverine flooding so far, with increasing urban flood events in recent years a new approach coupling 1D channel network with 2D surface flood flows started has come forward. A review of the urban flood model presented few models capable of simulating floods in an urban environment.

The modeling of the urban flood is a very complex phenomenon as human interference has changed the land use and hydrologic characteristics in an urban environment. Furthermore, urbanization and climate change have produced a substantial effect on the rainfall pattern. A full-fledged model capable of handling urban changes is rare and very costly. Again the output deliverables of these models are compromised with the nonavailability of precise raw dataset. Therefore, a methodology needs to be developed by combining different modeling techniques to produce accurate results with the available dataset.

References

Agilan, V., & Umamahesh, N. V. (2016). What are the best covariates for developing non-stationary rainfall intensity – duration frequency relationship? *Advances in Water Resources, 101*, 11–22. https://doi.org/10.1016/j.jhydrol.2016.08.052

Bayazit, M. (2015). Nonstationarity of hydrological records and recent trends in trend analysis: A state-of-the-art review. *Environmental Processes, 2*, 527–542. https://doi.org/10.1007/s40710-015-0081-7

Cavanaugh, N. R., Gershunov, A., Panorska, A. K., & Kozubowski, T. J. (2015). The probability distribution of intense daily precipitation. *Geophysical Research Letters, 42*, 1560.

Chatterjee, C., Forster, S., & Bronstert, A. (2008). Comparison of hydrodynamic models of different complexities to model floods with emergency storage areas. *Hydrological Processes, 22*(24), 4695–4709. https://doi.org/10.1002/hyp.7079

Crawford, H. H., & Linsley, R. K. (1966). *Digital Simulation in Hydrology: Stanford Watershed Model IV*, Technical Report No. 39, Department of Civil and Environmental Engineering, Stanford University, Stanford.

Dey, A. K., & Kamioka, S. (2007). An integrated modeling approach to predict flooding on urban basin. *Water Science and Technology, 55*(4), 19–29. https://doi.org/10.2166/wst.2007.091

Douglas, I., Alam, K., Maghenda, M., Mcdonnell, Y., Mclean, L., & Campbell, J. (2008). Unjust waters: Climate change, flooding and the urban poor in Africa. *Environment and Urbanization, 20*(1), 187–205. https://doi.org/10.1177/0956247808089156

Dutta, D., & Herath, S. (2004). Trend of floods in Asia and flood risk management with integrated river basin approach. In *Proceedings of the 2nd international conference of Asia-Pacific hydrology and water resources Association, Singapore* (Vol. 1).

Elliott, A. H., & Trowsdale, S. A. (2007). A review of models for low impact urban stormwater drainage. *Environmental Modelling and Software, 22*(3), 394–405. https://doi.org/10.1016/j.envsoft.2005.12.005

Gupta, K. (2020). Challenges in developing urban flood resilience in India. *Philosophical Transactions of the Royal Society A, 378*(2168), 20190211. https://doi.org/10.1098/rsta.2019.0211

Graham, D. N., & Butts, M. (2005). *Flexible, integrated watershed modelling with MIKE SHE* (pp. 245–272).

HEC. (2016). 'HEC-RAS, reference manual; USACE', version 5.0. (US Army Corps of Engineers, CPD-68, 2016).

Mitchell, V. G., Duncan, H., Inman, M., Rahilly, M., Stewart, J., Vieritz, A., Holt, P., Grant, A., Fletcher, T. D., Coleman, J., Maheepala, S., Sharma, A., Deletic, A., & Breen, P. (2007). State of the art review of integrated urban water Models. In *NOVATECH 2007, sixth international conference on sustainable techniques and strategies in urban water management*. 25–28 June 2007, Lyon, Rhone-Alpes, France.

Mondal, A., & Mujumdar, P. P. (2017). Hydrologic extremes under climate change: Non-stationarity and uncertainty. In E. Kolokytha, S. Oishi, & R. Teegavarapu (Eds.), *Sustainable water resources planning and management under climate change*. Springer. https://doi.org/10.1007/978-981-10-2051-3_2

Nguyen, H. Q., Degener, J., & Kappas, M. (2015). Flash flood prediction by coupling KINEROS2 and HEC-RAS models for tropical regions of Northern Vietnam. *Hydrology, 2*, 242–265. https://doi.org/10.3390/hydrology2040242

Pina, R. D., Simoes, N. E., Marques, J. A., & Sousa, J. O. (2011). Floodplain delineation with Free and Open Source Software. In *12th International Conference on Urban Drainage, Porto Alegre/Brazil*, 11–16 September 2011.

Prasad, R. K. (2014). Urban floods—a review. *International Journal of Innovative Research in Science, Engineering and Technology (IJIRSET), 5*(6), 2319–8753.

Ramachandraiah, C. (2011). Coping with urban flooding: A study of the 2009 Kurnool floods, India. *Environment and Urbanization, 23*(2), 431–446. https://doi.org/10.1177/0956247811418733

Ramos, H. M., Perez-Sanchez, M., Franco, A. B., & Lopez-Jimenez, P. A. (2017). Urban floods adaptation and sustainable drainage measures. *Fluids*. https://doi.org/10.3390/fluids2040061

Rossman, L. A. (2010). *Storm water management model user's manual*. USEPA Software.

Stelling, G. S., & Verwey, A. (2005). *Numerical flood simulation*. John Wiley and Sons, Ltd. https://doi.org/10.1002/0470848944.hsa025a

Shrestha, S., Fang, G., Rahman, A., & Aziz, K. (2014). Application of artificial neural networks in regional flood frequency analysis: A case study for Australia. *Stochastic Environmental Research and Risk Assessment, 28*(3), 541–554.

Todini, E. (2007). Hydrological catchment modeling: Past, present and future. *Hydrology and Earth System Sciences, 11*(1), 468–482.

Vanderkimpen, P., Melger, E., & Peeters, P. (2009). Flood modeling for risk evaluation – A MIKE FLOOD vs. SOBEK 1D2D benchmark study. In Samuels et al. (Eds.), *Flood risk management: Research and practice* (pp. 77–84). Taylor and Francis Group.

Xu, H., & Luo, Y. (2015). Climate change and its impacts on river discharge in two climate regions in China. *Hydrology and Earth System Sciences, 19*(11), 4609–4618.

Yerramilli, S. (2012). A hybrid approach of integrating HEC-RAS and GIS towards the identification and assessment of flood risk vulnerability in the City of Jackson, MS. *American Journal of Geographic Information System, 1*(1), 7–16. https://doi.org/10.5923/j.ajgis.20120101.02

Zoppou, C. (2001). Review of urban storm water models. *Environmental Modelling and Software, 16*, 195–231. https://doi.org/10.1016/S1364-8152(00)00084-0

Chapter 7
Technologies and Initiatives for Addressing Key Challenges and Achieving SDG 11 in India

Priyanka Kumaravel, R. Velkennedy, and Venkatesh Baskaran

Abstract Sustainable Development Goal 11 is one of 17 global goals set by the United Nations in 2015 as part of the 2030 Agenda for Sustainable Development. The main goal is to build more inclusive, resilient, and sustainable cities and human settlements, guaranteeing a higher quality of life for all citizens while reducing negative environmental consequences and fostering economic development as cities worldwide grapple with challenges such as pollution, congestion, and resource depletion. EcoTech urbanism represents a holistic approach to urban development that recognizes the interconnectedness of social, economic, and environmental factors by harnessing innovative technologies to create cities that are not only livable but also environmentally sustainable. This article investigates the important issues and obstacles faced by Indian cities in achieving Sustainable Development Goal 11. Urban areas face several challenges, including providing affordable housing, controlling air and water pollution, improving public transportation, solid waste management, energy efficiency, climate change mitigation, and implementing significant expenditures. This technology involves designing and constructing buildings with materials and technologies that minimize energy consumption, reduce carbon emissions, and enhance indoor air quality. By leveraging digital technologies such as GIS, intelligent transport systems and cities can optimize the management of resources such as housing, waste management, and transportation. This overview makes an important addition to talks about urban planning in India by analyzing the barriers to urbanization and providing remedies. This article underlines the importance of government operations in promoting programs to achieve SDG 11, which aligns with the SDG 2030 Agenda.

P. Kumaravel (✉) · R. Velkennedy
Department of Civil Engineering, Thiagarajar College of Engineering, Madurai, Tamil Nadu, India

V. Baskaran
Department of Civil Engineering, Alagappa Chettiar Government College of Engineering and Technology, Karaikudi, Tamil Nadu, India

V. Baskaran et al. (eds.), *EcoTech Urbanism*, The Urban Book Series, https://doi.org/10.1007/978-3-031-83833-0_7

Keywords Environmental sustainability · Sustainable cities · SDG-11 · Urbanization · Affordable housing

7.1 Introduction

As India's urban landscape continues evolving at an unprecedented pace, pursuing urban sustainability is a paramount challenge. It is anticipated that India's urban population will experience tremendous growth, reaching 600 million by the year 2036. By the year 2050, seven out of ten people may be living in metropolitan regions (Roy et al., 2023b). Cities are crucial to economic growth, contributing over 80% of global GDP and offering sustainable investment opportunities (Küfeoğlu, 2022). India stands at a crucial crossroads where the choices made today will profoundly shape the cities of tomorrow. With rising urbanization nationwide, sustainable urban planning is more important than ever. "The Road to Sustainable Cities: Addressing Key Issues and Challenges for India's Urban Planning" embarks on a journey to unravel the complexities inherent in India's urban landscape and chart a course toward a future where cities thrive in harmony with their inhabitants and the environment (da Silva et al., 2023). The expansion of cities leads to the consumption of agricultural lands, densification within cities, and migration from rural areas. The document discusses the impact of urban center's structure on travel behavior, emphasizing the need for efficient public transportation systems to address traffic issues. Furthermore, the pursuit of urban sustainability must prioritize inclusivity and equity to ensure that the benefits of development are accessible to all segments of society (Keivani, 2010). Addressing disparities in access to basic services, health care, education, and housing is indispensable for fostering socially cohesive and inclusive cities. Against this backdrop, this article endeavors to elucidate the key challenges confronting urban sustainability in India and delineate strategies to realize the vision of SDG 11 (Nagabhatla & Brahmbhatt, 2020). India can create dynamic, resilient, and inclusive cities that drive sustainable development in the twenty-first century by embracing innovation, collaboration, and sustainability. It also highlights the challenges Indian cities face in achieving Sustainable Development Goal 11, focusing on affordable housing, pollution control, public transport, waste management, and climate change mitigation (Kellison, 2022). EcoTech urbanism is a cutting-edge method that integrates ecological principles and technological technology to establish sustainable urban environments. Integrating EcoTech urbanism can have a crucial impact on reaching Sustainable Development Goal 11 (SDG 11) in the context of India's growing urbanization. SDG 11 aspires to create cities that are inclusive, safe, resilient, and sustainable (Vaidya & Chatterji, 2020). This chapter examines the fundamental elements of EcoTech urbanism and presents solutions for incorporating these ideas into Indian cities to tackle significant urban sustainability issues.

7.2 Sustainable Development Goal 11

In India, the primary objective of SDG 11 is to foster inclusive, safe, resilient, and sustainable urban development of rapid urbanization issues including inadequate housing, basic amenities, and infrastructure while promoting affordable housing programs like the Pradhan Mantri Awas Yojana (PMAY) and smart cities under the Smart Cities Mission is necessary (Brito, 2012). Moreover, SDG 11 emphasizes enhancing urban resilience to environmental and climate-related risks, improving public transport systems and infrastructure, and promoting sustainable consumption and waste management practices to ensure the well-being and quality of life of all residents, thereby contributing to India's overall sustainable development objectives (Tasaki et al., 2010).

Cities represent global life's future. The global population is predicted to approach eight billion in 2022, with over half residing in cities. This percentage is only anticipated to increase, with 70% of the population predicted to reside in cities by 2050 (Ruchira & Kansal, n.d.). Currently, around 1.1 billion people live in slums or slum-like conditions in cities, with an additional two billion predicted over the next 30 years (Roy et al., 2023a). However, many of these cities are unprepared for this fast urbanization, which has outpaced the construction of housing, infrastructure, and services, increasing slums or slum-like conditions (Dahl, 2012). The SDG 11 targets are briefly outlined in the Fig. 7.1.

Sustainable development requires a substantial transformation in how urban places are created and maintained (Lami et al., 2023). SDG 11 recognizes cities as economic, social, and cultural centers, but they also confront substantial difficulties in terms of sustainability and inclusion. By addressing these issues, SDG 11 strives to create livable, inclusive, and ecologically sustainable cities and human settlements that enhance quality of life and sustainable development (Gavaldà et al., 2023). SDG 11 seeks to convert cities into catalysts for sustainable development and fair growth by tackling the particular requirements and difficulties linked to urbanization. The specified targets and indicators establish an entire structure for assessing progress and directing endeavors to establish cities that improve the quality of life for all inhabitants.

7.3 Navigating India's Urbanization Odyssey: Toward Sustainable, Equitable Cities

India's urbanization history is characterized by sophisticated complexity, characterized by rapid population increase, dynamic migratory patterns, and massive urban expansion. Prominent to the subject matter is a country experiencing a significant change, with millions of people being attracted to the prospect of city living in search of economic development, education, and an improved standard of living. India is seeing an unprecedented rate of urban expansion, with the urban population

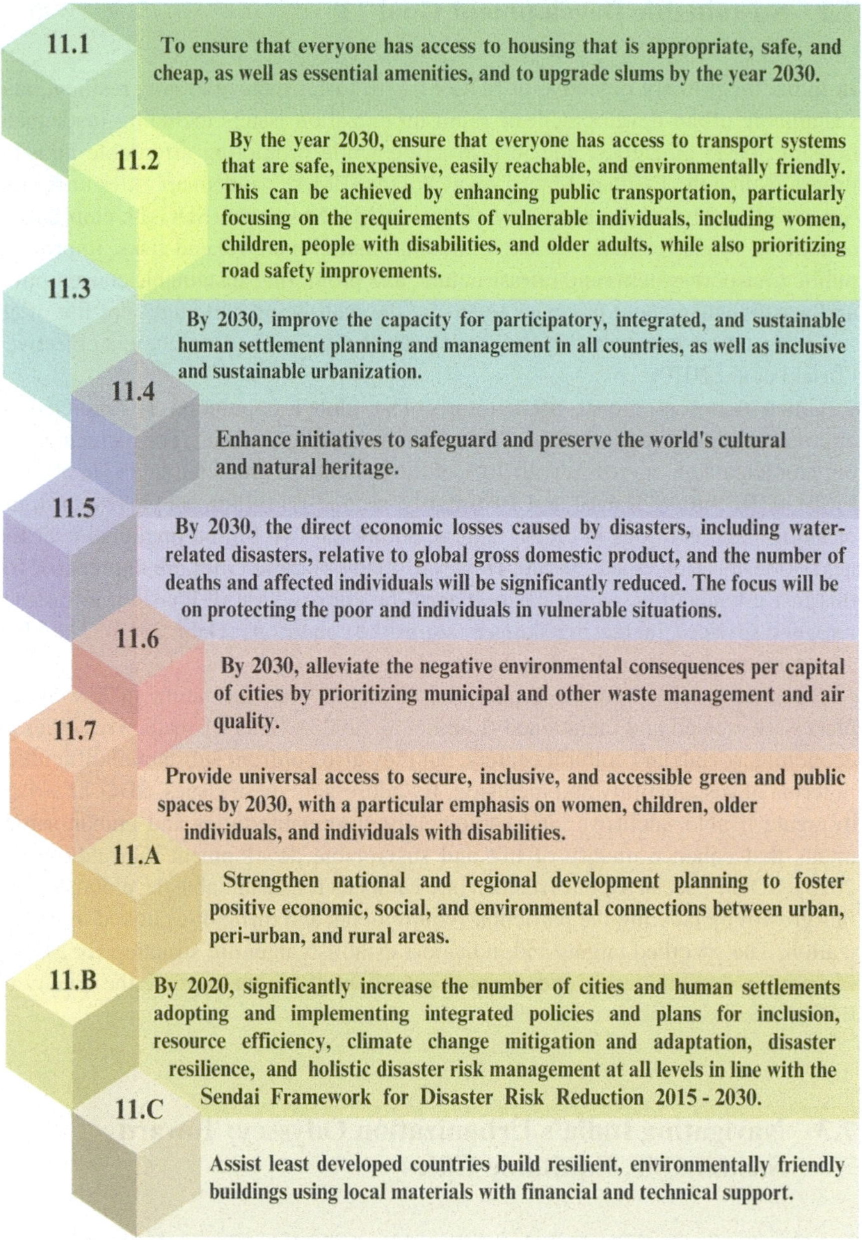

Fig. 7.1 Overview of SDG 11 targets

projected to exceed 600 million by 2030, making up a substantial share of the country's overall population (Flores, 2023). This demographic transition is not solely quantifiable; it signifies a fundamental restructuring of India's social, economic, and spatial environments, with significant consequences for governance, infrastructural advancement, and sustainability. As urban populations increase and cities get larger, the difficulties of ensuring sufficient housing, transportation, water supply, and sanitation become increasingly urgent (Karjalainen & Juhola, 2021). The scarcity of housing in metropolitan regions, which is approximately 19 million units, highlights the pressing necessity for comprehensive approaches to tackle the increasing requirement for inexpensive and sustainable housing (Kandpal & Okitasari, 2023). Furthermore, the pressure on urban infrastructure is evident, as transportation networks are failing to handle growing congestion, water supply systems are strained by increased demand, and waste management systems are being pushed to their maximum capacity. Within this framework, the notion of sustainability arises as a fundamental principle for urban development, requiring a comprehensive approach that harmonizes economic progress with environmental preservation and social fairness. India's process of urbanization is marked by the development of large cities such as Mumbai, Delhi, and Bangalore. These cities act as catalysts for economic expansion and creativity, drawing in skilled individuals and investments from both inside India and internationally. However, in addition to these busy and crowded cities, there is another story unfolding in the form of secondary cities like Pune, Hyderabad, and Ahmedabad (Ghosh & Kansal, 2014). These cities are undergoing fast growth and development, driven by the expansion of industries and the emergence of the IT industry. These cities serve as strategic points of opportunity, presenting a more equitable and evenly distributed approach to urban expansion that can relieve strain on large cities and foster regional progress. However, as India's urban area grows, it also faces the negative effects of fast urbanization, such as environmental decline, socioeconomic inequalities, and difficulties in governance (Zhang et al., 2024). The decline in air and water quality, worsened by industrial pollution and emissions from vehicles, presents significant risks to public health and overall welfare, especially in highly populated urban areas. Furthermore, the rapid increase in informal settlements and slums, which accommodate a large number of impoverished urban residents, highlights the immediate requirement for inclusive urban design and methods to reduce poverty. The Smart Cities Mission and the Atal Mission for Rejuvenation and Urban Transformation (AMRUT) are important initiatives that contribute significantly to the promotion of sustainable urban development. These initiatives utilize technology and innovation to enhance infrastructure and service delivery, while also empowering local communities and encouraging citizen participation. However, achieving the goal of sustainable urbanization goes beyond simply implementing technology solutions. It necessitates a profound reconsideration of urban governance structures, policy frameworks, and institutional capacity. Improving urban governance, fostering interagency collaboration, and advocating transparency and accountability are crucial measures for constructing resilient and inclusive cities capable of withstanding the challenges of the twenty-first century (Chang et al., 2023).

Figure 7.2 illustrates the state-level urbanization in India as of 2011. Furthermore, it is crucial to empower underprivileged people, such as women, children, and the elderly, to ensure that the advantages of urban growth are spread fairly and promote

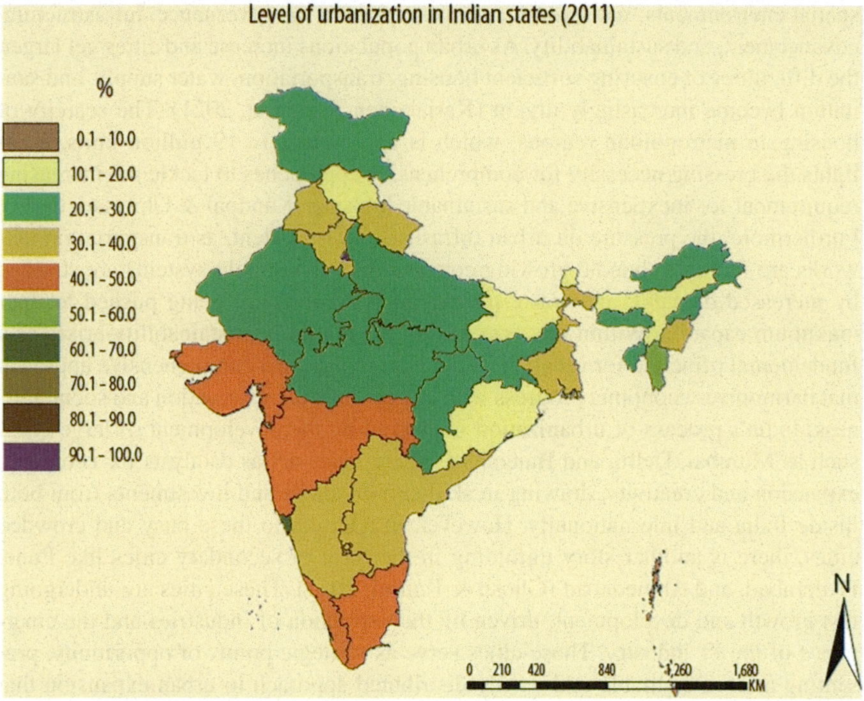

Fig. 7.2 State-level urbanization in India in 2011. (Source: Census Commissioner and Registrar General, 2011)

social inclusivity. As India progresses in its urbanization process, the decisions made now will determine the future of its cities. These choices will determine whether the cities become dynamic centers of opportunity and innovation or places marked by inequity and environmental decline. At this crucial juncture, it is essential to wholeheartedly adopt the values of sustainability, fairness, and resilience, and to set a direction toward a future where India's cities not only prosper but also become exemplars of urban excellence globally.

7.3.1 *Significance of Achieving SDG 11 in India*

In India, fulfilling Sustainable Development Goal (SDG) 11 is crucial, especially for EcoTech urbanism, which combines ecological principles with technological advancements to promote sustainable urban development (Kumar et al., 2023). India, one of the fastest-growing economies and urbanizing nations presents significant issues in sustainable urbanization. Livable, resilient communities are needed more than ever when the urban population approaches 600 million by 2030 (Pradhan et al., 2023). SDG 11, which seeks to make cities and human settlements inclusive, safe, resilient, and sustainable, provides a framework for addressing these issues and using EcoTech urbanism to promote change. EcoTech urbanism emphasizes the interdependence of urban and natural ecosystems to reduce environmental impact, maximize resource efficiency, and promote social fairness. This technique is significant in India because the country's natural landscapes range from crowded metropolitan centers to vulnerable ecosystems including wetlands, woods, and coastal regions. India can use its biodiversity and cultural heritage to create environmentally sustainable, culturally vibrant, and socially inclusive cities by adopting EcoTech urbanism.

SDG 11 in India offers a chance to address environmental issues such as air and water pollution, deforestation, and biodiversity loss. India's natural resources and ecosystems have been stressed by rapid urbanization and industrialization, causing environmental degradation and public health risks (Griggs et al., 2013). India can reduce these impacts and build healthier, more resilient communities by fostering EcoTech urbanism, which uses green infrastructure, renewable energy, and sustainable mobility. Urban planning with green spaces can reduce the urban heat island effect, enhance air quality, and offer citizens enjoyment. Clean energy technologies and energy-efficient buildings can reduce greenhouse gas emissions and help India meet its Paris Agreement goals. SDG 11 is essential for social equity and inclusive development in India, given increased urbanization and rising income inequality (Brito, 2012). Access to essential amenities, jobs, and housing in Indian cities is unequal, aggravating social tensions and marginalizing vulnerable groups including slum dwellers, informal laborers, and women. EcoTech urbanism stresses social inclusion and community empowerment to provide all people with basic services, facilities, and economic possibilities. Participatory planning that involves local communities in decision-making can improve marginalized groups' needs and preferences, resulting in more fair and inclusive outcomes. Investments in affordable housing, health care, and education infrastructure can increase social cohesion and quality of life for urban people of all socioeconomic backgrounds.

SDG 11 is also linked to India's broader sustainable development strategy, which includes poverty reduction, health and well-being, gender equality, and climate action. India, the world's second-most populous nation and economic leader, will impact global sustainability efforts with its sustainable urban development endeavors (Nabil et al., 2023). India can inspire other developing nations facing urbanization issues by adopting EcoTech urbanism and promoting environmental stewardship, social fairness, and economic development. India can expedite its transition to sustainable urbanization and help achieve the SDGs globally by using international collaborations and information exchange. SDG 11 is essential to EcoTech urbanism and sustainable, egalitarian cities in India. By adopting new technologies and using its natural and cultural assets, India can improve its urban population's future and inspire the world (Larsson & Hatzigeorgiou, 2022). Collective responsibility, inclusive governance, and transformative partnerships may help India build resilient, livable, innovative, creative, and opportunity-rich cities for SDG 11.

7.4 Key Issues and Challenges for Urban Planning

Urban planning faces numerous key issues and challenges, especially in rapid urbanization. These include inadequate infrastructure and basic services, such as water, sanitation, and electricity, leading to slums and informal settlements (Envisaging the Future of Cities, n.d.). Additionally, urban sprawl and inefficient land use contribute to congestion, pollution, and environmental degradation. Social inequality and exclusion are also prevalent, with marginalized communities often lacking access to essential services and facing displacement due to gentrification. Furthermore, the impacts of climate change, including extreme weather events and rising sea levels, pose significant risks to urban areas, necessitating adaptation and resilience strategies (Obe et al., 2024). Effective urban planning must address these multifaceted challenges by promoting sustainable and inclusive development, integrating land use and transportation planning, enhancing resilience to climate change, and fostering community participation in decision-making processes (Chang et al., 2023). These challenges encompass a spectrum of issues ranging from inadequate infrastructure and housing deficits to environmental degradation and socioeconomic disparities (Long et al., 2023). The key issues and challenges for urban planning to achieve sustainable development goal 11 in India are as follows

7.4.1 *Affordable Housing*

India faces enormous issues in urban planning for inexpensive housing, which are made worse by the country's fast urbanization and population expansion. With more than 31% of the population living in cities, and estimates showing that number will climb to over 40% by 2030, there is an unparalleled need for affordable housing. According to the Ministry of Housing and Urban Affairs (MoHUA), the urban housing deficit in India was approximately 18.78 million units in 2012. The scarcity predominantly impacts individuals from the economically weaker sections (EWS) and lower-income groups (LIG). The housing gap is marked by the widespread presence of informal settlements and slums, where millions of impoverished urban residents live in inadequate conditions without access to essential services. However, there are still a lot of obstacles to overcome, like the high cost of land, which accounts for a large amount of the costs associated with developing new homes (Nijman, 2012). Comprehensive urban planning techniques are essential since many metropolitan areas lack basic infrastructure and services, especially in informal settlements. Affordable housing initiatives also face challenges in navigating intricate legal and administrative frameworks and managing financial constraints (Shekhar & Ravi, 2023). The Government of India has implemented various initiatives to tackle the issue of affordable housing. These include the Pradhan Mantri Awas Yojana (Urban)—PMAY (U) (2015), which aims to provide housing for all by 2022 by offering subsidies to the urban poor and slum dwellers. Another scheme is the Credit-Linked Subsidy Scheme (CLSS) (2015), which delivers interest subsidies on housing loans for economically weaker sections (EWS), low-income groups (LIG), and middle-income groups (MIG) under PMAY (U). The Real Estate Regulatory Authority (RERA) (2016) has also been established to safeguard homebuyers and promote transparency in real estate investments. Lastly, the Affordable Rental Housing Complexes (ARHCs) (2020) have been introduced to offer affordable rental housing to urban migrants and the impoverished, particularly those affected by the COVID-19 pandemic. However, to ensure that cheap housing becomes a reality for millions of urban people across India, this ambitious aim demands creative solutions, community engagement, and sustainable design principles. Prefabrication and modular building can make affordable housing projects more feasible by reducing costs and construction time. Add smart city solutions and sustainable practices to these housing developments to improve living circumstances.

7.4.2 Air and Water Pollution

India has numerous instances of issues related with air pollution, which puts the country's citizens' health at significant danger. The country's enormous urbanization and industrialization, combined with automobile emissions and industrial pollutants, have made urban air and water quality a major concern (Mangweta et al., 2022). Recent statistics show that some Indian cities frequently rank among the world's most polluted, with PM2.5 and PM10 levels exceeding WHO guidelines. Alarmingly high concentrations of air pollutants, including particulate matter (PM2.5 and PM10), nitrogen dioxide (NO2), sulfur dioxide (SO2), and ground-level ozone (O3), are frequently measured in major cities like Delhi, Mumbai, Kolkata, and Chennai. These cities have high air pollution, which causes respiratory disorders, cardiovascular diseases, and untimely deaths. Untreated sewage, industrial effluents, and agricultural runoff pollute waterways and cause waterborne diseases and environmental degradation. To reduce air and water pollution, urban design must address problems like inadequate infrastructure, poor waste management, and environmental regulation (Rauber & Gonçalves, 2014). This requires increasing public transit, renewable energy, waste management, and wastewater treatment facilities (Yailymova et al., 2021). Ensuring environmental compliance and promoting sustainability requires public awareness and engagement (Gupta, 2023). Addressing these difficulties and emphasizing air and water quality will help Indian urban planners build healthier, more sustainable communities emphasizing citizens' well-being and protecting the environment for future generations. India is dedicated to 14 of the 20 most polluted cities in the world, with Delhi consistently ranking as the most contaminated, according to the World Health Organization's Global Urban Ambient Air Pollution Database. Many Indian cities have "hazardous" Air Quality Index (AQI) scores above 300. According to the Central Pollution Control Board (CPCB), approximately 70% of India's surface water is polluted, with the Ganga and Yamuna rivers being the worst (Shruti et al., 2019). The Economic Survey of India 2020–21 found that pollution-related diseases kill 12.5% of Indians, with air pollution being the third-leading cause. The Swachh Bharat Mission improved sanitation and waste management in 2014, while the Namami Gange Programme rejuvenated the Ganga River by reducing pollution and fostering sustainable development. In 2019, the National Clean Air Programme (NCAP) set city-specific action plans and tougher emission regulations to reduce PM2.5 and PM10 levels by 20–30% by 2024. Delhi-NCR also implemented the Graded Response Action Plan (GRAP) to address air pollution levels. Bharat Stage VI (BS-VI) pollution rules began in 2020, reducing vehicle emissions and encouraging greener fuels like CNG and electric vehicles. These extensive initiatives show India's dedication to environmental protection and a healthier, more sustainable future.

7.4.3 Public Transport

Urban planning in public transport in India faces many obstacles to effective and sustainable transportation networks. The overreliance on private vehicles due to poor public transport is a major issue. To accomplish SDG 11.2 in India, a comprehensive strategy is necessary to tackle deficiencies in infrastructure, overcome financial obstacles, address safety issues, and promote inclusivity and sustainability. Many steps forward have been taken, especially in developing metro trains and making electric vehicles easier to obtain. However, there is still a lot of work to be done. India's public transportation system needs to be changed to meet the needs of its growing urban population and contribute to sustainable development. This is feasible only with continued investment, policy support, and innovative concepts. Metro train developments in Delhi, Mumbai, and Bangalore have improved infrastructure and services, but public transport is still underdeveloped and inaccessible in many locations (Sharifi et al., 2024). In India, just 18% of urban trips are performed by public transit, much lower than in other emerging nations. Private vehicles cause traffic congestion, air pollution, and greenhouse gas emissions, worsening environmental degradation and public health problems (Verma et al., 2011). The lack of last-mile connectivity and integration between forms of transport makes it difficult for commuters to navigate urban regions. Uneven distribution of public transport services disproportionately impacts underprivileged communities, who have limited mobility options and must use more expensive and less reliable means. Urban planners must invest in public transport infrastructure, increase service quality and dependability, and encourage modal shifts away from private vehicles to address these issues. India can improve urban livability, reduce congestion, and lessen transportation's environmental and public health implications by encouraging sustainable mobility options and fair transit access.

7.4.4 Solid Waste Management

India's urban solid waste management planning is complex, where rapid urbanization and population growth have increased trash creation exponentially. Municipal solid waste production in the country exceeds 62 million metric tons per year and is expected to rise. The waste material is predominantly composed of organic matter (40–60%), recyclable materials such as plastics, paper, and glass (20–30%), and inert substances like construction debris (10–20%). Indore and Pune, which are frequently praised as an exemplary city for waste management in India, have successfully attained impressive levels of garbage segregation and cleanliness by actively involving the community and implementing effective management strategies. The city has successfully established garbage segregation and recycling programs by partnering with cooperatives of garbage pickers. However, inadequate garbage collection, segregation, treatment, and disposal infrastructure and systems cause environmental degradation, public health risks, and social inequalities (Kandpal & Okitasari, 2023). Open dumping and burning of rubbish, especially in informal settlements and peri-urban regions, pollutes the air and water and endangers adjacent residents. Waste management issues like resource depletion, landfill overflows, and soil and water contamination are compounded by improper source segregation and ineffective collection systems. The informal sector, which recycles and manages waste, typically works in unsafe and filthy conditions, exposing workers to health risks and abuse. Many Indian cities fail to follow waste management rules, despite the Swachh Bharat Mission's goal of universal sanitation and sustainable solid waste management. The Solid Waste Management Rules 2016 prioritize the practice of separating garbage at its source, processing waste in a decentralized manner, and holding waste generators accountable for their actions. Extended producer responsibility (EPR) laws are designed to hold producers accountable for the proper disposal of their products at the end of their shelf life. Smart bins, waste-to-energy technologies, and automatic systems for sorting trash are some of the new ideas that are being examined. Technological advances, community participation, and policy reforms are needed to address these concerns. India can overcome urban planning solid waste management difficulties and progress toward a more sustainable and resilient future by investing in decentralized waste management systems, increasing source segregation and recycling, and encouraging public knowledge and engagement.

7.4.5 Energy

Urban planning in energy management is difficult in India, where rising urbanization and industrialization have increased energy demand and environmental issues. The country consumes the third-most energy globally, mostly in cities. However, obsolete and inefficient energy infrastructure causes supply shortages, grid instability, and environmental deterioration. The significant use of fossil fuels for energy generation leads to air pollution, greenhouse gas emissions, and India's vulnerability to variable oil prices and supply interruptions. The lack of energy-efficient building rules and urban design standards strains the energy grid and limits renewable energy integration. India has established ambitious targets for increasing its renewable energy capacity. The nation's objective is to attain a total renewable energy capacity of 175 GW by the year 2022. This target comprises 100 GW from solar energy, 60 GW from wind energy, 10 GW from bio-power, and 5 GW from small hydro-power. Although there has been notable advancement toward the 2022 target, the objective has been expanded to 500 GW by 2030. India ranks as the fourth-largest global generator of wind power, boasting an installed capacity of nearly 40 GW. In addition, underprivileged urban communities like slum dwellers and informal settlements typically lack energy, exacerbating socioeconomic inequalities and hampering inclusive urban development. The Indian government has enacted a variety of measures to promote the growth of renewable energy, such as providing tax benefits, financial assistance, and requirements for purchasing renewable energy. The Jawaharlal Nehru National Solar Mission and the Green Energy Corridor programs aim to promote the construction of infrastructure and the integration of renewable energy into the power grid. Urban planning must prioritize energy efficiency, renewable energy adoption, and equal access to clean energy services to address these concerns. India can overcome urban planning energy management issues and move to a more sustainable and resilient energy future by investing in smart grid technologies, promoting energy-efficient building design and retrofitting, and enacting inclusive energy policies.

7.4.6 Climate Change

Climate change is a major urban planning issue. As they grow, cities contribute greatly to global greenhouse gas emissions and climate change. Urban areas use 70% of world energy and exceed 60% of carbon dioxide emissions. In the face of climate change, urban planning must reduce fossil fuel use and switch to renewable energy. This involves rigorous energy efficiency, renewable energy integration, and transportation and building carbon reduction. It also encourages public transit, walking, cycling, and vehicle reduction. Urban planning in climate change must handle the rising frequency and intensity of extreme weather occurrences. Cities are vulnerable to climate change-related heat waves, floods, and storms. Green infrastructure, floodproofing buildings, and emergency response strategies must be included in urban planning to reduce risk and boost resilience to these disasters. Indian urban planning is complicated by high population density and fast urbanization. India is the second-most populated nation, with 34% of its population living in cities, according to the World Bank. India has signed the Paris Agreement and has committed to decrease the emissions intensity of its GDP by 33–35% from the levels in 2005 by the year 2030. In addition, India's objective is to attain approximately

40% of its total installed electric power capacity from energy sources that are not derived from fossil fuels. This strains resources and infrastructure, making sustainable and climate-resilient city planning harder. With some of the world's most polluted cities and catastrophic weather occurrences including floods, droughts, and cyclones, India is vulnerable to climate change. Indian urban design must prioritize climate change adaptation and mitigation. In 2008, India launched the National Action Plan on Climate Change, and in 2011, it developed the National Urban Climate Change Mitigation and Adaptation Strategy. However, additional work is needed to address India's climate change urban planning difficulties.

7.4.7 Investment

India, a booming global economy, is currently at a crucial point in the pursuit of long-term economic growth and advancement. Financial investment is crucial in this path, including both domestic and foreign capital in different areas. This study analyzes the present condition of financial investment in India, emphasizing significant patterns, difficulties, and prospects that influence the country's economic environment. Urban planning encounters substantial obstacles in securing financial investment as a result of the growing influence of financial investors in determining urban redevelopment initiatives. The integration of financial markets with the urban built environment has resulted in a transformation where urban redevelopment projects are specifically designed to function as investment assets for financial investors, influencing policymaking and the physical appearance of metropolitan areas. India requires significant expenditures in urban infrastructure, amounting to an estimated $840 billion over the next 15 years. This underscores the vital role that private entities play in funding such projects. The Indian government has enacted many policy reforms to incentivize foreign investment, such as relaxing foreign direct investment regulations in areas such as military, insurance, and retail. The objectives of initiatives such as "Make in India" and the production-linked incentive (PLI) scheme are to enhance manufacturing and export capacities. Nevertheless, the main problem is the inadequate funding of infrastructure, despite the increasing demand caused by urbanization. This highlights the need to enhance the financial resources of urban local bodies (ULBs) to attract external funding and close the infrastructure deficit. These problems highlight the complex connection between financial investment and urban planning, requiring innovative methods to tackle financing shortfalls and guarantee long-lasting urban growth.

7.5 Driving Sustainable Cities: Government Initiatives in India

India has made many steps to achieve Sustainable Development Goal (SDG) 11, which aims to make human settlements and cities equitable, secure, durable, and sustainable. Some key initiatives include:

Pradhan Mantri Awas Yojana (PMAY), launched in 2015, is a flagship affordable housing scheme that aims to provide "Housing for All" by 2022. It comprises two components: Pradhan Mantri Awas Yojana (Urban) for urban areas and Pradhan Mantri Awas Yojana (Gramin) for rural. The urban component builds affordable homes for economically deprived, low-income, and middle-income urbanites to relieve the housing deficit and promote inclusive urbanization.

Smart Cities Mission was initiated in 2015 and is designed to convert specific cities into sustainable and citizen-friendly urban centers by utilizing technology, infrastructure development, and citizen engagement. The mission focuses on improving urban infrastructure, enhancing service delivery, promoting sustainable mobility, and strengthening governance mechanisms in selected cities across the country.

AMRUT (Atal Mission for Rejuvenation and Urban Transformation), launched in 2015, aims to ensure basic infrastructure services such as water supply, sewerage, and urban transport to improve the quality of life for residents in urban areas. The mission focuses on providing these services in mission cities with a population of over 100,000 and other identified cities with a population between 50,000 and 100,000.

Swachh Bharat Mission (Urban), launched in 2014, endeavors to realize the vision of a Clean India by addressing issues related to sanitation and hygiene in urban areas. The mission's primary objective is to construct toilets, manage solid waste, and promote behavioural changes to ensure cleanliness and hygiene in cities.

National Urban Livelihoods Mission (NULM), launched in 2013, endeavors to mitigate poverty and vulnerability among impoverished urban households by providing them with opportunities for qualified wage employment and gainful self-employment, improving their livelihoods and living standards.

National Clean Air Programme NCAP (2019) is a national program that was introduced in 2019 to address air pollution. Using 2017 levels as a baseline, it seeks to lower PM2.5 and PM10 concentrations by 20–30% by 2024. The initiative includes creating action plans tailored to each community, keeping an eye on air quality, raising public awareness, and implementing more stringent emission standards for automobiles and industry. The NCAP places a strong emphasis on cooperation between state and federal governments, urban municipal governments, and pertinent parties.

Housing for All by 2022 Mission, the objective is to guarantee that every family in India has access to a pucca house that is equipped with water, sanitation facilities, and electricity. It encompasses various housing schemes and initiatives, including PMAY, to achieve the overarching goal of providing housing for all.

These initiatives reflect the government's commitment to promoting sustainable urban development, enhancing the quality of life in cities, and addressing the housing needs of the urban population. Through these programs, the Government of India aims to contribute significantly to achieving SDG 11 targets and fostering inclusive and sustainable urbanization across the country.

7.6 Conclusion

Indian cities are facing environmental degradation and resource constraints. Urbanization in India needs better population and resource management. The paper explores the challenges and potentials of urbanization in India, focusing on sustainability. In conclusion, to achieve urban sustainability in India and effectively address the challenges for the success of SDG 11, it is necessary to take a multifaceted approach that takes into consideration the diverse nature of Indian cities, the asynchronous progress in SDG implementation, the necessity of incorporating environmental concerns into development planning, and the critical importance of translating broader sustainability targets onto the contextual level. Technological advancements like geographic information systems (GIS), Internet of Things (IoT), and artificial intelligence (AI) are essential for improving urban planning, enhancing infrastructure efficiency, and assuring optimal resource management. Employing these technologies in smart city programs shows clear advantages for strengthening urban mobility, managing energy, and ensuring public safety. In addition, government-led programs such as the Pradhan Mantri Awas Yojana (PMAY), Atal Mission for Rejuvenation and Urban Transformation (AMRUT), and the Swachh Bharat Mission are making substantial progress in the areas of affordable housing, urban infrastructure improvement, and cleanliness and sanitation. These initiatives, underpinned by inventive funding structures and collaborations between the public and private sectors, are crucial for expanding solutions and safeguarding their sustainability over the long term. The findings highlight the significance of good governance, cooperation across different scales of governance, and the incorporation of sustainable development goal indicators as context-driven targets in city plans to promote successful implementation and overcome challenges in sustainable urban development. India has the potential to make progress toward a more sustainable and resilient urban society by the objectives of SDG 11 if it can solve these main problems and leverage the insights gained from national experiences and city-level assessments.

References

Brito, L. (2012). Analyzing sustainable development goals. *Science (New York, N.Y.), 336*(6087), 1396. https://doi.org/10.1126/SCIENCE.1224531

Chang, Y., Ouyang, X., Fei, X., Sun, Z., Li, S., Jiang, H., & Li, H. (2023). Comprehensive assessment of sustainable development goal 11 at the sub-city scale: A case study of Guilin City. *Remote Sensing, 15*(19). https://doi.org/10.3390/rs15194722

da Silva, R. G. P., Lima, C. L., Quinn, S. D., Afelt, A., Laques, A. E., & Saito, C. H. (2023). Urban green spaces in Brazil: Challenges and opportunities in the context of the COVID-19 pandemic. *Studia Ecologiae et Bioethicae, 21*(1), 69–88. https://doi.org/10.21697/seb.2023.06

Dahl, A. L. (2012). Achievements and gaps in indicators for sustainability. *Ecological Indicators, 17*, 14–19. https://doi.org/10.1016/j.ecolind.2011.04.032

Envisaging the Future of Cities. (n.d.).

Flores, E. P. (2023). Sustainability reporting in Cambodia: Challenges and prospects. *Journal of Accounting, Finance, Economics, and Social Sciences, 8*(2). https://doi.org/10.62458/jafess.160224.8(2)11-22

Gavaldà, O., Gibbs, C., & Eicker, U. (2023). A review of current evaluation urban sustainability indicator frameworks and a proposal for improvement. *Sustainability, 15*(21), 15425. https://doi.org/10.3390/su152115425

Ghosh, R., & Kansal, A. (2014). Urban challenges in India and the mission for a sustainable habitat. *INTER DISCIPLINA, 2*(2). https://doi.org/10.22201/ceiich.24485705e.2014.2.46530

Griggs, D., Stafford-Smith, M., Gaffney, O., Rockström, J., Öhman, M. C., Shyamsundar, P., Steffen, W., Glaser, G., Kanie, N., & Noble, I. (2013). Policy: Sustainable development goals for people and planet. *Nature, 495*(7441), 305–307. https://doi.org/10.1038/495305A

Gupta, J. (2023). *Statistical assessment of spatial autocorrelation on air quality in Bengaluru, India* (pp. 254–265). https://doi.org/10.1007/978-3-031-31164-2_21

Kandpal, R., & Okitasari, M. (2023). Governance transformation towards localisation of sustainable development goal 11 in India. *World Development Sustainability, 2*, 100069. https://doi.org/10.1016/j.wds.2023.100069

Karjalainen, L. E., & Juhola, S. (2021). Urban transportation sustainability assessments: A systematic review of literature. *Transport Reviews, 41*(5), 659–684. https://doi.org/10.1080/01441647.2021.1879309

Keivani, R. (2010). A review of the main challenges to urban sustainability. *International Journal of Urban Sustainable Development, 1*(1–2), 5–16. https://doi.org/10.1080/19463131003704213

Kellison, T. (2022). An overview of sustainable development goal 11. In *The Routledge handbook of sport and sustainable development* (pp. 261–275). Taylor and Francis. https://doi.org/10.4324/9781003023968-32

Küfeoğlu, S. (2022). *SDG-11: Sustainable cities and communities* (pp. 385–408). https://doi.org/10.1007/978-3-031-07127-0_13

Kumar, D., Shandilya, A. K., & Varghese, S. G. (2023). Mapping sustainable cities and communities (SDG 11) research: A bibliometric review †. *Engineering Proceedings, 59*(1). https://doi.org/10.3390/engproc2023059125

Lami, I. M., Abastante, F., Gaballo, M., Mecca, B., & Todella, E. (2023). Fostering sustainable cities through additional SDG11-related indicators. *Valori e Valutazioni, 32*(5), 45–61. https://doi.org/10.48264/VVSIEV-20233205.

Larsson, A., & Hatzigeorgiou, A. (2022). Sustainability. In *Designing smart and resilient cities for a post-pandemic world* (pp. 290–343). Routledge. https://doi.org/10.4324/9781003222583-14

Long, G., Censoro, J., & Rietig, K. (2023). The sustainable development goals: Governing by goals, targets and indicators. *International Environmental Agreements: Politics, Law and Economics, 23*(2), 149–156. https://doi.org/10.1007/s10784-023-09604-y

Mangweta, R., Mokoele, N. J., & Monama, S. A. (2022). Building sustainable cities to address urban sprawl: A reflective analysis towards achieving SDGs. *EUREKA: Social and Humanities, 6*, 72–78. https://doi.org/10.21303/2504-5571.2022.002704

Nabil, N., Haleem, A., & Khattab Khalil, A. I. (2023). Determining the indicators of sustainability to be achieved in new cities and urban communities. *International Journal of Scientific Research and Management (IJSRM), 11*(10), 932–943. https://doi.org/10.18535/ijsrm/v11i10.ec01

Nagabhatla, N., & Brahmbhatt, R. (2020). Geospatial assessment of water-migration scenarios in the context of sustainable development goals (SDGs) 6, 11, and 16. *Remote Sensing, 12*(9). https://doi.org/10.3390/RS12091376

Nijman, J. (2012). India's urban challenge. *Eurasian Geography and Economics, 53*(1), 7–20. https://doi.org/10.2747/1539-7216.53.1.7

Obe, A., Bolatito, S., & Madinah, N. (2024). An inclusive infrastructure, industrialization and innovation: A SDG 9 goal for Uganda. *IUIU Mbale Campus ONOMÁZEIN, 63*, 71–86.

Pradhan, B. K., Yadav, S., Ghosh, J., & Prashad, A. (2023). Achieving the sustainable development goals (SDGs) in the Indian State of Odisha: Challenges and opportunities. *World Development Sustainability, 3*, 100078. https://doi.org/10.1016/j.wds.2023.100078

Rauber, A., & Gonçalves, A. R. (2014). *Measuring urban sprawl and environmental sustainability*. https://www.researchgate.net/publication/262012798

Roy, A., Garai, N., & Biswas, J. K. (2023a). Exploration of urban sustainability in India through the lens of sustainable development goals. *Discover Sustainability, 4*(1). https://doi.org/10.1007/s43621-023-00158-2

Roy, A., Garai, N., & Kumar Biswas, J. (2023b). *Exploration of urban sustainability of India through the lens of sustainable development goals*. https://doi.org/10.21203/rs.3.rs-2830577/v1

Ruchira, G., & Kansal, A. (n.d.). *Urban challenges in India and the mission for a sustainable habitat*. https://www.researchgate.net/publication/299538034

Sharifi, A., Allam, Z., Bibri, S. E., & Khavarian-Garmsir, A. R. (2024). Smart cities and sustainable development goals (SDGs): A systematic literature review of co-benefits and trade-offs. *Cities, 146*. https://doi.org/10.1016/j.cities.2023.104659

Shekhar, S., & Ravi, K. (2023). Characterising the slum environment from space for achieving SDGS. *International Archives of the Photogrammetry, Remote Sensing and Spatial Information Sciences – ISPRS Archives, 48*(M-1-2023), 331–338. https://doi.org/10.5194/isprs-archives-XLVIII-M-1-2023-331-2023

Shruti, Singh, P. K., & Ohri, A. (2019). Towards developing sustainable smart cities in India. *International Journal of Engineering and Advanced Technology, 9*(2), 2876–2880. https://doi.org/10.35940/ijeat.B3653.129219

Tasaki, T., Kameyama, Y., Seiji, H., Moriguchi, Y., & Hideo, H. (2010). A survey of national sustainable development indicators. *International Journal of Sustainable Development, 13*(4), 337–361. https://doi.org/10.1504/IJSD.2010.038173

Vaidya, H., & Chatterji, T. (2020). *SDG 11 sustainable cities and communities* (pp. 173–185). https://doi.org/10.1007/978-981-32-9927-6_12

Verma, A., Sreenivasulu, S., & Dash, N. (2011). Achieving sustainable transportation system for Indian cities-problems and issues. *Current Science, 100*(9), 1328–1339.

Yailymova, H., Yailymov, B., Kussul, N., & Shelestov, A. (2021). *Air quality estimation in Ukraine using SDG 11.6.2 indicator assessment*. https://doi.org/10.20944/preprints202110.0299.v1

Zhang, L., Sun, Y., Li, C., & Li, B. (2024). Promoting sustainable development in urban–rural areas: A new approach for evaluating the policies of characteristic towns in China. *Buildings, 14*(4), 1085. https://doi.org/10.3390/buildings14041085

Chapter 8
Integrating BIM with Green Building Certification System to Design Sustainable Building

G. S. Mahaaraja and S. M. Renuka

Abstract Global greenhouse gas emissions reached 49.4 billion tons of CO_2 equivalent, with approximately 17.5% of total energy-related emissions attributed to building operations. A pivotal strategy to curbing carbon emissions involves the development of sustainable buildings through the reduction of energy use intensity. This research draws upon insights derived from the smart city study conducted by Johnson Controls, which underscores the limited compliance of Indian buildings—merely 4%—with "green" prerequisites. The main aim of this study are to integrate Building Information Modeling (BIM) with energy analysis, which is performed by Autodesk Green Building Studio and green building rating system by evaluating the criteria of proposed model. To achieve this aim, the responses from consultants and architects were analyzed using SPSS software, which demonstrate the high importance of criteria that are easily constructible and affordable. The survey analysis shows that there are a total of 16 GRIHA criteria, 21 IGBC criteria, and 14 LEED criteria. The energy use intensity is 1318.6 MJ/m^2/year in the Base Model and 1187.2 MJ/m^2/year in the Proposed Model. Overall, compared to the Base Model for the same floor area, the Proposed Model exhibits greater energy efficiency, resulting in lower energy consumption and lower energy expenses.

Keywords Green building · BIM · Autodesk Revit · Energy use intensity · LEED · IGBC · GRIHA · GBS

G. S. Mahaaraja · S. M. Renuka (✉)
Division of Structural Engineering, Department of Civil Engineering, CEGC, Anna University, Chennai, Tamil Nadu, India
e-mail: renuka@annauniv.edu

V. Baskaran et al. (eds.), *EcoTech Urbanism*, The Urban Book Series,
https://doi.org/10.1007/978-3-031-83833-0_8

8.1 Introduction

In recent decades, greenhouse gases (GHGs), primarily in the form of CO_2 emissions, have surged to unprecedented levels due to global growth and resource consumption. To identify the main sectors responsible for these emissions, a graphic from Our World in Data utilizes 2016 data from Climate Watch and the World Resources Institute, revealing total emissions of 49.4 billion tons of CO_2 equivalents (CO_2e). Global GHG emissions can be categorized into four main sectors: energy, agriculture, industry, and waste. Notably, nearly three-quarters of GHG emissions stem from energy consumption. Buildings account for 17.5% of energy-related emissions, which aligns with the striking reality that cities consume 60–80% of the world's annual energy (Our World in Data, 2021). Green Building Rating Systems (GBRSs) are generally third-party standards that are voluntary and market-driven. They assess the sustainability of buildings through a multi-criteria evaluation and promote the integration of environmentally, socially, and economically sustainable practices in the design, construction, and operation of buildings or neighborhoods. The goal of green building is to guide and evaluate the project at every stage of its life cycle, in order to minimize its negative impact on the environment, the health and well-being of the building occupants, and even to reduce operational expenses (Marchi et al., 2021). The construction sector has shown increased interest in creating and building eco-friendly structures that can offer exceptional performance while also generating significant cost savings (Jalaei & Jrade, 2014). It is projected that by 2056, global economic activity will have multiplied by five, global population will have grown by more than 50%, global energy consumption will have risen nearly threefold, and global manufacturing activity will have increased at least threefold (Matthews, 2000; Ilha et al., 2009). Key decisions regarding the design of sustainable buildings are often made during the conceptual phase of the building's lifecycle. Energy analysis usually occurs after the architectural and engineering designs and associated documentation have been completed. Energy efficiency plays a crucial role in classifying building materials as environmentally friendly. The primary objective of utilizing energy-efficient materials is to decrease the reliance on externally generated power needed at the construction site (Jeong et al., 2010). As energy consumption rises, so do environmental challenges and issues related to energy supply. Consequently, the primary focus today is on creating energy-efficient buildings. Achieving this goal requires reducing energy consumption and minimizing the environmental impact of construction. The use of fossil fuels contributes significantly to serious environmental problems, with greenhouse gas emissions from their combustion being a major concern (Maglad et al., 2023). The adaptability and reliability of energy analysis conducted through BIM-based simulations, including tools like Revit, Insight, and Green Building Studio (Ferny Celina, 2022). BIM signifies one of the most significant transformations in the Architecture, Engineering, Construction, and Operation (AECO) industry. Projects utilizing BIM offer extensive data related to building geometry, materials, costs, envelope components, electrical systems, HVAC (heating, ventilation, and air-conditioning) systems, and the thermal characteristics of materials (Dubljević et al., 2023). The production of building materials requires energy, as does the construction process itself, along with the energy needed for heating, lighting, power, and ventilation in a finished building. Beyond energy use, the building industry is also a significant source of environmental pollution (Kukadia & Hall, 2004; Zimmermann et al., 2005; Pitt et al., 2009; Yahya & Boussabaine, 2010). A significant portion of raw materials is consumed, totaling three billion tons each year, which accounts for 40% of global usage (Zimmermann et al., 2005; Ding, 2008) and produces a huge amount of waste (Burgan & Sansom, 2006; Osmani et al., 2008). BIM tools enabled designers to explore various design alternatives prior to finalizing the design solution, resulting in cost and time savings while also promoting more energy-efficient building designs (Ferny Celina, 2022).

Green building rating systems cater to various project types, ranging from single-family homes and commercial structures to entire neighborhoods. These systems are available for new construction, emphasizing planning, design, and construction practices, as well as for existing buildings, which concentrate on operational and maintenance aspects throughout their lifespan. In India, prominent green rating systems include LEED, IGBC, and GRIHA (IGBC, 2019; A GRIHA Council, 2021; U.S. Green Building Council, 2023) are shown in Fig. 8.1, Tables 8.1, and 8.2, respectively.

Previously, numerous researchers have assessed the green criteria provided by LEED, IGBC, and GRIHA (IGBC, 2019; A GRIHA Council, 2021; U.S. Green Building Council, 2023). The primary aim of this study is to obtain the minimum certification for G+3 residential building by achieving the required minimum green points through a Delphi technique survey (Norouzi et al., 2021) conducted among consultants and architects. The goal is to identify a cost-effective and easily constructible approach to meet the green certification requirements by incorporating green-certified products into a 3D BIM proposed model and performimg energy analysis to reduce energy consumption.

Fig. 8.1 LEED rating system

Table 8.1 IGBC rating system

Rating	Points	Recognition
Certified	40–49	Good practices
Silver	50–59	Best practices
Gold	60–74	Outstanding performance
Platinum	75–89	National excellence
Super-platinum	90–100	Global leadership

Table 8.2 GRIHA rating system

GRIHA rating threshold	Rating
25–40 points	☆
41–55 points	☆☆
56–70 points	☆☆☆
71–85 points	☆☆☆☆
86 or more points	☆☆☆☆☆

From the past study, the review of the literature is presented below.

Farooqui (Azhar et al., 2009) Building Information Modeling (BIM) can facilitate intricate analyses of building performance, contributing to the development of an optimized sustainable design. This exploratory study assesses three software tools for building performance analysis—Ecotect™, Green Building Studio™ (GBS), and Virtual Environment™—to evaluate their effectiveness for sustainability analysis within a BIM framework (Jrade & Jalaei, 2013; Jalaei & Jrade, 2014). The integration of BIM and energy analysis tools with a green building certification system is essential for sustainable conceptual building design. It highlights the significance of the conceptual phase in supporting decision-makers, even when information may be limited in accuracy. Guor (Mishra & Gour, 2018) examined green building rating systems in India, emphasizing their growing popularity and their contribution to enhancing sustainability in building design and performance. It recommended tackling key challenges present in current rating systems and introduced the idea of a revised version. Prakash (Jadhav, 2019) investigated BIM-based energy analysis, highlighting its importance in meeting sustainability standards and creating options for energy harvesting while minimizing environmental impact. Palaniappan (Varma & Palaniappan, 2019) conducted a comprehensive comparison of ten green building rating systems revealed that energy-related indicators are often given top priority across many of these schemes, followed by considerations of environmental and social factors. Celina (2022) researched performance assessment through simulation using Revit Insight and Green Building Studio, concentrating on elements that influence energy consumption in buildings and highlighting the significance of optimizing building envelopes for improved energy efficiency. Khahro et al. (2021) investigated the effective use of energy in the construction sector by leveraging BIM-based decision-making, highlighting the advantages of implementing BIM for sustainable design, enhancing energy efficiency, and optimizing overall building performance. Guo et al. (2021) introduced a BIM-based approach for evaluating and optimizing green buildings was introduced, highlighting its capacity to assess the performance of green buildings and achieve energy savings through renovation strategies. Hesam (Norouzi et al., 2021) examined the criteria for green hospitals in Fars Province, Iran, utilizing the Delphi method, with an emphasis on eco-friendly hospital design and energy efficiency. Prada Hanga-Fărcaş et al. (2023) conducted a case study focusing on the University of Oradea, highlighting how participation in green campus ratings and building certifications can promote a sustainable development strategy. By registering in the UI Green Metric World University Rankings, the University of Oradea aims to enhance its position in the rankings. This initiative has led to improvements in green criteria and a greater commitment to sustainable development within the university campus. Maglad et al. (2023) examined BIM-based energy analysis and optimization, highlighting the significance of energy-efficient design decisions and strategies for enhancing energy efficiency in building projects. Ur Rehman et al. (2023) proposed a comprehensive BIM-based strategy for green building design, leveraging the LEED system to enhance energy and water conservation while fostering sustainable design practices. The implementation of Building Information Modeling (BIM) in existing academic buildings has proven effective in decreasing energy consumption and costs. Tools such as Autodesk Revit, Autodesk Insight, and Autodesk GBS were used for energy simulations, while the LEED, IGBC, and GRIHA rating systems helped identify and apply sustainable measures. By incorporating BIM into the design process, the LEED certification can be achieved more efficiently, leading to considerable savings in time and resources. Moreover, this BIM approach supports the generation of renewable energy, thereby mitigating environmental impact. Overall, using BIM can successfully meet energy and cost reduction targets while simplifying the LEED certification process. The Delphi technique (Norouzi et al., 2021) was applied to prioritize criteria by conducting three rounds of questionnaires to identify high-impact criteria.

8.2 Methodology

To achieve the objective of reducing energy consumption and increasing cost savings through the implementation of easily constructible criteria with minimal expenses, a comprehensive methodology was employed. The process involved the creation and distribution of a questionnaire to consultants and architects, utilizing the Delphi technique (Norouzi et al., 2021) in three rounds to gather expert opinions. Subsequently, both the base model and proposed model were developed using Autodesk Revit (Autodesk, 2023), creating 3D Building Information Models (BIM). Energy analysis was conducted using the Green Building Studio, a cloud-based analysis tool, to assess the energy performance of the models (Autodesk Green Building Studio, 2023). The following steps were undertaken in the methodology is shown in Fig. 8.2.

To lay the foundation for the study, a comprehensive literature review is conducted, focusing on the integration of Building Information Modeling (BIM) with energy analysis tools (Autodesk Green Building Studio, 2023) and green building systems. Emphasis will be placed on studies pertaining to residential buildings, and notable software applications like Green Building Studio (Autodesk Green Building Studio, 2023) and Revit will be explored. A suitable residential building will be carefully selected for the study, and the specific site location for the energy analysis will be identified to ensure the accuracy and relevance of the study's findings. To guide the study and establish benchmarks for green building practices, the green building criteria from renowned certification systems such as LEED, IGBC, and

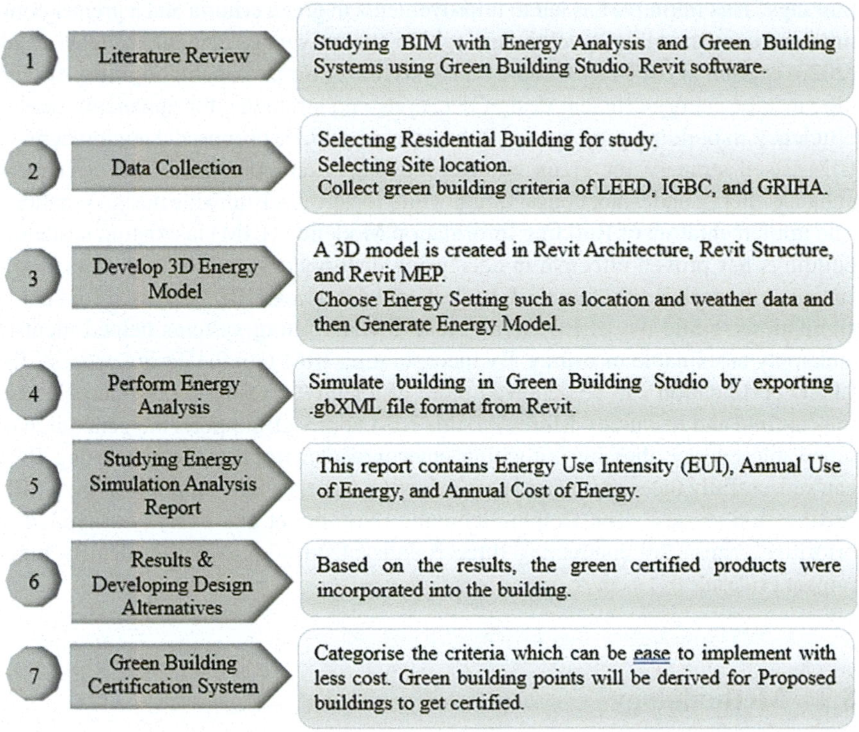

Fig. 8.2 Proposed methodology

GRIHA will be collected and utilized as essential guidelines (IGBC, 2019; A GRIHA Council, 2021; U.S. Green Building Council, 2023). A meticulous data collection process will be undertaken to gather relevant information about the selected residential building, including architectural plans, building materials, and energy consumption patterns. To create a comprehensive and accurate representation of the building, a detailed 3D model will be developed using Revit Architecture (Autodesk, 2023). The energy model will be fine-tuned by selecting appropriate energy settings, including location and weather data, to generate a precise and reliable energy model. Through the utilization of Green Building Studio (Autodesk Green Building Studio, 2023), energy simulations will be performed by exporting the Revit model in .gbXML file format (Autodesk, 2023), allowing for a thorough analysis of the building's energy performance. The results of the energy simulation will be analyzed in depth, with a focus on key metrics such as energy use intensity (EUI), annual energy consumption, and annual energy cost, to gain valuable insights into the building's energy efficiency. Based on the findings from the energy analysis, green-certified products and design alternatives will be integrated into the building to enhance its energy efficiency and overall sustainability. Green building criteria that can be easily implemented with minimal cost and effort will be meticulously identified and categorized, offering valuable insights for future green building projects. By quantifying the green building points for the proposed design alternatives, the study will aim to achieve certification under the chosen green building certification system, demonstrating the building's commitment to sustainable practices. The study will culminate in the presentation of a comprehensive methodology, detailing the step-by-step approach taken to achieve green building certification for the proposed residential building. This methodology will encompass the integration of BIM, energy analysis, and adherence to green building criteria, highlighting the building's commitment to sustainability and environmental responsibility.

8.2.1 Data Collection

The criteria for the multifamily residential building, as referenced from the LEED, IGBC, and GRIHA guidelines (IGBC, 2019; A GRIHA Council, 2021; U.S. Green Building Council, 2023), have been compiled and assigned unique criterion IDs for each specific criterion. These criteria checklist are detailed in Appendix A, Appendix B, and Appendix C, corresponding to LEED, IGBC, and GRIHA guidelines (IGBC, 2019; A GRIHA Council, 2021; U.S. Green Building Council, 2023), respectively. The compilation of listed criteria in these appendices serves as a comprehensive reference for the design and assessment of energy-efficient and sustainable multifamily residential buildings. The questionnaire used for evaluating each criterion of LEED BD+C Multi-family, IGBC, and GRIHA follows a 5-point Likert scale. Each criterion is presented with a specific question regarding its feasibility for easy construction with minimal costs. Participants then provide their responses using the 5-point Likert scale. The high RII importance criteria, which is easily constructible with less cost is obtained from questionnaire conducted by using three rounds of Delphi technique. Through the utilization of the Delphi technique, a questionnaire was administered to 23 experts comprising green building certificate consultants and architects. In general, the sample size for Delphi technique should be in the range of 10–50 response. In the initial round, participants provided their responses on the listed criteria using a 5-point Likert scale, ranging from "1—strongly disagree" to "5—strongly agree." Following the first round, a summary of the results was presented to the experts, who were then requested to reevaluate their ratings considering the feedback provided. This process was repeated for a total of three rounds to gather comprehensive and refined insights from the experts. Based on the experts' assessments, a compilation of green points criteria, focusing on ease of implementation and cost-effectiveness, was derived and presented in Table 8.4. Subsequently, the Relative Importance Index (RII) was computed using the SPSS software to quantify the significance of each criterion, with the RII scale indicated in Table 8.3 (Rooshdi et al., 2018). This systematic approach allowed for the identification and prioritization of criteria that are both feasible and economically viable for implementation in green building projects.

Table 8.3 RII scale

Values	Importance
$0.8 \leq RII \leq 1$	High
$0.6 \leq RII \leq 0.8$	Medium-high
$0.4 \leq RII \leq 0.6$	Medium
$0.2 \leq RII \leq 0.4$	Low-medium
$0 \leq RII \leq 0.2$	Low

The Relative Importance Index is calculated by using the formula, shown in Eq. 8.1 (Rooshdi et al., 2018):

$$\text{Relative Importance Index} = \frac{5n_5 + 4n_4 + 3n_3 + 2n_2 + 1n_1}{A * N} \tag{8.1}$$

n_5 = Total number of respondents for Strongly Agree
n_4 = Total number of respondents for Agree
n_3 = Total number of respondents for Neutral
n_2 = Total number of respondents for Disagree
n_1 = Total number of respondents for Strongly Disagree
A (Highest Weight) = 5
N (Total Number of Response) = 23

8.3 Base Model Creation

The project details pertain to a stilt+3 floor residential apartment proposed in the Kodambakkam, Chennai, Tamil Nadu is shown in Fig. 8.3. The apartment type is residential, with a plot size of 18 × 31 m^2. There are 3 units of 2BHK and 3 units of 3BHK, spread across 3 floors. The floor areas are 94 m^2 for 2BHK units and 111 m^2 for 3BHK units. The architectural plans are available in a .dwg file extension, which provides the necessary floor plans for the project.

Table 8.4 High importance criteria with ID and RII values from the survey

LEED	IGBC	GRIHA
IP (Integrative Process) (0.9391)	SA 2 (Site Preservation) (0.9478)	SSP 1 (Green Infrastructure) (0.9043)
LT 7 (Reduced Parking Footprint) (0.8609)	SSP 1 (Basic Amenities) (0.9130)	SSP 2 (Low Impact Design) (0.8783)
SS 4 (Rainwater Management) (0.8435)	SSP 2 (Proximity to Public Transport) (0.9043)	SSP 3 (Design to Mitigate UHIE) (0.9217)
SS 5 (Heat Island Reduction) (0.8696)	SSP 4 (Natural Topography or Vegetation) (0.8087)	CM 1 (Air and Soil Pollution Control) (0.9304)
WE 1 (Outdoor Water Use Reduction) (0.8261)	SSP 5 (Preservation or Transplantation) (0.8087)	CM 2 (Top Soil Preservation) (0.8957)
WE 2 (Indoor Water Use Reduction) (0.8261)	SSP 6 (Heat Island Reduction, Non-roof) (0.8522)	CM 3(Construction Management Practices) (0.8522)
WE 4 (Water Metering) (0.8696)	SSP 7 (Heat Island Reduction, Roof) (0.8522)	EE 1 (Energy Optimization) (0.9130)
EA 1 (Enhanced Commissioning) (0.8348)	SSP 9 (Universal Design) (0.9391)	OC 1 (Visual Comfort) (0.9043)
EA 2 (Optimize Energy Performance) (0.8870)	SSP 10 (Basic Facilities for Construction) (0.9043)	WM 1 (Water Demand Reduction) (0.8609)
MR 2 (Environmentally Preferable Products) (0.9217)	SSP 11 (Green Building Guidelines) 0.9304)	WM 3 (Rainwater Management) (0.9304)
MR 3 (C&D Waste Management) (0.8870)	WC 2 (Management of Irrigation Systems) 0.9391)	SBM 1 (Utilization of Alternative Materials in Building) (0.8783)
EQ 2 (No Environmental Tobacco Smoke) (0.9130)	WC 3 (Rainwater Harvesting, Roof & Non-roof) (0.8957)	SES 1 (Safety and Sanitation for Construction Workers) (0.9304)
IN 2 (LEED Accredited Professional) (0.9043)	WC 6 (Water Metering) (0.8783)	SES 2 (Universal Accessibility) (0.8957)
RP (Regional Priority) (0.9217)	EE 2 (Enhanced Energy Efficiency) (0.8261)	SES 3 (Dedicated Facilities for Service Staff) (0.9130)
	EE 5 (Commissioning, Post-Installation) (0.8870)	SES 4 (Positive Social Impact) (0.8957)
	EE 6 (Energy Metering and Management) (0.8783)	PMM 2 (Smart Metering and Monitoring) (0.8870)
	BMR 1 (Sustainable Building Materials) (0.8696)	
	BMR 3 (Handling of Waste Materials, during construction) (0.8957)	
	BMR 4 (Use of Certified Green Materials) (0.9043)	
	IEQ 3 (Outdoor Views) (0.8783)	
	ID 4 (IGBC Accredited Professional) (0.9478)	

*The above criteria is in the form of Criterion ID (criterion name) (RII value)

Fig. 8.3 Project location

The project location for the study is located at second street of Akbarabad, kodambakkam, Chennai, Tamil Nadu. The basic amenities like hospitals, hotels, supermarkets, bus stand, railway station, schools, colleges, ATMs, and parks are located around radius of 1 km distance. The base model of Revit first floor consists of 2BHK and 3BHK with staircase and lift facility is shown in Fig. 8.4 and it is typical till third floor. The 3D base model for the study is shown in Fig. 8.5.

Fig. 8.4 First floor 2BHK and 3BHK plan view

Fig. 8.5 3D BIM base model

8.4 Results and Discussion

This section encompasses the utilization of SPSS for survey analysis, aiming to identify criteria of high importance that are both cost-effective and easily implementable. Subsequently, a model based on these criteria is developed for evaluation. Energy analysis are conducted for both the baseline and proposed models. The ultimate objective is to assess the viability of the proposed model, ensuring it meets the requisite minimum criteria points set forth by LEED, IGBC, and GRIHA certifications.

8.4.1 Survey Analysis Using SPSS

The RII values which is in the range of 0.8–1 is said to be High is shown in Table 8.3, which is easy to implement with less cost. The Fig. 8.6 represents the rating values assigned to various sustainability certification systems—LEED, IGBC, and GRIHA—based on the Range of Influence Index (RII) is shown in the form of green rating matrix is shown in Fig. 8.6. For a high level of influence, LEED has a value of 14, IGBC has a value of 21, and GRIHA has a value of 16, with the condition $0.8 \leq \text{RII} \leq 1$. For a medium-high level of influence, LEED has a value of 20, IGBC has a value of 17, and GRIHA has a value of 9, with the condition $0.6 \leq \text{RII} \leq 0.8$. For a medium level of influence, LEED has a value of 3, IGBC has a value of 4, and GRIHA has a value of 3, with the condition $0.4 \leq \text{RII} \leq 0.6$. For a low-medium level of influence, LEED has a value of 3, IGBC has no value, and GRIHA has no value, with the condition $0.2 \leq \text{RII} \leq 0.4$. For a low level of influence, LEED, IGBC, and GRIHA all have values of 0, with the condition $0 \leq \text{RII} \leq 0.2$.

	LEED	IGBC	GRIHA	
High	14	21	16	0.8 ≤ RII ≤ 1
Medium-High	20	17	9	0.6 ≤ RII ≤ 0.8
Medium	3	4	3	0.4 ≤ RII ≤ 0.6
Low-Medium	3	0	0	0.2 ≤ RII ≤ 0.4
Low	0	0	0	0 ≤ RII ≤ 0.2

Fig. 8.6 Green rating matrix

8.4.2 *Green Points Criteria to Be Implemented in Model from Survey*

From the questionnaire analysis using SPSS software (IBM, 2020), response collected from the consultants and architects, the Table 8.4 shows the high importance of criteria, which is easy to implement with less cost. It shows that totally 14 criteria from LEED (U.S. Green Building Council, 2023), 21 criteria from IGBC (2019), and 16 criteria from GRIHA (A GRIHA Council, 2021).

8.4.3 *Proposed Model Creation for Main Study*

Based on the survey response, the proposed model for the study is shown in Fig. 8.7, it contains green-certified products like AAC blocks from JK smart block, green grid pavers, and double glazing unit from Saint-Gobain, which are incorporated into the Revit for analyses in GBS.

The green grid pavers in the non-roof area of ground floor, as shown in Fig. 8.8, are used to reduce surface runoff volume and improve groundwater quality.

The building input specification for both base model and proposed model are shown in Table 8.5. The materials used in proposed model are green-certified products, which allow to create the Revit family by incorporating green products specification.

8.4.4 *Energy Analysis of Base and Proposed Model*

The energy analysis of both base model and proposed model is performed in cloud-based analysis of GBS (Autodesk Green Building Studio, 2023) after exporting the model from Revit in .gbXML format. Before exporting the model, the energy setting is given to both model, as shown in Table 8.6. The energy model of proposed project is shown in Fig. 8.9. The analysis is done by uploading the .gbXML format in GBS after creating the project defaults.

Fig. 8.7 Proposed 3D BIM model

Fig. 8.8 Green grid pavers

Table 8.5 Building input specification

	Base model	Proposed model
Wall	Brick masonry, **U-Value = 2.6567**	JK smart block, AAC Block, **U-Value = 0.6481**
Window, Slide Door	1/4 in Pilkington single glazing, **U-Value = 6.7, VLT = 0.9, SHGC = 0.89**	Double glazing, **U-Value = 2.85, VLT = 0.68, SHGC = 0.49**
Roof	Cast in-situ Concrete, **U-Value = 6.9733**	Cast in-situ concrete with white roofing tile, **U-Value = 0.6**

Table 8.6 Energy setting for both model

Parameter	Value
Ground plane	GF
Project phase	New construction
Building type	Multifamily
Building operating schedule	24/7 facility
Export category	Spaces

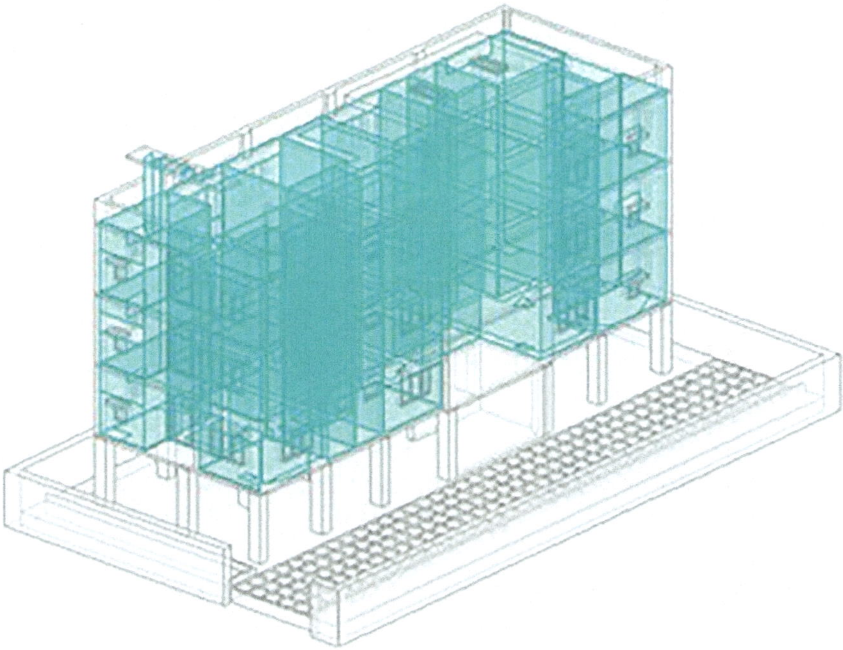

Fig. 8.9 Energy model of proposed project

After the run has created, the energy analysis report of both base and proposed model is shown in Fig. 8.10. The detailed analysis is shown in Table 8.7.

Table 8.7 compares the base model and proposed model of the building in terms of various parameters. The floor area remains constant at 509 m^2 for both models. The proposed model exhibits reduced energy use intensity, with 1187.2 MJ/m^2/year, compared to 1318.6 MJ/m^2/year in the base model. The electric cost per kilowatt-hour remains unchanged at Rs. 6.5 for both models. Moreover, the proposed model demonstrates a decrease in total annual electric energy consumption, with 92,756 Kwh, as opposed to 1,11,333 Kwh in the base model. The total annual electric cost in the proposed model is reduced to Rs 6,02,913 from Rs 7,23,661 in the base model. Similarly, the proposed model also exhibits a reduced total energy cost of Rs. 6,02,913 compared to Rs 7,23,661 in the base model. In conclusion, the proposed model showcases improved energy efficiency, leading to lower energy consumption and reduced energy costs in comparison to the base model, despite having the same floor area.

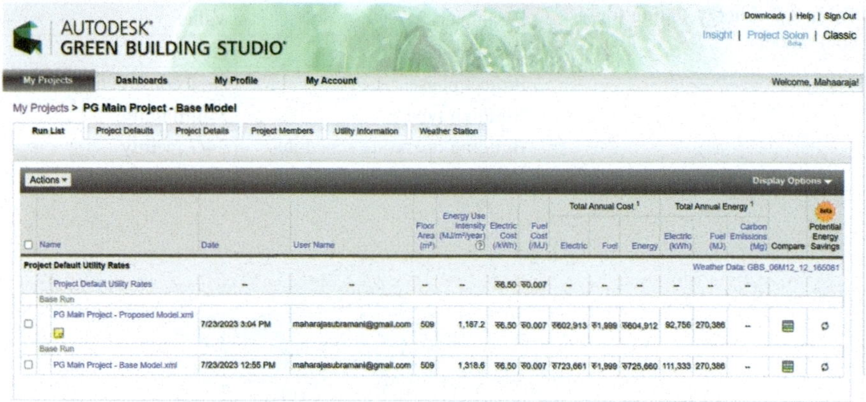

Fig. 8.10 Energy analysis report from GBS for both model

Table 8.7 Detailed energy analysis report for both model

Name of building	Base model	Proposed model
Floor area (m^2)	509	509
Energy use intensity ($MJ/m^2/year$)	1318.6	1187.2
Electric cost (/Kwh)	Rs 6.5	Rs 6.5
Total annual electric energy (Kwh)	1,11,333	92,756
Total annual electric cost	Rs 7,23,661	Rs 6,02,913
Total energy cost	**Rs 7,23,661**	**Rs 6,02,913**

8.4.5 Evaluation of LEED BD+C Multi-Family Criteria

The LEED criteria collected from the survey, shown in Table 8.4, are implemented and evaluated using the LEED BD+C Multi-Family reference guide (U.S. Green Building Council, 2023), which is provided in the USGBC portal. The points of LEED criteria evaluation is shown in Table 8.8.

The LT 3 (high-priority site) criteria is not included in the survey response because the experts says it is site-specific and not applicable to all locations. This criterion is only partially achieved for the selected project location, as it depends on the specific site and zone conditions. From the survey results, it is evident that 14 LEED criteria hold high importance and have been successfully achieved. The proposed residential building has attained 40 points out of the maximum 53 points in the LEED survey, resulting in its certification as a green building, as illustrated in Fig. 8.11.

Table 8.8 LEED criteria evaluation

Criteria ID	Maximum points	Intent obtained	Achieved points
IP	1	Early analysis of building is done to interrelate the high performance, cost-effective, equitable project outcomes	1
LT 7	1	By selling parking separately from all property sales or leases, that is, unbundling parking	1
SS 4	3	10% of the site is planted with trees, and some of the area is paved with green grid pavers, which reduces the surface runoff volume and thereby improves water quality	3
SS 5	2	Non-roof measures: Tactically situating trees and vegetation on the exterior of buildings can effectively obstruct sunlight, thereby reducing the extent to which buildings heat up and require cooling	2
WE 1	2	The project reduced its landscape water requirement by 35% from the calculated baseline for the site's peak watering month. Reductions are achieved by plant species selection and installing an irrigation system with drip irrigation	2
WE 2	6	The building installed a 6 lpf (liter per flush) tank in toilets, 3.8 lpf for urinals, 8.3 lpm at 415 kPa for kitchen faucets, and 9.5 lpm at 550 kPa for shower stalls to reduce indoor potable water consumption by 45%	5
WE 4	1	Meter water systems serve water for more than 80% of the irrigated landscaped area	1
EA 1	6	This is achieved by verifying the design documents and as-built drawing	4
EA 2	18	The proposed project's energy performance has an EUI of 1187.1 MJ/m^2/year, which is reduced by 131.4 MJ/m^2/year and below the ASHRAE 90.1 benchmark	10
MR 2	2	The project uses environmentally preferable products, such as green-certified products like AAC blocks, high-performance double glaze units, and roof coating—high-performance emulsion for roof tiles white	2
MR 3	2	The project follows a waste management plan and diverts 50% of the total construction and demolition materials from landfills and incineration facilities	2
EQ 2	2	Smoking inside the building is prohibited, and no-smoke policies are communicated to occupants, with no-smoke signage placed in and around the building	1
IN 2	1	At least one principal participant of the project team must be a LEED accredited professional (AP) with a specialty appropriate for the project	1
RP	4	This is the automatic achievement point that happens through LEED online based on the normal credits attempted and achieved	4
LT 3	2	Historic district: The project location is in Kodambakkam, Chennai	1

Fig. 8.11 LEED achieved points and rating

8.4.6 Evaluation of IGBC Residential Criteria

The IGBC criteria collected from the survey, which is shown in Table 8.4 is implemented and evaluated using the IGBC residential reference guide (IGBC, 2019), which is provided in the IGBC portal. The points of IGBC criteria evaluation is shown in Table 8.9.

From the survey, the high-importance criteria of 21 IGBC criteria is achieved. The proposed residential building achieved 51 points out of the maximum 56 points in the IGBC survey and certified as GOOD PRACTICES, as shown in Fig. 8.12.

8.4.7 Evaluation of GRIHA Residential Criteria

The GRIHA criteria collected from the survey, which is shown in Table 8.4 is implemented and evaluated using the GRIHA residential reference guide (A GRIHA Council, 2021), which is provided in the GRIHA portal. The points of GRIHA criteria evaluation is shown in Table 8.10.

From the survey, the high-importance criteria of 16 GRIHA criteria were achieved. The proposed residential building achieved 48 points out of the maximum 60 points in the GRIHA survey and was rated as a two-star rating, as shown in Fig. 8.13.

Table 8.9 IGBC criteria evaluation

Criteria ID	Maximum points	Intent obtained	Achieved points
SA 2	2	10% of the landscape is undisturbed, and 75% of the trees on the site are used in the building design	2
SSP 1	1	ATM, parks, schools, restaurants, and supermarkets are available in and around 1 km radius from the site	1
SSP 2	1	Kodambakkam railway station and a bus stop are located within an 800 m walking distance from the building	1
SSP 4	2	20% of the area is left for vegetation purpose	2
SSP 5	1	Totally 6 trees are preserved and 7 new trees are planted	1
SSP 6	2	75% of the parking are under covered	2
SSP 7	2	95% of the roof area is covered with high reflective material	2
SSP 9	1	Seating areas are provided near lift lobbies, and Braille and audio assistance in lifts are available for visually impaired people	1
SSP 10	1	Basic facilities for the construction workforce are provided, including sanitary facilities, PPE, first-aid kits, drinking water, and a site emergency alarm	1
SSP 11	1	Project-specific green building guidelines are provided to help building occupants implement and utilize green features post-occupancy	1
WC 2	1	Soil moisture sensors and drip irrigation are installed to use water efficiently	1
WC 3	4	Kodambakkam attains 191 mm of average peak month rainfall, in which the one day rainfall is considered to be 18% of the average peak month rainfall	4
WC 6	2	Brass analog residential water meters 25 mm, Multijet class A, are installed to measure the total potable water use for the building and associated grounds	2
EE 2	15	10% of EUI is reduced and 10% cot related to energy is saved. Five-star-rated fans, as well as pumps and motors, are installed. The building envelope includes U-value and SHGC	12
EE 5	2	The building equipment and systems commissioned in the design stage are verified to achieve the owner's requirements	2
EE 6	2	Building energy performance is monitored by installing energy meters which is followed by submeters	2
BMR 1	8	30% of the material used are locally available, and 10% are recycled material	6
BMR 3	1	75% of the waste generated is planned to recycle or reuse	1

(continued)

Table 8.9 (continued)

Criteria ID	Maximum points	Intent obtained	Achieved points
BMR 4	5	Products used include “JK Smart Blox (AAC block), high-performance double glaze units, roof coating—high-performance emulsion for roof tiles white, exterior wall primer.”	5
IEQ 3	1	The building occupants must not have any obstruction of views at least 8 m (26.2 feet) from the exterior vision glazing	1
ID 4	1	One principal participant of the project team shall be an IGBC accredited professional	1

Fig. 8.12 IGBC achieved points and rating

The points distribution among the LEED, IGBC, and GRIHA is shown in Fig. 8.14. It is represented in graphical form, with maximum points and achieved points on the horizontal axis, and green points on the vertical axis.

The data from Fig. 8.14 represents the performance of a building project in three different sustainability rating systems: LEED, IGBC, and GRIHA. The building project earned 40 points out of a maximum of 53 points available in the LEED (Leadership in Energy and Environmental Design) rating system. With this score, the building qualifies for the “Certified” level, indicating that it has met the basic sustainability requirements set by LEED. In the IGBC (Indian Green Building Council) rating system, the building project achieved an impressive score of 51 points out of a maximum of 56 points. This performance corresponds to the “Good

Table 8.10 GRIHA criteria evaluation

Criteria ID	Maximum points	Intent obtained	Achieved points
SSP 1	5	Minimum of one tree for every 80 m^2 of site area is planted and maintained in order to increase vegetation on the site	4
		Per capita gross area for the building is 19.6 m^2/person (range = 12.5–50 m^2/person)	
		Basic amenities are available near the site of 800 m radius	
SSP 2	5	AAC blocks, green grid pavers, high reflective roof tiles are used in the project	3
SSP 3	2	The reduction in predicted hourly average air temperature is 2.5 °C less than the base model	2
CM 1	1	Water were sprinkled on unpaved pathways on the site using non-potable water. The speed of vehicular movement on-site is limited to 10 km/h. It ensured that vehicles carrying waste materials out of the site were covered	1
CM 2	1	Topsoil from disturbed areas on the site were preserved, stabilized, and its fertility were maintained throughout the construction period. Moreover, all soil needed for landscaping, including roof gardens, which met through the utilization of this conserved soil	1
CM 3	1	Employ gunny bags, the ponding technique, or curing compounds, and incorporate water-reducing admixtures into the concrete mix	1
EE 1	12	10% of EUI is reduced and 10% cot related to energy is saved. Five-star-rated fans, as well as pumps and motors are installed. Building envelope includes U-value and SHGC	6
OC 1	4	Daylight is provided in the building to ensure visual comfort	4
WM 1	4	The landscape water demand of the project is reduced by 80%	2
WM 3	5	Green grid pavers are paved in 40% of the site, which reduces the surface runoff volume by more than 75%	4
SBM 1	5	AAC blocks is used in the building, PPC is used for masonry mortar and plasters, 5% of M-sand is used by weight of fine aggregate in structural concrete	5
SES 1	1	Arrange a minimum of two events throughout the construction phase to foster environmental awareness among the construction workforce	1
SES 2	2	Seating area are provided near lift lobbies. Braille and audio assistance in lifts are available for visually impaired people	2
SES 3	2	The provision of dedicated rooms for resting for service staff on-site is ensured, along with the provision of on-site toilets for service staff	2
SES 4	3	The environmental awareness is provided to the occupants	3
PMM 2	7	Digital metering is installed to monitor the energy and water consumption	7

Fig. 8.13 GRIHA achieved points and rating

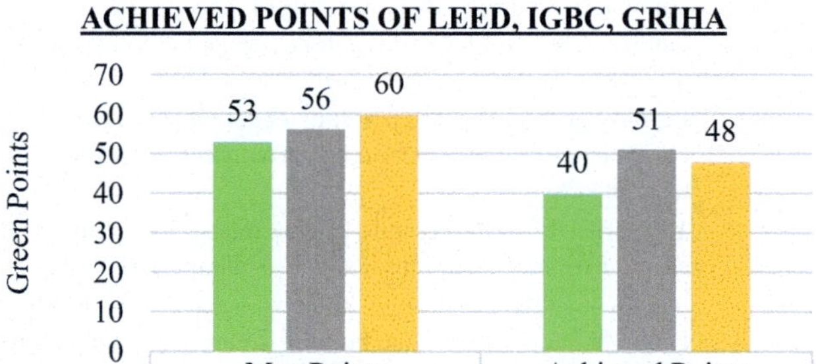

	Max Points	Achieved Points
LEED	53	40
IGBC	56	51
GRIHA	60	48

LEED IGBC GRIHA

Fig. 8.14 Points distribution among the LEED, IGBC, and GRIHA

Practices" level, which signifies a high standard of green building practices and environmental consciousness. For GRIHA (Green Rating for Integrated Habitat Assessment), the building project received 48 points out of a maximum of 60 points. This score qualifies the building for a "two-star rating," indicating that it has successfully implemented sustainable design and construction strategies, earning a notable level of recognition in environmental performance.

Overall, the building project has demonstrated a commendable commitment to sustainability, achieving respectable ratings in all three rating systems—LEED Certified, IGBC Good Practices, and GRIHA 2-Star Rating.

8.5 Conclusions

As per Climate Watch, the World Resources Institute 2020, the global GHG emissions contributes about 17.5% for buildings. In order to reduce the energy-related emissions, this study has been carried out. The aim of the study is to integrate Building Information Modeling (BIM) with energy analysis and established green building certification systems represents a pivotal approach for the purposeful design of environmentally sustainable structures. To address this aim, the following methodology is carried out to streamline the selection of criteria from prominent certifications such as LEED, IGBC, and GRIHA, focusing on those seamlessly executable within the construction phase while incurring minimal costs. Additionally, a comprehensive energy performance analysis is conducted to gauge the building's operational efficiency. Through systematic assessment, this research contributes to evaluating the building's adherence to the fundamental requisites for obtaining a baseline green building certification. To optimize energy consumption, factors like energy use intensity (EUI), U-value, and solar heat gain coefficient (SHGC) values were taken into account, resulting in a reduction in energy consumption within the building. A questionnaire, designed using the Delphi technique, was then administered in three rounds, with 23 respondents each. Criteria with a Relative Importance Index (RII) value greater than 0.8 were categorized as high-importance criteria. Consequently, the high-importance criteria for LEED, IGBC, and GRIHA were identified, comprising 14, 21, and 16 criteria, respectively is shown in Table 8.4. For further evaluation, a 3D base model and a proposed model were created using Revit Architecture, and their energy analysis was carried out using Green Building Studio (GBS). The proposed model was carefully assessed to meet the necessary criteria points for obtaining green building certification. The primary goal was to achieve certification with the minimum number of green points while demonstrating a remarkable 10% improvement in energy efficiency, resulting in reduced energy consumption and substantial energy cost savings of Rs 1,20,748 compared to the base model, despite both models having the same floor area of 502 m^2. Currently, the Industries and Commerce Department offers a significant 25% subsidy on the total fixed capital investment for buildings that achieve green ratings, excluding certain costs like land, land development, preliminary expenses, preoperative expenses, and consultancy fees. This incentive is applicable to both MSME and large industries. Given the growing energy consumption in residential buildings, it is strongly recommended that the government further encourages green building certification by providing an additional 10% subsidy on the total fixed capital investment for such projects. This measure would not only raise awareness about sustainable construction practices but also foster a significant uptake of green building initiatives and contribute to a greener and more energy-efficient future.

Appendix 8A

Rating System: LEED Residential BD+C: Multi-Family

The LEED BD+C Residential: Multifamily criteria are divided into seven categories, which include Location and Transportation, Sustainable Sites, Water Efficiency, Energy and Atmosphere, Materials and Resources, Indoor Environmental Quality, and Innovation. From the (U.S. Green Building Council, 2023), the total possible credit for LEED BD+C Residential: Multifamily is 110 points, in which every criteria is given with criterion ID. The LEED checklist and the possible credits used in this study is available in U.S. Green Building Council 2023 for the system LEED Residential BD+C: Multi-Family.

Appendix 8B

The Indian Green Building Council (IGBC) has developed a checklist to evaluate the environmental performance of buildings. The checklist includes various categories and credits that a building can achieve, based on its design and operational aspects. The IGBC 2019 criteria is given with criterian ID. From the IGBC 2019, the total possible credit for IGBC Residential for tenant-occupied buildings is 100 points. If required, the IGBC checklist and the total possible credits used in this study are available in IGBC 2019 webpage.

Appendix 8C

The LEED, IGBC, and GRIHA checklist consists of several criteria, each of which is assigned a certain number of credits. The total number of credits available varies depending on the type of building or community being rated. A GRIHA Council 2021 criteria is given with criterion ID. The total possible credit for GRIHA Residential is 105 points. If required, the GRIHA checklist and the total possible credits used in this study are available in A GRIHA Council 2021 webpage.

References

A GRIHA Council. (2021). *A GRIHA Council Publication Volume 1 Introduction to National Rating System – GRIHA*.

Autodesk. (2023). *Revit 2023*.

Autodesk Green Building Studio. (2023). *Building Performance Analysis*.

Azhar, S., Brown, J., & Farooqui, R. (2009). BIM-based sustainability analysis: An evaluation of building performance analysis software. In *Proceedings of the 45th ASC Annual Conference*.

Burgan, B. A., & Sansom, M. R. (2006). Sustainable steel construction. *Journal of Constructional Steel Research, 62*(11), 1178–1183.

Celina, F. (2022). *Energy Analysis with Revit Insight and Green Building Studio | Autodesk University*.

Ding, G. K. (2008). Sustainable construction—The role of environmental assessment tools. *Journal of Environmental Management, 86*(3), 451–464.

Dubljević, S., Tepavčević, B., Markoski, B., & Anđelković, A. S. (2023). Computational BIM tool for automated LEED certification process. *Energy and Buildings, 292*, 113168.

Guo, K., Li, Q., Zhang, L., & Wu, X. (2021). BIM-based green building evaluation and optimization: A case study. *Journal of Cleaner Production, 320*, 128824.

IBM. (2020). *SPSS Statistics, IBM*.

IGBC. (2019). *Green Homes Rating System – Version 3.0 for Multi-dwelling Residential Units*.

Ilha, M. S. d. O., de Oliveira, L. H., & Gonçalves, O. M. (2009). Environmental assessment of residential buildings with an emphasis on water conservation. *Building Services Engineering Research and Technology, 30*(1), 15–26.

Jadhav, P. P. (2019). *Assesment of building energy by performing simulation with BIM* (pp. 49–53). Faculty of Engineering Hunedoara.

Jalaei, F., & Jrade, A. (2014). Integrating Building Information Modeling (BIM) and energy analysis tools with green building certification system to conceptually design sustainable buildings. *Journal of Information Technology in Construction (ITcon), 19*, 494–519.

Jeong, K.-S., Lee, K.-W., & Lim, H.-K. (2010). Risk assessment on hazards for decommissioning safety of a nuclear facility. *Annals of Nuclear Energy, 37*(12), 1751–1762.

Jrade, A., & Jalaei, F. (2013). Integrating building information modelling with sustainability to design building projects at the conceptual stage. *Building Simulation, 6*(4), 429–444.

Khahro, S. H., Kumar, D., Siddiqui, F. H., Ali, T. H., Raza, M. S., & Khoso, A. R. (2021). Optimizing energy use, cost and carbon emission through BUILDING information modelling and a sustainability approach: A case-study of a hospital building. *Sustainability, 13*(7), 3675.

Kukadia, V., & Hall, D. (2004). *Improving air quality in urban environments: Guidance for the construction industry*. BRE Bookshop.

Maglad, A. M., Houda, M., Alrowais, R., Khan, A. M., Jameel, M., Rehman, S. K. U., Khan, H., Javed, M. F., & Rehman, M. F. (2023). Bim-based energy analysis and optimization using insight 360 (case study). *Case Studies in Construction Materials, 18*, e01755.

Marchi, L., Antonini, E., & Politi, S. (2021). Green Building Rating Systems (GBRSs). *Encyclopedia, 1*(4), 998–1009.

Matthews, E. (Ed.). (2000). *The weight of nations: material outflows from industrial economies*. World Resources Institute.

Mishra, P., & Gour, S. (2018). Green building rating system. *International Journal of Research in Engineering, Science and Management, 1*.

Norouzi, D., Vahdat, S., & Hesam, S. (2021). Investigating Green Hospital Criteria Using Delphi Method for Fars Province, Southwest of Iran, 2019. *Journal of Community Health Research*.

Osmani, M., Glass, J., & Price, A. D. (2008). Architects' perspectives on construction waste reduction by design. *Waste Management, 28*(7), 1147–1158.

Our World in Data. (2021). *A global breakdown of greenhouse gas emissions by sector, visual capitalist*.

Pitt, M., Tucker, M., Riley, M., & Longden, J. (2009). Towards sustainable construction: promotion and best practices. *Construction Innovation, 9*(2), 201–224.

Prada Hanga-Fărcaş, I. F., Bungău, C. C., Scurt, A. A., Cristea, M., & Prada, M. F. (2023). The building certification system – A tool of sustainable development of university campuses. *Journal of Applied Engineering Sciences, 13*(1), 105–112.

Rooshdi, R. R. R. M., Majid, M. Z. A., Sahamir, S. R., & Ismail, N. A. A. (2018). Relative importance index of sustainable design and construction activities criteria for green highway. *Chemical Engineering Transactions, 63*, 151–156.

U.S. Green Building Council. (2023). *LEED v4.1 | Building Design and Construction*.

Ur Rehman, H. S., Raza, M. A., Masood, R., Khan, M. A., Alamgir, S., Javed, M. A., Roy, K., & Lim, J. B. P. (2023). A multi-facet BIM based approach for Green Building design of new multi-family residential building using LEED system. *International Journal of Construction Management, 23*(12), 2024–2038.

Varma, C. R. S., & Palaniappan, S. (2019). Comparision of green building rating schemes used in North America, Europe and Asia. *Habitat International, 89*, 101989.

Yahya, K., & Boussabaine, H. (2010). Quantifying environmental impacts and eco-costs from brick waste. *Architectural Engineering and Design Management, 6*(3), 189–206.

Zimmermann, M., Althaus, H.-J., & Haas, A. (2005). Benchmarks for sustainable construction: A contribution to develop a standard. *Energy and Buildings, 37*(11), 1147–1157.

Chapter 9
Regenerative and Carbon Neutral Construction: A Sustainable Building Practices

M. Naveenkumar, Aparna Anil Kumar, S. R. Preethiga, and K. Senthilkumar

Abstract Sustainable building practices are an essential approach to constructing buildings that minimize negative impacts on the environment and promote long-term ecological balance. These practices involve the use of environmentally friendly materials, energy-efficient systems, and sustainable design principles. Sustainable building materials, such as reclaimed wood, bamboo, and recycled steel, reduce waste and conserve natural resources. Energy-efficient systems, such as solar panels, wind turbines, and geothermal heating and cooling, significantly reduce energy consumption and greenhouse gas emissions. Sustainable design principles, such as passive solar heating, natural ventilation, and daylighting, maximize the use of natural resources and reduce the need for artificial lighting and heating. Additionally, sustainable building practices promote the use of green spaces, water-efficient landscaping, and rainwater harvesting, which help to conserve water resources. These practices also prioritize the health and well-being of building occupants, promoting indoor air quality, thermal comfort, and acoustic comfort. They contribute to the creation of healthy, productive, and comfortable indoor environments, which can lead to increased productivity and reduced absenteeism. Furthermore, these practices can lead to cost savings in the long term, as energy-efficient systems and sustainable materials can reduce operating costs and maintenance expenses. Additionally, they can increase property values and attract tenants who prioritize sustainability. Sustainable building practices are crucial for creating environmentally responsible, energy-efficient, and healthy buildings that benefit both the environment and building occupants. By incorporating sustainable materials, energy-efficient systems, and sustainable design principles, we can reduce our carbon footprint, conserve natural resources, and promote long-term ecological balance. This chapter focuses on the utilization of energy-efficient design for resilient infrastructure, highlighting the principles and strategies that drive the transformative potential of sustainable construction.

M. Naveenkumar (✉) · A. A. Kumar · S. R. Preethiga
Department of Civil Engineering, Easwari Engineering College, Chennai, Tamil Nadu, India

K. Senthilkumar
Department of Food Technology, Kongu Engineering College, Erode, Tamil Nadu, India

V. Baskaran et al. (eds.), *EcoTech Urbanism*, The Urban Book Series,
https://doi.org/10.1007/978-3-031-83833-0_9

Keywords Sustainable building practices · Environmental sustainability · Energy-efficient systems · Sustainable materials · Cost savings · Ecological balance · Comfortable indoor environments

9.1 Introduction

The construction industry can greatly affect the environment through air pollution, water pollution, noise pollution, waste generation, fuel consumption, and material extraction and manufacturing. These effects can be reduced by adopting sustainable construction practices like designing eco-friendly structures, installing energy efficient systems, and embracing strategies for reducing waste. The approach to regenerative architecture is about thinking ahead of how we can design buildings that heal the earth instead of leaving it in a worse condition than it was before. The shift from sustainability to regeneration means that architects have to figure out how buildings can work as parts of larger systems that produce and distribute resources such as clean water, energy, or food. Regenerative buildings incorporate natural processes, thereby establishing a mutually beneficial relationship between a building and its surrounding, thus contributing toward making the world better.

Regenerative and carbon-neutral construction are two critical components of sustainable building practices that aim to transform the built environment into a more eco-friendly and resilient space. These approaches focus on not only reducing the environmental impact of buildings but also actively contributing to the health and well-being of occupants while promoting ecological health. Regenerative construction involves designing and building structures that not only minimize their environmental footprint but also actively contribute to the health and well-being of occupants and the surrounding ecosystem. This approach emphasizes the integration of natural systems and the use of materials that sequester and store carbon, transforming buildings into carbon sinks.

Carbon-neutral construction, on the other hand, focuses on achieving net-zero energy consumption in buildings. This involves the strategic use of renewable energy sources, energy-efficient systems, and smart controls to minimize energy consumption. Additionally, carbon-neutral design incorporates carbon offsetting and negative carbon technologies to achieve net negative carbon emissions, surpassing carbon neutrality requirements. Key strategies for regenerative and carbon-neutral construction include integrating renewable energy sources, employing smart controls and automation, exploring carbon offsetting and negative carbon technologies, optimizing building materials and construction, considering lifecycle management, monitoring and evaluating energy performance, and raising awareness through education and training. By adopting these strategies, the construction industry can move toward a more sustainable future, transforming buildings into carbon sinks that not only reduce their environmental impact but also actively contribute to the health and well-being of occupants and the surrounding ecosystem.

9.2 Understanding Sustainability in Building Practices

9.2.1 Definition and Principles of Sustainable Construction

Sustainable construction means building structures that minimize their negative effects on the environment, the society, in general, and the economy. It consists of using renewable and recyclable materials, cutting down on energy use and waste, and protecting the natural environment. The main aim of sustainable construction is to reduce its environmental impact so that at the end of such a building's life cycle it would have left least traces on planet earth (Kitek Kuzman et al., 2014).

Designing buildings to minimize adverse environmental impact and consider environmental, social, and economic impacts, as shown in Fig. 9.1.

- Using construction techniques and robust materials that can withstand the test of time to reduce the need for frequent replacements and maintenance.
- Implementing design techniques and strategies to reduce energy consumption during planning, building, and maintenance phases.
- Implementing strategies and processes to minimize waste generation during construction and promoting the use of recycled materials.

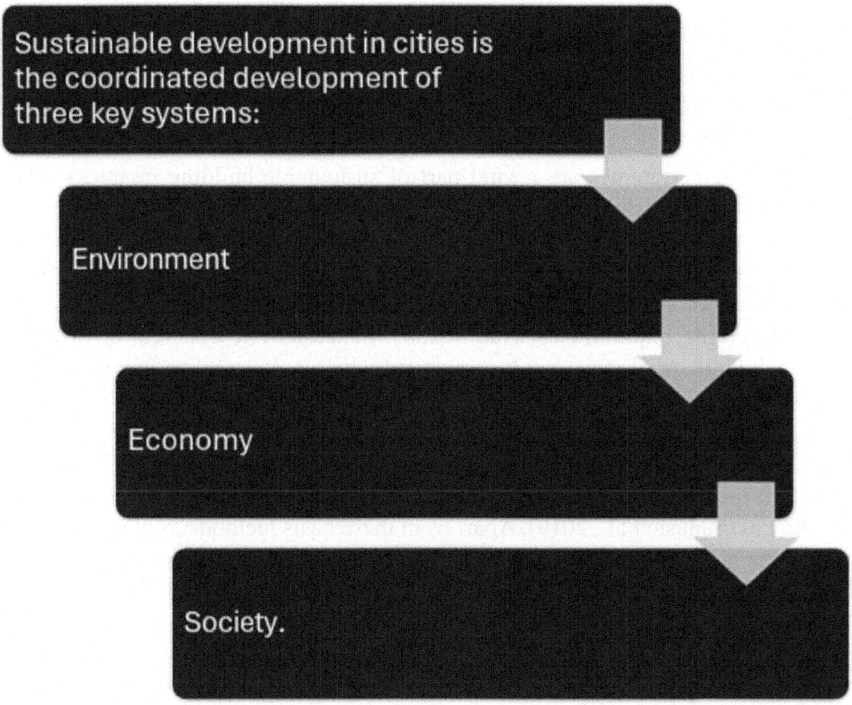

Fig. 9.1 Schematic representation of sustainable types

Designing buildings to conserve water effectively through various strategies like rainwater harvesting and efficient irrigation.

9.2.2 Benefits of Sustainable Building Practices

- Cost reduction: Sustainable construction can lead to lower building maintenance and operation costs by using durable and sustainable materials, reducing energy consumption, and incorporating water conservation measures.
- Increased productivity: Environmentally friendly workspaces in sustainable buildings can enhance employee performance and reduce absenteeism, creating a conducive environment for improved productivity (Tramontin et al., 2012).
- Environmental benefits: Sustainable building practices reduce energy consumption, promote cleaner air and water, and decrease the carbon footprint, leading to a healthier environment and a more sustainable future.
- Economic growth: The green building industry contributes significantly to economic growth by creating jobs, boosting the economy, and increasing demand for skilled labor in sustainable construction practices.

9.3 Sustainable Design Principles

9.3.1 Integration of Passive Design Strategies

Passive design strategies are a vital part of sustainable building practices. These strategies aim to cut down on energy use and reduce negative effects on the environment by using natural elements and smart design tactics. For instance, the orientation of a construction project should be optimized for maximum exposure to the sun during winter months, while minimizing it in summer months, thus decreasing heating and cooling requirements. Additionally, materials like concrete or brick that possess high thermal mass can be employed which help stabilize indoor temperatures through absorbing extra heat during daytime then releasing it at night. Adequate ventilation is also necessary, whereby cross-ventilation is promoted through placement of windows and openings in strategic locations while stack ventilation is created by alternating high and low vents to establish a natural cycle of air movement (Sanusi et al., 2019). Apart from these basis methods, roofs with vegetation coverings or reflective coatings may significantly minimize thermal loads gain, as well as improve insulation properties against heat transmission into conditioned spaces below them. Moreover, thinking adaptably while planning enables easy changes over time for different uses, thus ensuring sustainability throughout the lifetime of a building. Adoption of rainwater harvesting systems, coupled with graywater recycling systems, also helps meet non-potable needs conserve freshwater resources, sustainably manage demand, reduce costs related to treatment, transportation, and disposal, and bring about self-sufficiency.

9.3.2 Use of Renewable Energy Sources

Construction sites are responsible for approximately 39% of global carbon dioxide emissions resulting from energy use and industrial processes. The adoption of sustainable building practices requires the use of renewable energy sources. To solve this problem, the field of construction has been looking to different forms of renewable energy such as solar, wind, hydroelectric, and hydrogen power. Among the most used renewable energies on construction sites is solar energy. Lighting systems, tools, and other equipment can be powered by electricity generated from the capture of sunlight on rooftops, scaffolding, or even ground-mounted solar panels. Wind power also presents itself as viable especially in areas not connected to the grid where construction companies can associate with wind farms. Additionally, construction workers can tap into hydropower plants nearby which are said to be the cleanest source of energy available globally for electricity production during their activities as well. Another type of renewable energy that may be used at construction sites is bioenergy. This involves converting organic waste like wood chips and sawdust produced through construction processes into heat or electricity using methods such as anaerobic digestion and biomass combustion, respectively.

9.3.3 Efficient Water Management Systems

Efficient water management systems in sustainable construction and buildings are essential for reducing water consumption, promoting water conservation, and minimizing environmental impact. These systems encompass a range of strategies, including the use of low-flow fixtures, graywater recycling, rainwater harvesting, and efficient water management technologies. By implementing these practices, construction projects can optimize water use, reduce reliance on external water sources, and create self-sustaining water systems within buildings. Efficient water management not only contributes to environmental sustainability but also helps in achieving cost savings, promoting resource efficiency, and enhancing the overall sustainability of construction projects (Morganti et al., 2019). Strategies for efficient water management include installing low-flow fixtures, recycling greywater, harvesting rainwater, detecting and monitoring leaks, using water-efficient landscaping, and implementing water-efficient construction practices. These strategies can help reduce water consumption, lower water bills, and minimize wastewater discharge. For example, the Bullitt Center in Seattle uses a closed-loop water system, recycling graywater and rainwater for toilet flushing and irrigation, reducing potable water use by 83%. Similarly, the Dockside Green Development in Victoria,

Canada, incorporates a district-scale water reclamation system, recycling wastewater for toilet flushing and irrigation, reducing potable water use by 65%.

9.4 Energy Efficiency Measures

9.4.1 Insulation and Thermal Performance

Using insulation materials that have low heat conductivity such as walls and roofs can help save a lot of energy while improving temperatures within living spaces. This means that during the life cycle of a building, less power will be needed for cooling or heating due to better thermal insulation provided by materials with low thermal conductivity. By doing this, not only are operating costs reduced but also greenhouse gas emissions that result from using energy are cut down, making it environmentally friendly. Traditional examples of insulation materials are expanded polystyrene (EPS) and rockwool, while newer ones include aerogel panels and thermo-rock boards.

9.4.2 Energy-Efficient Lighting and Appliances

The use of LED lighting, which consumes only a fraction of the energy compared to traditional incandescent bulbs, is becoming increasingly prevalent in both residential and commercial buildings. LED lights offer superior performance, a longer lifespan, and greater reliability, making them a cost-effective and sustainable choice for indoor and outdoor lighting applications. In addition to efficient lighting, the selection of energy-efficient appliances is essential for creating a green home. High-performance appliances, such as front-loading washing machines and Energy Star-certified refrigerators and dishwashers, can reduce energy use by up to 50% compared to conventional models. These appliances are often built with higher-quality materials and superior craftsmanship, resulting in a longer lifespan and fewer repairs, further enhancing their economic and environmental benefits (Appleby et al., 2011). When designing a self-built green home, it is crucial to consider the site's orientation to maximize natural resources like sunlight and wind for energy generation, reducing the overall energy requirements of the building. Sustainable and eco-friendly building materials with low volatile organic compound (VOC) emissions should be prioritized, along with the expertise of architects and designers specializing in green home construction, as shown in Fig. 9.2.

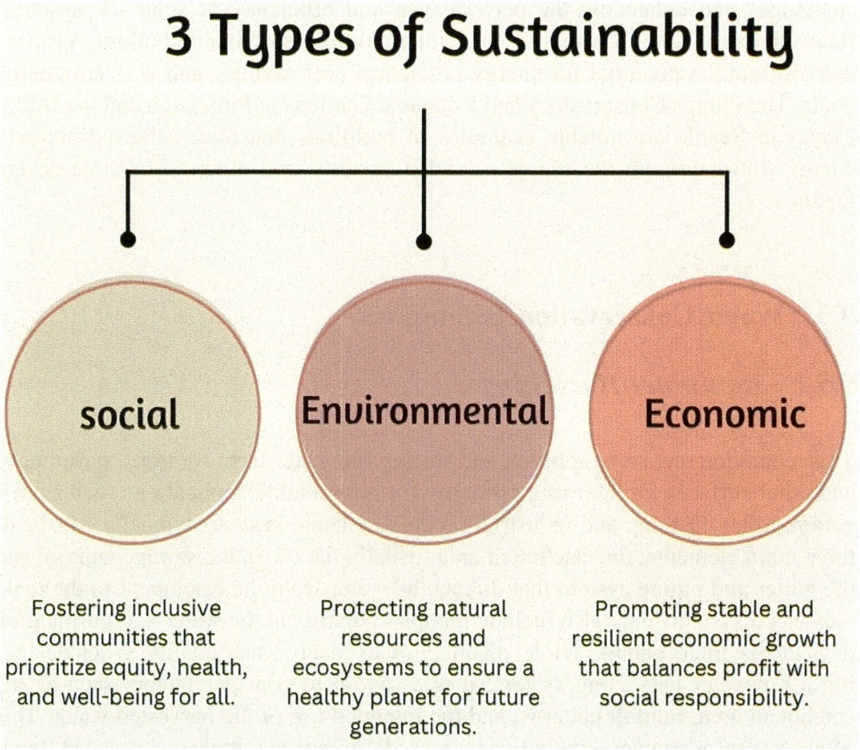

Fig. 9.2 Representation of sustainable types

9.4.3 Solar Photovoltaic Systems

Solar photovoltaic (PV) systems play a crucial role in sustainable building practices, converting sunlight into electricity and reducing reliance on fossil fuels. These systems can be installed on rooftops, facades, or even integrated into building materials like glass, generating renewable energy for lighting, heating, and cooling systems. The use of solar PV systems in sustainable buildings offers numerous benefits, including reduced energy consumption, cost savings on utility bills, and the potential for net-zero energy buildings that generate as much energy as they consume. Solar PV systems can be designed to optimize energy efficiency and sustainability, considering factors such as building orientation, solar panel placement, and integration with the building's overall architecture and energy systems (Guenther et al., 2007). The integration of solar PV systems with smart building systems can further enhance energy efficiency and cost savings by allowing for greater control and optimization of energy usage. Despite the benefits, solar PV systems face challenges such as high initial costs, integration with existing infrastructure, and the availability of suitable green building materials. However, advancements in solar technology, such as bifacial solar panels and perovskite solar cells, are addressing these challenges and enhancing the performance and efficiency of solar PV systems. Real-life examples of successful solar integration in sustainable building projects demonstrate the potential for energy efficiency, cost savings, and net-zero energy goals. The Phipps Conservatory and Botanical Gardens in Pittsburgh and the Bullitt Center in Seattle are notable examples of buildings that have achieved net-zero energy status through the use of solar PV systems and other sustainable design features.

9.5 Water Conservation Techniques

9.5.1 Rainwater Harvesting

This technique involves capturing and storing rainwater from rooftops, pavements, and other surfaces for later use, primarily for non-potable applications such as irrigation, toilet flushing, and industrial processes. These systems typically consist of three main elements: the catchment area (usually the roof), the storage tank(s), and the gutter and piping system that directs the water from the catchment to the tank. Advanced systems may also include pumps to distribute the water and purification devices like filters and ultraviolet disinfection to ensure water quality. When designing a rainwater harvesting system, it is essential to consider factors such as the catchment area, rainfall patterns, and the intended use of the harvested water. This offers several advantages, including increased groundwater recharge, reduced flooding, and lower water bills for homeowners. The stored rainwater can be used for various purposes, such as gardening, toilet flushing, and even drinking water if adequately treated. However, it is crucial to ensure that the rainwater is collected from clean surfaces and that the storage tanks are properly maintained to prevent contamination.

9.5.2 Graywater Recycling

Graywater recycling involves treating wastewater from appliances like showers, baths, and sinks to be reused for non-potable purposes such as toilet flushing and irrigation. The process begins with collecting wastewater from appliances and passing it through a collection unit for contaminant removal using biological, chemical, and physical actions (Guenther, 2008). The treated water undergoes ultrafiltration to remove particles, bacteria, and viruses before being stored in a tank for reuse. Graywater recycling systems are beneficial for businesses, particularly those requiring a constant water supply for baths, basins, and toilets, as they can reduce water consumption by up to 40%. The installation costs of graywater recycling systems vary, with payback periods ranging from 2.5 to 5 years, depending on the system type. Businesses investing in graywater recycling systems can benefit from tax relief and reduced water bills, contributing to water conservation and sustainability efforts.

9.6 Waste Reduction and Management

9.6.1 Construction Waste Minimization

One of the key strategies for construction waste minimization is to incorporate waste reduction considerations into the design process. This can be achieved by estimating and quantifying potential waste generation during the early stages of design, allowing for informed decision-making and the implementation of waste-reducing measures. Design professionals can specify materials with recycled content, modular dimensions, and standardized sizes to minimize waste. Additionally, designing for deconstruction and incorporating prefabrication or off-site construction techniques can significantly reduce waste generation. Material selection is another critical factor in construction waste minimization. Effective waste management strategies, such as on-site sorting, recycling, and proper disposal of hazardous materials, are also essential for minimizing the environmental impact of construction waste. Successful construction waste minimization requires collaboration among all stakeholders, including designers, contractors, suppliers, and waste management professionals.

9.6.2 Recycling and Reuse Practices

Recycling and reuse practices are becoming increasingly important in the construction industry as a means of reducing waste, conserving resources, and promoting sustainability. One of the key strategies for recycling and reuse is designing buildings for adaptability, disassembly, and reuse. This approach involves using simple structural systems, modular components, and mechanical fasteners instead of adhesives, making it easier to dismantle and reuse materials at the end of a building's life assessment. Deconstruction, the process of carefully dismantling a building to salvage usable materials, is another effective way to reduce waste and increase reuse. Recycled materials can be used in various applications, such as crushed concrete as aggregate in new concrete mixes, wood for landscaping, and metals for new products. Manufacturers are also incorporating recycled content into their products, reducing the demand for virgin materials. To promote recycling and reuse, construction professionals can plan ahead by creating a waste reduction or recycling plan, communicate it to all parties involved, and establish a system for handling materials efficiently. They can also look for materials on exchanges or reuse centers, use standard dimensions to reduce cutoff waste, and prioritize materials that are durable, reusable, or recyclable.

9.6.3 Implementation of Circular Economy Principles

A circular economy aims to keep materials in circulation at their highest value for as long as possible, minimizing waste and environmental impact. One of the key steps in implementing a circular economy is adapting business models to prioritize circularity. This involves rethinking traditional linear models of "take-make-use-dispose" and transitioning toward a circular approach of "take-make-use-recover" (Bokalders et al., 2009). Businesses can achieve this by designing products for longevity, repairability, and recyclability, ensuring that materials can be easily recovered and reused at the end of a product's life cycle. Another essential aspect of implementing a circular economy is collaboration across the value chain. This requires effective communication and coordination among stakeholders, including suppliers, manufacturers, retailers, and consumers, to ensure that materials are properly recovered and reused. Initiatives such as product take-back programs and reverse logistics can facilitate the recovery of materials and promote a circular flow of resources. Governments also play a crucial role in supporting the transition to a circular economy through policies, regulations, and incentives. These measures can include extended producer responsibility schemes, landfill bans, and tax incentives for businesses that adopt circular practices. Implementing a circular economy requires a holistic approach that considers the entire life cycle of products and materials. This includes design decisions, manufacturing processes, distribution channels, and end-of-life management.

9.7 Sustainable Materials and Resources

9.7.1 Selection Criteria for Eco-Friendly Materials

When evaluating the sustainability of eco-friendly materials, several criteria should be considered to ensure they have minimal environmental impact throughout their life cycle. Firstly, renewability: the material should come from renewable sources, such as plants or recycled materials, the rate of resource replenishment should match or exceed the rate of consumption. Secondly, resource efficiency: efficient use of raw materials and energy in the production process and minimal waste generation during manufacturing. Thirdly, low environmental impact: low greenhouse gas emissions during production and transportation, minimal pollution and toxic by-products during manufacturing and disposal. Next is biodegradability: ability to decompose naturally without causing harm to the environment and to breakdown into nontoxic components that can reintegrate into the ecosystem (Muller Brook, 2018). Recyclability: capability of being recycled into new products at the end of its life cycle, with established infrastructure and processes for recycling the material. Durability and longevity: long-lasting and resistant to wear and tear, reducing the need for frequent replacement. Maintains functionality and performance over an extended period. Energy efficiency: low energy consumption during production, transportation, and disposal. Use of renewable energy sources in the manufacturing process. Nontoxicity: free from harmful chemicals and substances that can affect human health and the environment. Safe for use and handling throughout its life cycle. Next, social responsibility: ethical sourcing of raw materials, ensuring fair labor practices and community welfare and positive impact on local economies and societies.

9.7.2 Use of Recycled and Salvaged Materials

Using recycled and salvaged materials is an important strategy in sustainable development, contributing to reduced waste, conservation of natural resources, and minimized environmental impact. Here are some key aspects and benefits: Benefits of using recycled and salvaged materials, waste reduction, divert materials from landfills, reducing the environmental burden of waste disposal. Extends the life cycle of existing materials, promoting a circular economy. Resource conservation, decreases the need for virgin raw materials, preserving natural resources, lowers the environmental impact associated with resource extraction and processing. Energy savings, often requires less energy to process recycled materials compared to producing new ones from raw material, reducing the carbon footprint associated with production and transportation. Cost efficiency: can be more cost-effective than purchasing new materials, especially if salvaged locally. Reduces costs associated with waste disposal and landfill use.

9.7.3 Certification and Labeling Systems

Establishing a robust labeling system for eco-friendly materials involves several key criteria to ensure transparency, credibility, and comprehensiveness. Here are the essential criteria for an effective labeling system: clarity and transparency, clear information: labels should provide easily understandable information about the material's eco-friendly attributes (Majerska-Palubicka et al., 2014). Transparency discloses all relevant details about the material's sourcing, production, and environmental impact. Standardization and consistency uniform standards: use consistent criteria and metrics across all products to enable easy comparison. Certifications: align with recognized standards and certifications (e.g., LEED, cradle to cradle, FSC). Third-party verification independent audits: ensure that claims are verified by

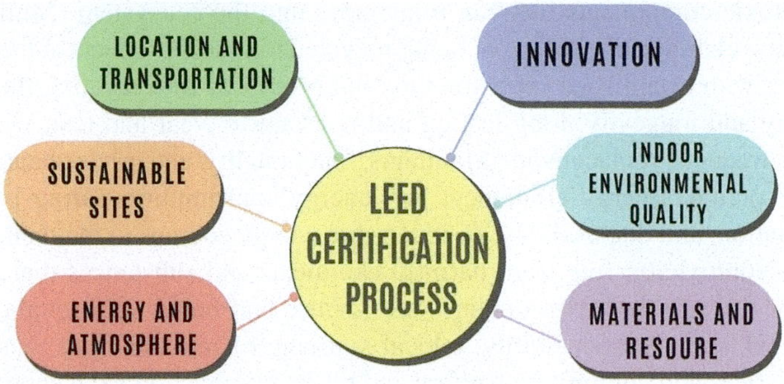

Fig. 9.3 Representation of points to be earned for LEED certification

independent, third-party organization. Credibility: increase trust in the label by using reputable certifiers.

9.8 Green Buildings Certification and Standards

9.8.1 LEED (Leadership in Energy and Environmental Design)

Leadership in Energy and Environmental Design (LEED) is a globally recognized green building certification system developed by the U.S. Green Building Council (USGBC). LEED provides a framework for healthy, efficient, and cost-saving green buildings and is applicable to virtually all building types, as shown in Fig. 9.3 (Todeschi et al., 2021).

9.8.2 BREEAM (Building Research Establishment Environmental Assessment Method)

The Building Research Establishment Environmental Assessment Method (BREEAM) is one of the world's leading sustainability assessment methods for master planning projects, infrastructure, and buildings (Peng et al., 2012) developed by the Building Research Establishment (BRE) in the UK. BREEAM sets standards for best practices in sustainable building design, construction, and operation, as shown in Fig. 9.4.

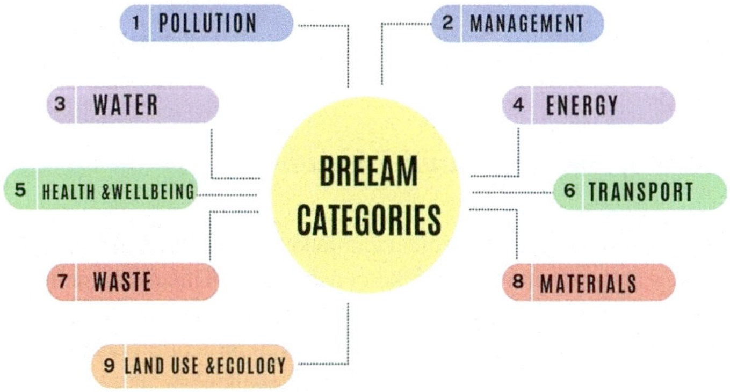

Fig. 9.4 Representation of points to be earned for BREEAM certification

9.8.3 Passive House Standard

The passive house standard is a rigorous, voluntary standard for energy efficiency in buildings, significantly reducing their ecological footprint. It results in ultralow energy buildings that require little energy for space heating or cooling. The concept, developed by the Passive House Institute (PHI) in Germany, focuses on superior energy performance and comfort.

9.9 Challenges and Barriers to Sustainable Construction

9.9.1 Cost Consideration

Cost considerations for a sustainable building project encompass several critical factors aimed at balancing environmental benefits with financial feasibility. Firstly, initial construction costs can be higher due to the use of eco-friendly materials and technologies, such as energy-efficient windows, solar panels, and green insulation. These upfront investments, however, are often offset by long-term savings in operational costs, primarily through reduced energy consumption and lower utility bills. Additionally, incorporating sustainable practices can lead to decreased maintenance costs over the building's life cycle due to the durability and longevity of green materials. Another significant consideration is the potential for financial incentives and subsidies. Governments and institutions frequently offer tax credits, grants, and rebates to support sustainable construction, which can substantially mitigate initial expenditures. Furthermore, sustainable buildings often achieve higher property values and rental incomes, given the increasing market demand for environmentally responsible living and working spaces. Lifecycle cost analysis (LCCA) is a crucial tool in assessing the overall cost-effectiveness of sustainable projects. This approach evaluates the total cost of ownership, including initial investment, operational costs, maintenance, and end-of-life disposal or recycling (Khoshnaw et al., 2019).

9.9.2 Lack of Awareness and Education

Lack of awareness in a building project can lead to significant problems, undermining the success and sustainability of the construction. This issue manifests in various ways throughout the project life cycle, from planning and design to execution and completion. At the planning stage, lack of awareness often results in inadequate feasibility studies and risk assessments. Project managers may overlook essential factors such as environmental impact, local regulations, and community needs. This oversight can lead to legal complications, increased costs, and delays. For instance, failing to consider environmental regulations may result in fines or the need to halt construction to address compliance issues, causing substantial financial and time losses. During the design phase, insufficient awareness can result in poor architectural and engineering decisions. Designers may not fully understand the local climate, leading to structures that are not energy-efficient or resilient to local weather conditions.

9.9.3 Regulatory and Policy Issues

Building projects are subject to a complex array of regulatory and policy issues that can significantly impact their planning, development, and completion. Key among these are zoning laws, which dictate land use and determine what types of structures can be built in specific areas. Zoning regulations can affect building height, density, and the intended use of the property, whether residential, commercial, or industrial (Daugelaite et al., 2021). Building codes and standards, which ensure that structures meet minimum safety and health requirements, also play a crucial role. These codes cover aspects such as electrical systems, plumbing, fire safety, and structural integrity. Environmental regulations are another critical factor, requiring builders to assess and mitigate the environmental impact of their projects. This can include issues like waste management, water quality, and the preservation of natural habitats. Compliance with the National Environmental Policy Act (NEPA) or similar local legislation often necessitates detailed environmental impact assessments and public consultations.

9.10 Future Trends in Sustainable Buildings Practices

9.10.1 Advancements in Green Technologies

One significant development is the integration of renewable energy sources, such as solar panels and wind turbines, directly into building designs. These technologies not only reduce reliance on fossil fuels but also contribute to a building's energy independence. Energy-efficient systems, like advanced HVAC (heating, ventilation, and air conditioning) units and LED lighting, are also becoming standard. These systems often incorporate smart technology, allowing for real-time monitoring and adjustments to optimize energy usage. Additionally, the use of sustainable building materials has gained prominence (Li et al., 2014). Materials such as cross-laminated timber, recycled steel, and low-VOC (volatile organic compounds) paints reduce the carbon footprint and improve indoor air quality. Green roofs and walls, which are covered with vegetation, provide insulation, reduce urban heat islands, and improve air quality. Water conservation technologies are also critical in sustainable building projects. Systems for rainwater harvesting, graywater recycling, and low-flow fixtures ensure that water use is minimized. Furthermore, advancements in building automation systems (BAS) enable comprehensive control over a building's operations, enhancing efficiency and reducing waste. Passive design strategies, which leverage natural light, ventilation, and thermal mass, reduce energy consumption by optimizing the building's orientation and structure.

9.10.2 Adoption of Net-Zero and Carbon-Neutral Building

Net-zero and carbon-neutral buildings are a huge thing when it comes to sustainable building projects because their aim is to eliminate the carbon footprint that comes with construction and operation. Net-zero structures are created in such a way that they can generate more power than what is necessary for them over a year by mainly using sources like wind turbines, solar panels, and geothermal systems. To reduce the amount of energy these buildings need, their design also includes things such as high-efficiency HVAC units, LED lighting, and advanced insulation materials. Carbon-neutral buildings take this further by compensating for any remaining emissions through methods like purchasing offsets for carbon or afforestation schemes where trees are planted to absorb CO_2, as well as integrating carbon capture technology into their design. These types of buildings are built with materials that can be easily sustained over time; for example, environmentally friendly pieces like recycled steel, reclaimed wood, and low carbon concrete help cut down on embodied carbon which refers to the emissions released during production and transportation of building materials too. In addition, rainwater harvesting systems, among other water conservation methods, may also be used to cut down consumption levels while also reducing power costs from pumping water around premises. They are referred to as smart based on how well they can achieve their own energy needs without relying much, if at all, on an external supply (either electrical grid or gas supply lines).

9.10.3 Importance of Resilient Design

Resilience is a critical aspect of sustainable building projects, especially in the face of increasing environmental challenges such as extreme weather events, natural disasters, and climate change impacts. Building resilience involves designing structures and systems that can withstand and adapt to these challenges, ensuring the long-term viability and functionality of the built environment. One key aspect of resilience in sustainable building projects is robustness in construction. This involves using durable materials and construction techniques that can withstand high winds, floods, earthquakes, and other hazards. Another important aspect of resilience is redundancy and flexibility in building systems. This means designing buildings with backup systems for essential services such as power, water, and communications, ensuring that they can continue to operate even during disruptions. Flexible design allows buildings to adapt to changing conditions over time, such as shifts in climate patterns or evolving user needs. Furthermore, sustainable building projects often prioritize location and site selection to minimize exposure to environmental risks. This includes avoiding floodplains, wildfire-prone areas, and other high-risk zones, as well as maximizing access to public transportation and other essential services to reduce vulnerability to disruptions. Resilience in sustainable building projects also extends beyond the individual building level to consider broader community resilience. This involves integrating buildings into their surrounding environment in a way that supports ecosystem health, enhances community connectivity, and fosters social cohesion. By promoting community resilience, sustainable buildings contribute to overall societal well-being and help communities bounce back stronger from adverse events (Al-Kodmany et al., 2016).

9.11 Conclusion

In summary, the sustainable building project is a completely new way of doing things when it comes to construction; it takes into account the environment, uses resources effectively, and is fair to society as a whole. Future buildings will need to do better than this one environmentally because we have used systems that are very advanced. This means that the project uses things like solar panels, rainwater harvesting systems, and insulation that saves energy. Additionally, it has been constructed using materials found within the area and methods that do not harm the environment, thus creating jobs locally while engaging communities around it. Besides, such kind of houses are meant to improve people's health while living in them; hence, they should have enough natural light, good ventilation, and places where they can relax. In general terms, sustainable building serves as an example of creativity and motivation, in that design should consider the surroundings to ensure that life continues to thrive even amidst challenges posed by climate change.

References

Al-Kodmany, K. M. (2016). Sustainable tall buildings: Cases from the global south. *International Journal of Architectural Research: Arch Net-IJAR, 10*.

Appleby, P. (2011). *Integrated sustainable design of buildings*. Earthscan.

Bokalders, V. (2009). *The whole building handbook: How to design healthy, efficient, and sustainable buildings*. Earthscan.

Daugelaite, A., & Grazuleviciute-Vileniske, I. (2021). The relationship between ethics and aesthetics in sustainable architecture of the Baltic Sea region. *Sustainability, 13*.

Guenther, R. (2007). *Sustainable healthcare architecture*. John Wiley & Sons, Inc.

Guenther, R. (2008). *Sustainable healthcare architecture*. John Wiley & Sons.

Khoshnaw, R. (2019). Sustainable construction in Kurdish vernacular Architecture. *Periodica Polytechnica Architecture, 50*, 178.

Kuzman, K., Manja, & Kutnar, A. (2014). Sustainable buildings. In *Contemporary Slovenian timber architecture for sustainability* (pp. 39–56). Springer International Publishing.

Li, G., Qiao, Q. Q., & Zhou, X. M. (2014). The engineering examples of sustainable architecture. *Advanced Materials Research*, 1008–1009.

Majerska-Palubicka, B. (2014). Sustainable landscape between buildings. *Advanced Engineering Forum, 12*, 171.

Morganti, M., Habib, E., Currà, E., & Cecere, C. (2019). Traditional buildings Back to the future. Adaptive energy efficiency in reuse. In *Sustainable vernacular architecture* (pp. 409–425). Springer International Publishing.

Muller, B. (2018). New horizons for sustainable architecture. *Nature and Culture, 13*, 189.

Peng, H., Liu, Y., Le Zhang, Y., Tan, M. Q., & Gan, C. (2012). A case study of integrated and human-oriented sustainable architectural design. *Advanced Materials Research, 368-373*, 3706.

Sanusi, A. N. Z., Azmin, A. K., Abdullah, F., & Kassim, M. H. (2019). Climatic adaptations of colonial school buildings in Malaysia. In *Sustainable vernacular architecture* (pp. 275–292). Springer International Publishing.

Todeschi, V., Beltramino, S., El Jamous, B., & Mutani, G. (2021). Low-energy architecture for sustainable neighborhoods. *Tecnica Italiana-Italian Journal of Engineering Science, 65*, 83.

Tramontin, V., Loggia, C., Basciu, M., & Saddi, M. (2012). Adaptive buildings for sustainable architecture. In *Sustainability in energy and buildings* (pp. 275–284). Springer Berlin Heidelberg.

Chapter 10
Innovative Approach to Sustainable Urban Road Construction: Enhancing Pervious Concrete with GGBS and HDPE Fiber Reinforcement

T. Seethalakshmi and A. Ciola Rixy Winslet

Abstract Pervious concrete is made up of a layer of mortar, coarse aggregate, and bonding that creates a structure that resembles a honeycomb. The fact that pervious paving has holes in it that let water seep through to the underlying materials underneath is one of its key features. Additionally, it lessens water pollution and peak stormwater flow while encouraging groundwater recharging. The primary goal of this study is to replace cement with GGBS, which makes it environmentally friendly, while also aiming to strike the ideal balance between strength and permeability. Utilizing a by-product like GGBS is a sustainable construction practices, which results in lower CO_2 emissions. In this paper, 10%, 20%, and 30% of GGBS have been replaced to cement. HDPE plastic fiber with length 18 ± 2 mm, which is obtained from the used HDPE cement bags, is used as a reinforcing material. Different proportion of plastic fibers 1%, 1.5%, and 2% was added in pervious concrete trial. The specimens were casted and examinations like the compressive strength test, infiltration test, flexure strength test, and impact test were conducted to determine its mechanical properties. When GGBS was optimized, the mechanical properties of pervious concrete were boosted. Based on the test results, the optimum dosages for GGBS are 20%. Compressive strength was 23.44% more than in the control group. Adding 1.5% HDPE fiber was the most beneficial to improve the flexure properties. It is 41.68% greater than control mix, but infiltration rate decreases with increase in fibers. Considering 1.5% HDPE fiber-reinforced concrete to ordinary concrete, it was found that the impact strength increased to more than 100% for the first fracture and 85.52% for the final failure, respectively. It has been demonstrated that the mix GH-7 (20% GGBS with 1.5% HDPE fiber) has better

T. Seethalakshmi (✉) · A. C. R. Winslet
Department of Civil Engineering, Government College of Engineering,
Tirunelveli, Tamilnadu, India
e-mail: seethalakshmi@gcetly.ac.in

V. Baskaran et al. (eds.), *EcoTech Urbanism*, The Urban Book Series,
https://doi.org/10.1007/978-3-031-83833-0_10

compressive strength, flexural strength, and impact strength, and thus can be utilized as a paving material.

Keywords Paving material · Environment friendly · Sustainable construction practices · GGBS · HDPE plastic bag strips

10.1 Introduction

Concrete is a material that is widely used in construction because of its affordability, strong compressive strength, and durability. Unfortunately, spalling, cracking, and water buildup are issues brought on by the limited permeability of conventional concrete. The invention of pervious concrete was an attempt to address this issue. Having little to no sand used in the design and production process results in a lot of interconnected voids, which is one of the advantages of pervious concrete over ordinary concrete (Yan et al., 2023). The percentage of void space in pervious concrete determines how it primarily varies from conventional concrete. Usually, the void spaces are between 15% and 25%. (Lee et al., 2022). Because of this characteristic, pervious concrete can offer a number of benefits to the environment, including as reducing runoff, increasing groundwater levels, reducing the effect of urban heat islands, treating wastewater, enhancing skid resistance, and absorbing noise from vehicles (Feri'c et al., 2023; Mitrosz et al., 2023; Oinam et al., 2022).

To influence the characteristics and functions of pervious concrete systems, supplementary cementitious materials (SCMs) have been introduced over time. In this study, GGBS is used in place of cement to make concrete more durable and less harmful to the environment (Lin et al., 2022). Concrete can significantly reduce its harmful environmental effects and help to protect natural resources for future generations by reducing the carbon dioxide emissions associated with it. It has been proposed that workability and compressive strength can be enhanced by partially replacing with SCMs (Saboo et al., 2019).

Fiber variations in concrete increase its resistance to impact, abrasion, and shattering. Fibers improve the properties of the material, but only to the extent that too much of them will be undesirable. When HDPE plastics are disposed of in a landfill, the chemicals they contain can seep into the ground, nearby water, and the air, raising emissions and pollution levels. It is also more economical and energy-efficient to reuse HDPE rather than using virgin plastic to create new items (Tamrin & Nurdiana, 2020).

Therefore, HDPE fiber from cement bags is chopped into thin strips for this investigation and added to pervious concrete. Given that HDPE is a nonbiodegradable waste material, it is a highly valuable disposable material for concrete projects due to its longevity. Waste utilization has been one of the considerations in the construction industry toward sustainability. The utilization of GGBS and HDPE in the creation of pervious concrete introduces a new component to concrete mix design. If implemented extensively, this approach might transform the industry of building by reducing construction expenses and minimizing the amount of waste materials that end up in landfills (El-Hassan et al., 2023).

10.2 Methodology

To the objective of this proposed research, the track as mentioned below was very effective. The collection of appropriate data about the quantity of HDPE wastes produced in construction industry is much essential. Review of literatures to use the HDPE fibers in pavement construction, the selection of materials, and the tests to be conducted on the chosen materials to validate the properties of those materials and proportion of each constituent material in preparing the pervious concrete were done according to the methodology (Fig. 10.1).

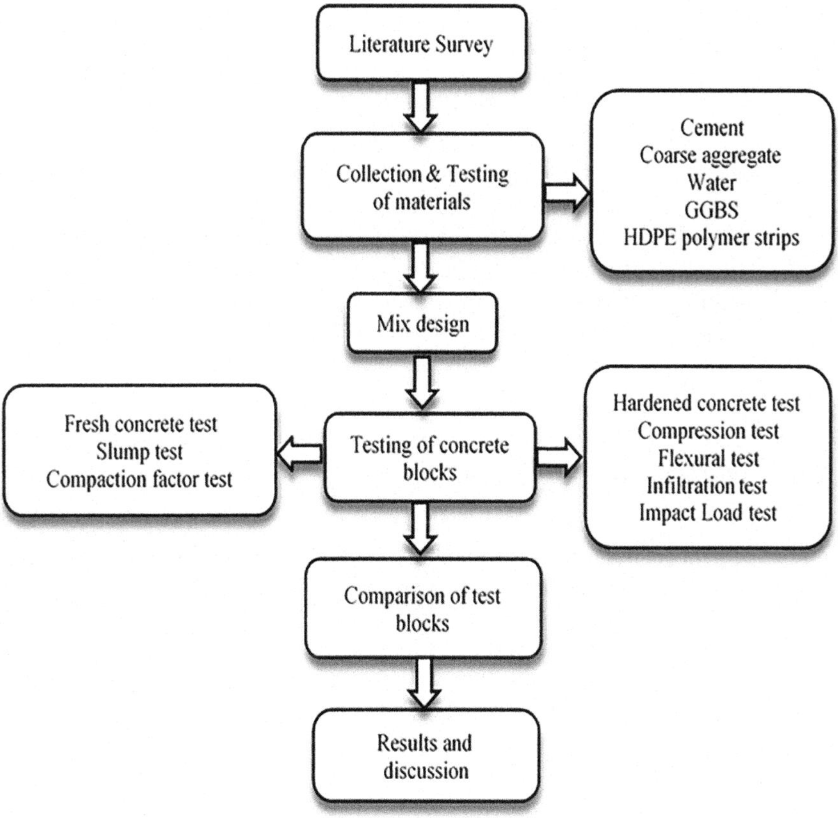

Fig. 10.1 Methodology

10.3 Materials Used

Prior to casting the samples, the materials that are utilized to make pervious concrete were examined. According to the guidelines in the applicable Indian Standard codes, the compounds were examined.

(a) *Cement*—The most common kind of cement used in the building sector is called Portland Pozzolana Cement (PPC). It is the essential ingredient of mortar and concrete. The binder, also referred to as cement gel, is primarily responsible for controlling most of the properties of the concrete. Portland Pozzolana cement (ChettinadSuper Grade) is what is utilized.
(b) *Water*—When using water for this mixing process, it must be pure and fresh, free of any organic or harmful substances that could weaken the mortar's properties.
(c) *Coarse aggregate*—Coarse aggregate is used to stabilize and strengthen concrete. The particles that make up coarse aggregates are bigger than 4.75 mm. In this analysis, uniform coarse aggregate that conformed with IS 383-1970 and had a length of 12.5 mm was used.
(d) *Ground granulated blast-furnace slag (GGBS)*—The primary factors influencing the composition of slag are the raw materials utilized in the production of iron. The usual chemical composition is composed of 40% calcium oxide, 13% alumina, 8% magnesia, and 35% silica (Deepa et al., 2023). Depending on the beginning temperature and the cooling methods employed, slags with a glass content of 90–100% are often suitable for blending with Portland cement.
(e) *High-density polyethylene (HDPE) bags*—Flexible, waterproof, transparent or waxy, affordable, easily processed using most methods, chemically resistant, and having good low temperature endurance (down to −60 °C). HDPE fiber from cement bags is chopped into thin strips of length 18 ± 2 mm.

10.4 Material Testing

The constituent materials locally available and used in this current study were tested to understand the physical properties and to proportion the ingredients of concrete to achieve a concrete grade of M40.

(a) Test for cement.

1. Standard consistent value—31.5%.
2. Initial setting time—60 min.
3. Final setting time—240 min.
4. Specific gravity—3.15.
5. Fineness (<90 microns)—3%.
6. Soundness (Le Chatlier's apparatus)—1.6 mm expansion.

(b) Test for coarse aggregate.

1. Specific gravity—2.79.
2. Fineness modulus—7.64.
3. Water absorption—0.35%.
4. Density—1670 kg/m^3.

(c) Test for GGBS.

1. Specific gravity—2.9.
2. Bulk density—1200 kg/m^3.
3. Fineness—350 m^2/kg.

(d) Test for HDPE.

1. Tensile strength—0.20–0.40 N/mm^2.
2. Thermal coefficient of expansion—100–220 × 10^{-6}.
3. Max cont use temp—65 °C.
4. Density—0.944–0.965 g/cm^3.

10.5 Mix Design

The mix proportion takes into consideration the factors, such as the strength, durability, workability, and desired appearance of the concrete, as well as the specific materials being used.The objective is to blend the different ingredients to produce concrete that is almost ideal for achieving a balance between the material's desired qualities at the most affordable price.The essential data obtained from the material testing are used in the mix design process to ensure the desired properties and performance of concrete. The mix design with designation grade M40 for this study was completed in accordance with the Indian Standard, IS:10262–2019 & IS 456:2000. Table 10.1 gives the final mix proportion of the control mix and Table 10.2 gives the various proportions of GGBS and HDPE fiber added in the study.

Table 10.1 Mix proportion

Grade	Cement (kg)	Fine aggregate (kg)	Coarse aggregate 12.5 mm (kg)	w/c ratio	Super plasticizer
M40	450	0	1524.73	0.35	1.2%

Table 10.2 Various proportions of GGBS and HDPE

Level	Factor	
	GGBS%	HDPE fibers%
1	10	1
2	20	1.5
3	30	2

10.6 Experimental Investigation

10.6.1 Fresh Concrete Test

As the concrete-making material of this study excludes fine aggregate, the behaviour of concrete in its fresh state is required to be studied.

(a) Slump cone test

A concrete slump test examines a batch's consistency to determine how smoothly the concrete will flow. The primary goal of the test is ensuring that identical concrete batches have consistent strength and quality. You may determine if a mix will have great workability or not and whether the water-to-cement ratio is excess by assessing the overall "slump" of the concrete. The workability of control mix is medium. By adding GGBS, the workability of concrete increased and reached highest at 20% GGBS. The workability of the slump with 1.5% HDPE fiber shows better workability.

(b) Compaction factor test

The workability of cement concrete is also assessed by the compacting factor test. The test determines how much the concrete weighs differently before and after compaction. The weight of the partially compacted concrete divided by the weight of the compacted concrete yields the compaction factor. The ease with which concrete may be compacted is indicated by the compaction factor value. Greater workability, or the ability to deal with and shape concrete more easily, is indicated by the higher compaction factor. The workability of the concrete reached highest in the G-H 7 mix (20% GGBS and 1.5% HDPE fiber).

10.6.2 Hardened Concrete Test

(a) Compressive test

The 7- and 28-day compressive strength of cube (150 × 150 × 150 mm) was evaluated in accordance with IS:516. The specimen, which was placed in the center of the bearing plate on the press, was subjected to a steady, consistent load. The purpose of the test was to determine the cubic specimens' compressive strengthin accordance with the instructions given in IS:516 (1959), titled "Method of test for strength of concrete." The loading setup along with the cube specimen is shown in Fig. 10.2a. The formula for calculating the compressive strength is given in Eq. (10.1).

$$f_c = F / A \tag{10.1}$$

where

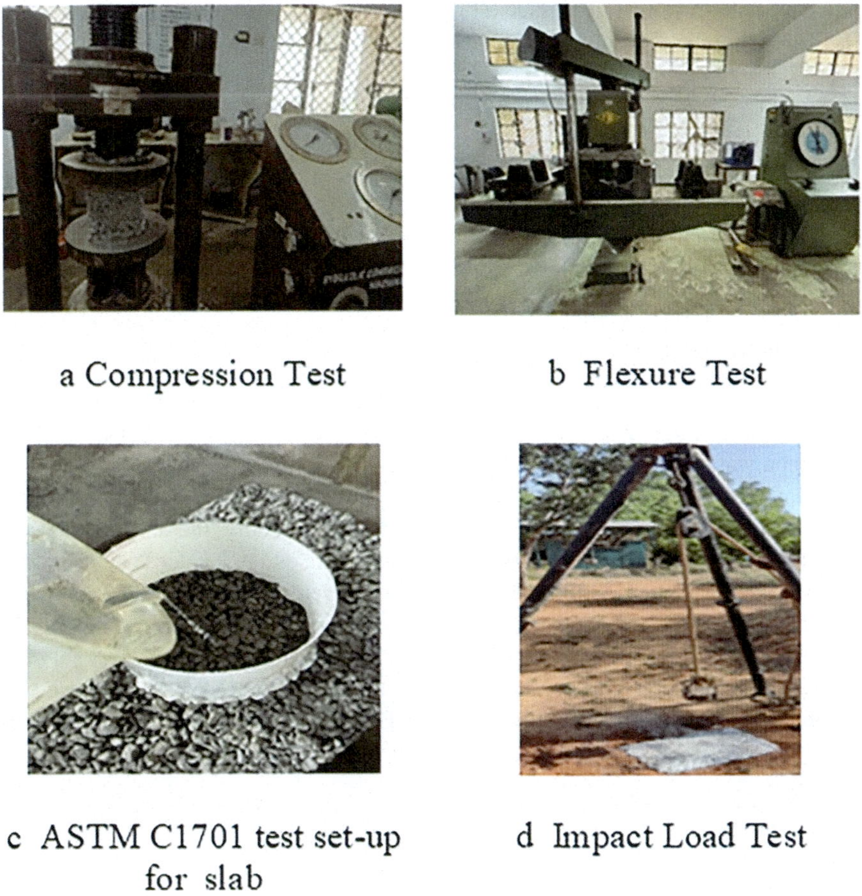

Fig. 10.2 Test setup

f_c stands for the cubic compressive strength (MPa),
F denotes the maximum load given to the specimen (N).
A is the area of the specimen to which the load is applied (mm^2).

(a) Flexure test

The specimens (600 mm × 450 mm × 60 mm) were evaluated in accordance with IS:516 (1959) in the leaf spring machine after 28 days of cure. This test's flexural behavior was examined, and a comparison between the typical mix and various GGBS and HDPE fiber percentages was made. A graph was created by plotting different blends against flexural strength. After 28 days of curing, slabs were manually tested under a 200 kN maximum capacity flexural testing machine with a 200 mm/min maximum straining speed. The line load was placed at the center of the slab, and it was simply supported along its whole effective length. The loading setup along with the specimen is shown in Fig. 10.2b. Flexural strength is calculated from the formula

$$\text{Flexural strength} = PL / bd^2 \ N / \text{mm}^2$$

where,

P = the ultimate load (N).
L = c/c distance between the supports (600 mm).
b = specimen breath (450 mm),
d = specimen depth (60 mm),

(b) Infiltration test

A plastic ring measuring 30.48 cm (12 in) in diameter served as the permeameter for the slab testing. A sealant was used to adhere the rings to the surface. Every specimen was positioned atop two metal supports. As per ASTM C1701, a prewetting test must be conducted, and within 2 min of the prewetting, an actual test must be conducted. Throughout the prewetting and test procedure, the amount of water required to pass through the specimen was controlled, and the water level inside the ring was kept between 1 and 2 cm. The slabs were prewet with 3.6 ± 0.05 L of water, and the time it took for the water to permeate through the specimen was recorded. The ring was filled with 18.0 ± 0.05 kg of water in total. The time interval between the water's contact with the pervious concrete surface and its disappearance was recorded, with accuracy to the nearest 0.1. The setup used in the current study is shown in Fig 10.2c. The following equation was used to calculate the infiltration rate (I):

$$I = KM / D^2 t \, \text{mm} / \text{h}$$

where,

M (kg) is the mass of infiltrated water.
D (mm) is inner diameter of the permeameter.
t (s) is time needed for a measured amount of water to infiltrate the specimen,
K (mm^3 s/kg/h) is constant of 4,583,666,000 [in SI units].

(c) Impact test

The measurement of the impact absorption capacity resulting from an external load is the aim of the impact test conducted on the concrete specimen. Slab specimens (600 mm × 450 mm × 60 mm) were constructed in order to assess impact strength. The impact testing apparatus used in the current study is depicted in Fig. 10.2d. The first crack and final failure needed multiple strikes, after which readings were recorded. Impact energy is found using the formula

$$\text{Impact energy} = W \times h \times n$$

where,

w is the hammer's weight (45.4 N),
h is the fall's height (0.457 m),
n is the quantity of strikes needed to produce the first crack or the last failure.

10.7 Results and Discussion

(a) Compressive strength test results.

- The 7- and 28 day compressive strength test results were displayed in Fig. 10.3.
- Compressive strength varies from 35.4–43.66 MPa. GBFS and HPMC fiber pervious concrete's 7- and 28-day cubic compressive strengths, respectively, obtained their greatest values of 14.84 MPa and 43.66 MPa, which had increased by 34.66% and 23.44%, respectively, when compared to the control group. Conversely, the lowest compressive strength value measured 40.1 MPa, 13.3% greater than the standard mix.
- Pervious concrete's mechanical qualities increased once GGBS was optimized (Lin et al., 2022). The test results indicate that a 20% dosage is the ideal for GGBS. Mix GH-7 (20% GGBS with 1.5% HDPE fiber) has a 23.44% higher compressive strength than the control group. GGBS above 20% results in a decrease in compressive strength. When compared to the 20% GGBS specimen, the compressive strength decreased by 4.1% when added at a ratio of 30% of GGBS (Pradhan et al., 2022). The addition of

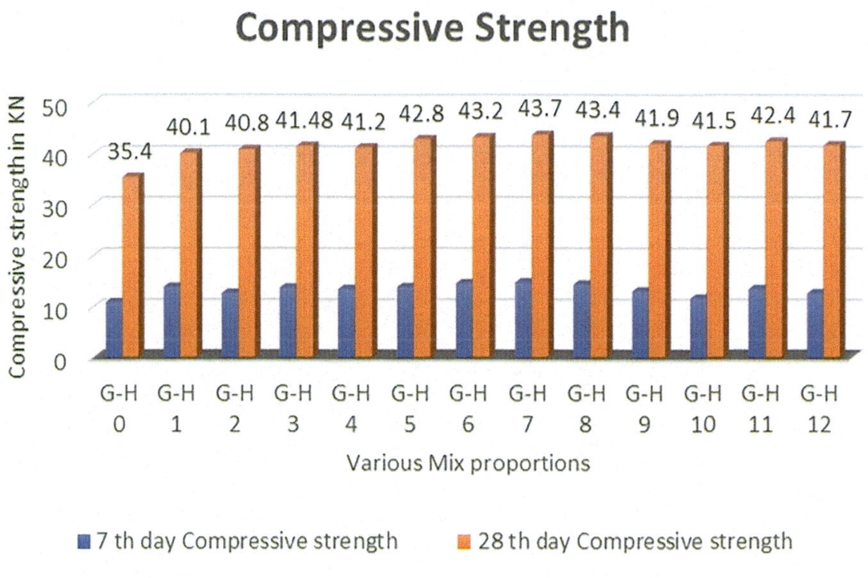

Fig. 10.3 Relation between various mix proportions and compressive strength

GGBS strengthens the pervious concrete up to a specific proportion, which accounts for the gain in compressive strength.

- It is evident that increasing the ratio of GGBS and HDPE fibers causes the 7- and 28-day compressive strengths to rise, producing the intended test results.

(b) Flexural strength test results

- The 28-day flexural strength test results were presented in Fig. 10.4.
- The flexural strength is varying from 4.42% to 7.58%. Maximum flexural strength was measured at GH-7 (20% GGBS with 1.5% HDPE fiber) mix and it is 41.68% greater than control mix. On the other hand, lowest value of flexural strength was measured at G-H 1 (10% GGBS) and it is 4.74 MPa which is 6.75% higher than the conventional mix.
- It is evident that the 28-day flexural strength first rises with fiber addition, and that the flexural strength falls as fiber addition exceeds 2%. 1.5% fibers produced the best improvement rates. The flexural strength of concrete has increased due to the incorporation of HDPE fibers. According to Abeer et al. (2024), they are crucial for preserving the even distribution of internal stresses, stopping cracks from forming, and strengthening existing fissures as they propagate. Nonetheless, there was a minor decrease in flexural strength with the high fiber rate of HDPE (2%) present.

(c) Infiltration test results

- Measuring the infiltration rate for a pervious concrete slab is highly essential as the concrete mixture includes no fine aggregate. The voids formed in any concrete specimen are duly filled by the fine aggregate only. As this is a no-fines concrete, the infiltration test findings at 28 days were presented in Fig. 10.5.

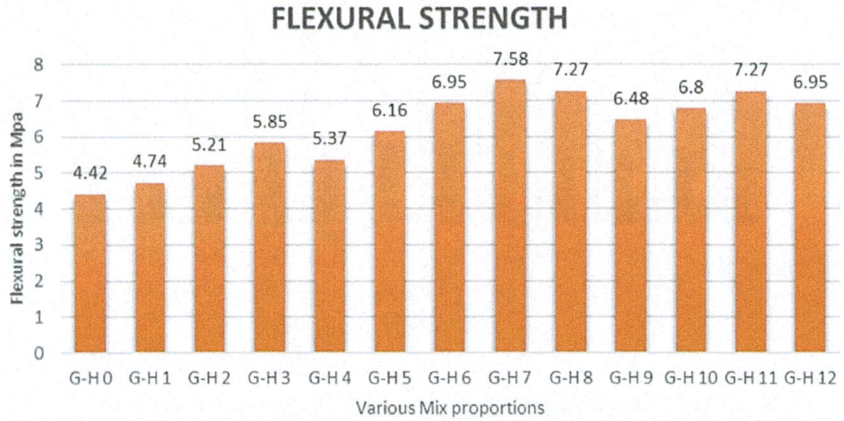

Fig. 10.4 Relation between various mix proportion and Flexural strength

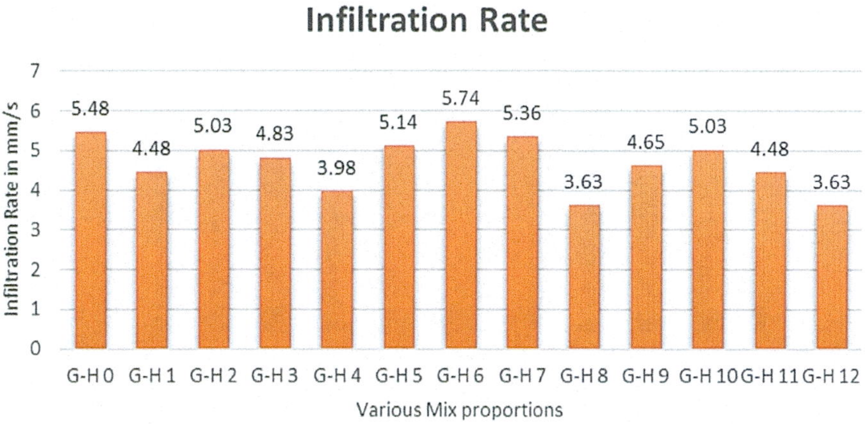

Fig. 10.5 Relation between various mix proportion and infiltration rate

- Particles in GGBS are usually finer than those in cement. The spaces between the cement particles are filled when GGBS is added to the concrete mixture. The mixture's packing density can be increased by this finer particle size distribution, which will also lessen the spacing between the particles and change the flow characteristics. Thus, infiltration rate first increases with addition of GGBS up to 20% and then decreases with addition of 30% of GGBS.
- Infiltration rate varies from 3.63–5.74 mm/s. Sample mix G-H 6(20% GGBS and 1% HDPE fibers) showed an increased infiltration rate of 5.74 mm/s, which is 4.74% higher than the control mix and a decreased infiltration rate of 3.63 mm/s of mix (G-H 8 and G-H 12 with 2% of HDPE fibers), which is 33.75% lesser than the control mix. Adding 1.5% HDPE fiber was the most beneficial to improve the flexure properties, but infiltration rate decreases with increase in fibers.
- It was discovered that a 1% fiber addition improved the pervious concrete's penetration. (Singh et al., 2022). However, with fiber content above 1.5% wt, the porosity began to decrease due to the excessive fiber content. This densification of the concrete structure contributes to reduced porosity and improved durability. Hence, in areas with heavy rainfall, 1% HDPE fibers can be used for better infiltration, and for moderate rainfall areas, 1.5% HDPE fibers can be used for better mechanical performance.

(d) Impact test results

- The test specimen was made at 28 days, and Figs. 10.6 and 10.7 show the variation in impact energy and number of blows for the first crack and final failure, respectively.
- When compared to the control mix, it was found that the impact strengths of 1%, 1.5%, and 2% HDPE fiber-reinforced concrete with 10% GGBS

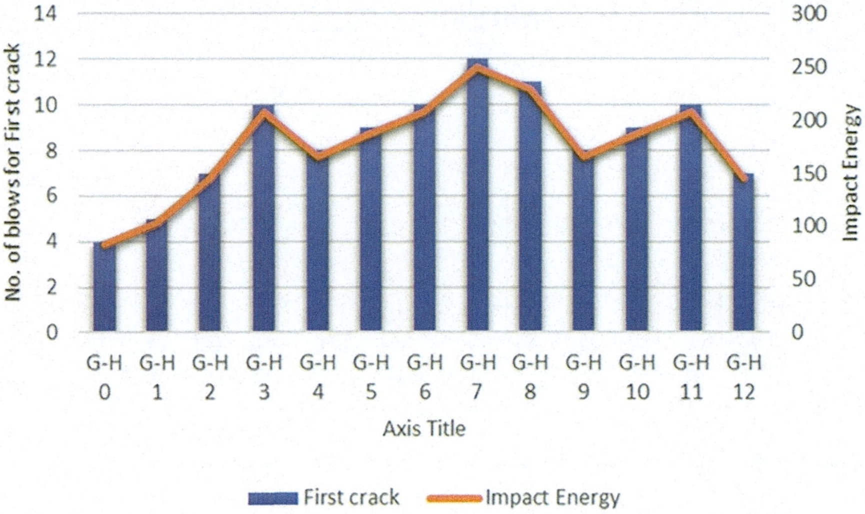

Fig. 10.6 Relation between various mix proportion and number of blows for first crack

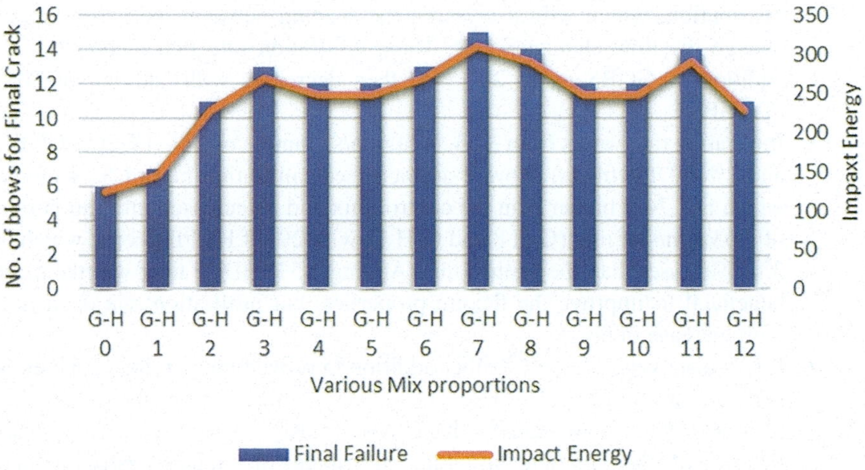

Fig. 10.7 Relation between various mix proportion and number of blows for final crack

increased to more than 58.84%, 73.67%, and 66.66% for the specimen's first crack and 58.84%, 73.67%, and 66.66% for the final failure, respectively. Because HDPE-fiber pervious concrete absorbs more energy, it takes more hits to cause the first crack and eventual failure (Yi Wu et al. 2022). The standard concrete samples' insufficient resilience to impact has caused them to shatter into several fragments.

- When compared to the control mix, it was found that 1%, 1.5%, and 2% HDPE fiber-reinforced concrete with 20% GGBS gave an improved impact

strength of more than 85.7%, 100%, and 93.35% for the specimen's first crack and 73.67%, 85.52%, and 80% for the final failure, respectively. Comparing 1%, 1.5%, and 2% HDPE fiber-reinforced concrete with 30% GGBS to the control mix, it was found that the impact strength increased to more than 76.9%, 85.41%, and 54.57% for the specimen's first crack and 66.66%, 80%, and 58.84% for the final failure.
- The results showed that adding 1.5% HDPE may greatly improve the impact resistance of the pervious slab specimen because the material's fibers promote sample particle binding and delay the failure process.

10.8 Conclusion

The performance of pervious concrete cubes and slabs casted by replacing cement with GGBS at 10%, 20%, and 30% with addition of HDPE fiber at 1%, 1.5%, and 2% in a M40 grade of concrete was studied to understand the behavior of GGBS and HDPE fibers, an eco-friendly sustainable alternative suggested for minimizing the disposal issues related to GGBS and HDPE wastes.

- When GGBS was added to concrete, the degree of workability improved at 20% GGBS and then decreased at 30% of GGBS.
- When the percentage of fiber in concrete increases, the density of the material increases as well (Deepa et al., 2023).
- The 28-day cubic compressive strength, flexural strength test, and infiltration test of GGBS & HDPE fiber-incorporated pervious concrete reached the highest values of 43.66 MPa, 7.58 MP, and 5.74 mm/s respectively. These values exceeded those of the control group by 23.44%, 41.68%, and 4.74%, respectively. Similarly, it was observed that 1.5% HDPE (G-H 7) fiber-reinforced concrete showed an increased impact resistance of more than 100%, for the first crack and 85.52% for the final failure on the specimens when compared to conventional concrete.
- It has been demonstrated that mix GH-7 (20% GGBS with 1.5% HDPE fiber) has better compressive strength, flexural strength, and impact strength, and thus can be utilized as a paving material.
- To make concrete more durable and less harmful to the environment, GGBS is used in place of cement. By lowering the carbon dioxide emissions linked to concrete, it can greatly lessen its negative effects on the environment and contribute to the preservation of natural resources for future generations.
- HDPE fibers are mostly extracted from cement bags, which means that no non-biodegradable waste will be generated on the construction site and that they will be extremely helpful for the durability of concrete projects (Brasileiro et al., 2024). The concrete's significantly increased flexural capability further supported the investigation (El-Hassan et al., 2023). Higher impact resistance was also observed for HDPE-added.
- Because of its special porous structure and low mechanical qualities, pervious concrete can be used in environmentally friendly construction for areas such as

basketball lanes, tennis courts, open car parking spaces, pedestrian walkways, and pavements (Deepa et al., 2023).

- It offers long-term use and sustainable benefits when utilized in residential building for features like patios and pathways. In many different kinds of commercial and industrial environments, where controlling runoff and preserving water quality are crucial factors, pervious concrete can be used.
- In order to promote sustainable landscaping techniques, it can be included into landscape designs to create surfaces and pathways that let water pass through. In places where groundwater recharge and flood management depend on rainwater infiltration and retention, pervious concrete works effectively.
- It can be incorporated into green infrastructure initiatives to lessen the impact of the urban heat island and assist with water management. High-strength pervious concrete, having been taken into consideration, is an adaptable substrate that may be used in many situations where finding a balance between water permeability and structural strength is essential.

According to the study, adding HDPE strips and partially replacing cement with GGBS significantly increases the mechanical strength of pervious concrete, which supports the development of sustainable infrastructure and urbanization.

Acknowledgments First and foremost, I want to express my gratitude to the All-Powerful God for his kind guidance during the course of this work.

I owe the greatest debt of gratitude to the Head of Civil Engineering Department for permitting us to utilize the facilities available in the laboratory for completing it successfully.

My profound gratitude goes out to my parents, who have made significant sacrifices for my welfare and education. I am grateful to everyone who helped me finish my project, either directly or indirectly.

References

Abeer, S. Z., et al. (2024). Abrasion and maintenance of high-strength fiber-reinforced pervious concrete. *Buildings, 14*(1), 127. https://doi.org/10.3390/buildings14010127

Brasileiro, K. P. T. V., et al. (2024). Influence of recycled aggregates and silica fume on the performance of pervious concrete, *82*, 10834. 1 April 2024.

Deepa, K., et al. (2023). Experimental research on pervious concrete performance with partial replacement of foundry sand nano and GGBS. *Materials Today: Proceedings, 2023*, 534. https://doi.org/10.1016/j.matpr.2023.10.046

El-Hassan, H., et al. (2023). Synergic effect of recycled aggregates, waste glass, and slag on the properties of pervious concrete. *Developments in the Built Environment, 15*, 100189. https://doi.org/10.1016/j.dibe.2023.100189

Feri'c, K., et al. (2023). Effect of aggregate size and compaction on the strength and hydraulic properties of pervious concrete. *Sustainability, 15*, 1146. https://doi.org/10.3390/su15021146

Lee, M.-G., et al. (2022). Mechanical properties of high—strength pervious concrete with steel fiber or glass fiber. *Buildings, 12*(5), 620. https://doi.org/10.3390/buildings12050620

Lin, K.-L., et al. (2022). Application of ternary cementless hybrid binders for pervious concrete. *Construction and Building Materials, 346*(5), 128497.

Mitrosz, O., et al. (2023). Influence of the addition of recycled aggregates and polymer fibers on the properties of pervious concrete. *Materials, 16*, 5222.

Oinam, Y., et al. (2022). Characteristics of GGBFS—based pervious concrete considering rheological properties of the binder. *International Journal of Concrete Structure Materials, 16*, 62.

Pradhan, S. K., et al. (2022). Performance assessment of pervious concrete road on strength and permeability by using silica fume. *Materials Today: Proceedings, 60*(Part 1), 559–568.

Saboo, N., et al. (2019). Effect of fly ash and metakaolin on pervious concrete properties. *Construction and Building Materials, 223*, 322–328.

Singh, A., et al. (2022). Glass and carbon fiber reinforced polymer composite wastes in pervious concrete: Material characterization and lifecycle assessment. *Resources, Conservation and Recycling, 182*, 106304.

Tamrin, & Nurdiana, J. (2020). The effect of recycled HDPE plastic additions on concrete performance. *Recycling, 2021*(6), 18.

Yan, X., et al. (2023). Performance analysis and admixture optimization of GBFS-HPMC/fiber pervious concrete. *Materials, 16*, 6455. https://doi.org/10.3390/ma16196455

Yi, W., et al. (2022). The strength and elastic modulus of pervious concrete considering pore and fiber during freeze–thaw cycles. *Sustainability, 14*(23), 16217. https://doi.org/10.3390/su142316217

Chapter 11
Sustainable Building Practices: The Path to a Resilient Future for Developing Countries

T. Chockalingam, C. Vijayaprabha, A. J. Jeya Arthi, and K. Saranya

Abstract The growing concentration of people in urban areas across developing countries resulted in a huge shortage of basic amenities, especially shelter, in terms of quantity and quality. As the population of developing countries is growing exponentially every year, there is an alarming situation to overcome the above-stated issue. Furthermore, the resources required to fulfil this gap also depleting. There is an immediate urgency to switch over to sustainable development; thereby, the depletion of available resources is minimal. The practices adopted among developing countries to achieve sustainable development such as design, durability, energy efficiency, waste reduction, indoor air quality, water conservation, and building materials are discussed with relevant case studies. The target status achieved for "sustainable development goal 9—build resilient infrastructure, promote inclusive and sustainable industrialization and foster innovation" was also discussed.

Keywords Sustainable building practices · Sustainable construction · Sustainable development goal

T. Chockalingam (✉)
Department of Civil Engineering, Ramco Institute of Technology, Rajapalayam, Tamil Nadu, India
e-mail: chockalingam@ritrjpm.ac.in

C. Vijayaprabha
Department of Civil Engineering, Alagappa Chettiar Government College of Engineering and Technology, Karaikudi, Tamil Nadu, India

A. J. Jeya Arthi
Department of Civil Engineering, Rajalakshmi Engineering College, Chennai, India

K. Saranya
Department of Civil Engineering, Sri Ramakrishna Engineering College, Coimbatore, India

V. Baskaran et al. (eds.), *EcoTech Urbanism*, The Urban Book Series,
https://doi.org/10.1007/978-3-031-83833-0_11

11.1 Introduction

Construction industry across the world is facing lot of problems due to rapid urbanization and population growth especially in developing countries. As the developing countries are striving hard for progress in economy and a better standard of living, the adoption of sustainable development goals has emerged as a key indicator for a sustainable future. This chapter explores the multifaceted role of sustainable building practices especially Indoor air quality, water conservation and building materials adopted by developing countries for constructing infrastructures to foster the long-term resilience.

The demand of constructing infrastructure due to industrial development and population growth in developing countries needs to be fulfilled. Approximately 39% of energy-related CO_2 emissions globally are due to construction sector. Out of 39% of CO_2 global emissions, a significant portion is attributed due to the activities happening in developing countries (*Global Status Report for Buildings and Construction: Beyond foundations—Mainstreaming sustainable solutions to cut emissions from the buildings sector*, 2024). At the same time, the built environment is susceptible to climate-related disasters, mandating the need for suitable construction approaches (UNDRR, 2023). A wide range of sustainable building practices needs to be adopted to minimize the environmental effects. Some of the practices include, but are not limited to:

- Indoor air quality (IAQ).
- Water conservation.
- Eco-friendly building materials.
- Industrial waste as building materials.

By adopting the abovementioned practices, the developing countries may get numerous advantages beyond protecting their environment. Buildings adopting sustainable practices may enhance the IAQ and leads to better comfort. Their operational costs tend to be reduced over their lifecycle cost, contributing to long-term economic sustainability (Finance Corporation, 2023).

This chapter will delve into suitable case studies from Indian country highlighting the achievements and long-term impacts after effective implementation of sustainable building practices. It will also address the industrial waste as building materials, their composition, and its impact on strength, leaching, and durability aspects. By providing an overview, this chapter aims to equip the readers with knowledge and insights required to drive the construction sector with sustainable building practices.

11.2 Sustainable Development Strategies

11.2.1 Indoor Air Quality

Sustainability solutions give a healthy and comfortable life to human beings. Maintaining good and healthy air quality in the workplace can enhance worker comfort, productivity, and overall well-being. It becomes a fundamental need to the people because of more vehicular usage, industrialization, and urbanization. Poor quality of air affects the children and elders severely. Many cities have become unsuitable for leading a healthy life, but still people want to continue over there due to their situation. Biological origin contaminants, such as dust mites, pets, cockroaches, and mice, and nonbiological origin contaminants, such as smoke and volatile organic compounds, have turned indoor air quality into the worst condition. A research conducted by Harvard University reported that the particulate matters with a diameter of 2.5 μm or lesser than that referred as PM2.5 causes the increase in mortality rate by 10–27% (Sung & Hsiao, 2021). The acceptance limit of the pollutants such as carbon dioxide, carbon monoxide, formaldehyde, PM2.5, PM10, ozone, and total volatile organic compounds are 1000 ppm, 10 ppm, 0.1 ppm, 12 $\mu g/m^3$, 15 $\mu g/m^3$, 0.5 ppm, and 3 ppm, respectively (Hashim et al., 2019).

The following methods are used to maintain indoor air quality within our houses and to maintain healthy life as well. Nature will always exist to cure every problem in our life. The role of plantation to maintain the indoor air quality is more vital. The indoor plantation will give both physiological and psychological benefits of purifying the air as well as maintaining a peaceful atmosphere inside the house (Kavathekar & Bantanur, 2022). One of the plants is Areca palm, potted, which is effective in removing acetone, CO_2, xylene, toluene, benzene, trichloroethylene, carbon monoxide, and formaldehyde pollutants from petroleum, paints, and wood (Bhargava et al., 2021). Dracaena, Golden pothos, Snake plant, Bird nest fern, Spider plant, English ivy, Aloe vera, Weeping fig, Dumb cane, and Peace lily are some of the plants, known by their common name, that can be potted inside the house to get rid of pollutants (Nisitha et al., 2023). Other than the pollutants, the indoor humidity control is also important to manage excess humidity, along with condensation which can lead to eye irritation, dryness of nose and throat fungal diseases, bacterial growth, allergy, and breathing problem to the occupants (Gulandaz et al., 2024). The use of humidifiers has been common in low indoor relative humidity places, especially during winter. A study on the utilization of humidification equipment reported that examined on testing 817 patients with allergies, a notable improvement was observed on 65% of the patients, and 30% improvement was reported on the utilization on home humidifiers (Arundel et al., 1986). Monitoring or testing indoor air quality (IAQ) is crucial to assess the presence of contaminants that can impact the comfort, productivity, and well-being of occupants. A very simple and energy-efficient method of improving the indoor air quality is air filtration technique using portable air cleaners, heating, ventilating, and air conditioning. Sometimes, these filters are multilayer filter system that consists of a prefilter, a carbon filter, and an antibacterial filter called a high-efficiency particulate air filter (HEPA). It can remove particles greater than or equal to 0.3 μm in diameter with 99.97% (Sublett, 2011). The Internet of Things (IOT) technology is also comfortably used for the indoor air quality. The indoor air quality can be analyzed by MATLAB simulation, fuzzy control applied to the data of fine particulate matters and CO_2 in the indoor environment. For this a logic base was predicted on the AQI data for further improvement of air quality in homes. Even though the system exists to assess the indoor air quality, the available system will analyze 24 h data and there is no hourly monitoring system. Indoor Air Quality Index (IAQI) is the total toxic level concentration ratio, which is fixed by the human tolerance limit. Using advanced algorithms, a control station can get the information from the sensors about the air quality and can intervein the ventilation or air conditioning system to get optimal level of indoor air quality (Amado & Cruz, 2018; De Capua et al., 2023). Indoor air quality is necessary to lead a healthy life which can be attained by providing proper ventilation, plantation, air purifier, air filters, and by maintaining the humidity above 30–40%.

11.2.2 Water Conservation

In the recent era, we all know the importance of saving water, which is essential to protect for the utilization of future generations. There are two factors on water utilization: one is need basis, another one is want basis. Drinking, cooking, and cleaning are the basic needs that essentially require water. Industrial production, food production, transportation, communication, waste processing, and other goods and services require water which is in large quantity. Some purposes, like recreation, leisure, and luxury goods, require water, which may be unnecessary but it is being practiced due to luxurious lifestyle of people (Bostanci, 2020; Gleick, 2003).

At the same time, the increase in population, industrialization, urbanization, global warming, climate change, depletion of sources of natural freshwaters due to less rainfall, and lack of responsibility among people to save water are all factors giving us a dangerous warning about water scarcity. To conserve water, the government should implement strict rules and regulations on people and to concerned authorities. The following measures should have been taken to protect the water sources:

1. The surface water to be properly stored to use draught seasons.
2. Rainwater harvesting method to be implemented in each house.
3. Making artificial recharge and percolation tanks.
4. Provision of catchment area protection.
5. Minimize the use of water by providing drip sprinkler irrigation.
6. Minimize the evaporation and evapotranspiration water loss.
7. Recycling the wastewater (Kurunthachalam, 2014).

Rainwater is a free source of water that is often missed for conservation. If it is conserved properly, it can be converted into a main water resource, as it is the purest form of water and can be used for potable uses like drinking, bathing, cooking, and dishwashing after some preliminary treatment. Other than potable use, non-potable utilization needs no treatment. Due to its high purity and with minimum treatment, the rainwater possibly meet out the water demand (Che-Ani et al., 2014). Sustainable water conservation also includes the recycling of waste water. Wastewater treatment consists of three phases: primary, secondary, and tertiary treatment. This treatment reduces the concentration of waste in the effluent and makes it possible to recycle in any other way. In biorefineries during the treatment process, the wastewater can be converted into biogas as energy recovery process. The organic maters in the wastewater that is difficult to treat using conventional methods can be effectively converted into burning renewable fuel. This method will make economic as well as sustainable development and reduces cost and lower the greenhouse gas emission (Li et al., 2023). Methane is the major part of the biogas, which can be a required source of energy to run the complete wastewater treatment plant (Al-Juaidi et al., 2014). Nowadays, wastewater is another sustainable resource to get valuable products such as clean water, fertilizer, nutrient, and alternate materials replacing natural resources. In such a way, the wastewater contributes to a circular economy and supporting environmental sustainability (Ashraf et al., 2021; Baawain et al., 2020; López-Morales & Rodríguez-Tapia, 2019).

11.2.3 Building Materials

Our earth consists of plenty of natural resources used by humankind. There are two types namely renewable and nonrenewable resources. Every time renewable resources can regenerate themselves—for example, animals, plants, water, and air. Also, the sun and wind fall under renewable resources. Nonrenewable resources, as specified by their name, cannot be brought up again once they are used up. They are generally nonliving things. Nonrenewable resources take more than a life span of a human to form, even taking millions of years (Chakrawarthi et al., 2021). Some metals are produced from minerals, also called nonrenewable resources. The sustainability of building construction involves the utilization of waste material, recycled material, energy-efficient materials, and water-efficient materials.

The construction activities require an enormous amount of renewable and nonrenewable resources for their successful completion. Apart from water, river sand is the most demandable natural resource categorized as a nonrenewable resource. Further, the sustainable building practice of using recycled metal reduces mining operations and minimizes the environmental impact. Steel and aluminum can be used without losing its original property. The recycled metal can be used in structural framing and roofing (Abera, 2024). Apart from water, river sand is the most demandable natural resource categorized as a nonrenewable resource. The material derived from industry or the material disposed of as waste is replaced with sand provided the material should induce the same property as river sand or contribute to the strength. Industrial wastes contribute to greenhouse gas emissions such as methane and carbon dioxide. Some hazardous waste contaminates the soil, water supply, and leads to the direct emission of harmful gases or gas particles into the air. Also, the storage and transportation of these wastes create business risks and affect the cost associated with these components.

The utilization of waste is directly linked with sustainability. Waste reduction and waste management need sustainable principles such as reuse, recover, recycle, and remanufacture. The increasing cost of waste disposal in combustion, landfills, depletion of natural resources, and the necessity for sustainable development have amplified the need to reuse the materials as substitutes for natural resources.

(Chakrawarthi et al., 2016; Chilamkurthy et al., 2016; Vijayaprabha et al., 2017). Volcanic ash (Játiva et al., 2021), seashells (Safi et al., 2015) and Coconut Shells (Prakash et al., 2021), plastic wastes (Al-Sinan & Bubshait, 2022), fly ash (Raju & Dharmar, 2020), silica fume, bottom ash, M sand (Aliabdo et al., 2014), quarry dust (Meisuh et al., 2018), copper slag (Vijayaprabha, 2020), steel slag, recycled aggregate, and demolition wastes are some of the evolving sustainable materials in building construction. Pervious concrete is one of the water-efficient building materials made of cement, water, coarse aggregate, and no-fine aggregate. It is a highly porous and permeable concrete structure, which is an alternative material for pavements that can reduce excessive runoff in urban areas (Chockalingam et al., 2023).

11.2.4 Industrial Wastes as Building Materials

The following section deals with some of the industrial waste materials which can be partially or fully replaced for conventional fine aggregate in order to reduce the dependency on river sand. Some of the industrial waste materials used in building construction are as follows:

- Copper slag.
- Steel slag.
- Ferrochrome slag.
- Ground-granulated blast furnace slag (GGBFS).

11.2.4.1 Manufacturing

Steel slag is obtained from steel manufacturing plants, which can be tamped to make finer particles as compared to river sand (Subathra Devi & Gnanavel, 2014). From molten slag, 100 kg/ton of steel is disposed of as waste. Copper slag is a waste by-product derived from the copper separation industry. Each ton of copper produces 1.2 tons of copper slag (Murari et al., 2015). Ferrochrome is the combination of chromium and iron oxide, which is produced from the smelting process of electric arc furnaces. For every ton of ferrochrome alloy production, there will be 1.2 tons of slag production (Panda et al., 2013). GGBS is the waste delivered from the manufacturing of pig iron. 0.5 tons at each ton of steel is disposed of as waste (Saranya et al., 2018).

Table 11.1 shows the physical properties of river sand and different supplementary materials to the river. Steel slag and copper slag show higher density than river sand. Hence, the unit weight of the concrete is more compared to normal concrete. The density tensed to increase by 15%. The next higher dense material is ground-granulated blast furnace slag (Manhas & Moohmend, 2018). The fineness modulus is very high for steel slag compared to all other wastes. So, the steel slag requires more water for good workability. Ferrochrome holds a porous

Table 11.1 Physical properties

Material	Specific gravity	Water absorption	Fineness modulus
River sand (Vijayaprabha & Brindha, 2020)	2.51	1.25	2.742
Steel slag (Nguyen et al., 2020)	3.56	2.1	3.0
Copper slag (Vijayaprabha, 2020)	3.56	0.15	2.99
Ferrochrome slag (Jena & Panigrahi, 2019)	2.63	2.01	2.33
Ground-granulated blast furnace slag (Saranya et al., 2020)	2.9	1.4	2.75

structure that absorbs more water. Even though the fineness modulus is lower than of sand, the water absorption is about 2.01%, leading to less water in the concrete mixture (Dash & Patro, 2021). The specific gravity of ferrochrome slag is slightly heavier than sand.

11.2.4.2 Chemical Composition

Steel slag includes C_2S, C_3S, and C_4AF, and free C_2S are the primary constituents that give it cementitious properties (Guo et al., 2019), whereas ferrochrome slag consists of SiO_3, Al_2O_3, and MgO. Also, ferrochrome slag was identified with 8.5% of chromium.

Shi et al. (2008) discussed the microstructure of copper slag. The metals present in copper slag are silica, alumina, lime, and magnesia, which are most stable in oxide and silicate forms. Basic patterns of Fe_3O_4 (magnetite) and Fe_2SiO_4 (fayalite) are the primary phases found in the slag. Among the total mass of chemical constituents, 95% of the masses are the oxides of silica-alumina, ferrous, and magnesium. The (ASTM C618, 1995) standards specify the minimum criteria to be a pozzolana. From Table 11.2, copper slag and GGBS prove to act as a pozzolanic material since the summation of primary oxide components such as SiO_2, Al_2O_3, Fe_2O_3, and CaO is beyond 75%.

11.2.4.3 Effect on Fresh Concrete Properties

Fresh concrete properties are as important as hardened concrete properties because the workability, bleeding, and segregation in fresh state influence the hardened state properties. The workability of concrete is measured by slump, and the loss of slump is used to compare fluidity of two different mix proportions. Slump decreases as the steel slag replacement increases even the water absorption is less than the steel slag. More than 50% replacement of steel slag shows much reduction in workability (Nguyen et al., 2020). Also, the percentage of increase in density was observed to be 10.2% (Yu et al., 2016). The roughness, high angularity, and irregular texture of steel slag give a negative impact on workability (Sezer & Gülderen, 2015).

Table 11.2 Chemical composition

Composition	Cement (Vijayaprabha et al., 2017)	Steel slag (Nguyen et al., 2020)	Copper slag (Chakrawarthi et al., 2016)	Ferrochrome slag (Dash & Patro, 2021)	GGBS (Gnanadurai et al., 2021)
SiO_2	20.85	9–17.5	33.05	26.68	29.54
Al_2O_3	4.78	6.3–12.2	2.79	23.41	13.44
Fe_2O_3	3.51	20.3–32	53.45	5.48	0.673
CaO	63.06	23.9–35	6.06	4.63	36.72
MgO	2.32	2.9–15	1.56	23.92	7.12
SO_3	2.48	0.1–0.6	1.89	0.67	2.1
Na_2O	0.24	–	0.28	0.101	–
K_2O	0.55	–	0.61	0.45	–
MnO	–	2.5–5.6	–	0.721	–
Mn_2O_3	–	–	–	0.81	–
TiO_2	–	0.5–0.8	0	0.66	–
$SiO_2 + Al_2O_3 + Fe_2O_3 + CaO$	92.2	59.5	95.8	60.2	80.37

The fresh concrete property of copper slag concrete is enhanced due to the low water absorption of copper slag particles. Hence, there will be enormous free water available in concrete making the concrete becomes more workable. Also, there is no need for high-range water reducers in concrete (Brindha & Nagan, 2011). Ferrochrome slag delays the initial and final setting time. Workability of concrete increases as the ferrochrome content increases because of its high fineness (Işil & Hasan, 2020).

11.2.4.4 Effect on Compressive Strength

Compressive strength of concrete is discussed with early age strength and later strength in case of introducing any waste material to concrete, because some materials accelerate the hydration process and enhance the early age strength, while some materials retard the hydration process to achieve the target strength beyond 28 days. Steel slag shows variability in the trendline of the compressive strength. Initially, the replacement of steel slag increases the compressive strength. Steel slag, when used as 20–60% replacement, shows a decreasing trend of compressive strength. However, with 80–100% replacement, the compressive strength increases when compared to control concrete. It is evident that the variation of compressive strength oscillates between negative to positive leads to unstable prediction over the compressive strength values. The bleeding may be the reason for the decrement of compressive strength (Gao et al., 2017). Also due to the roughness of steel slag, the cement required to coat the slag is high. Hence, there would be an insufficiency of cement paste resulting in decreasing the hydration rate when the steel slag is replaced in a larger percentage.

Ferrochrome slag concrete delivers an increase in compressive strength as the curing period increases. But the compressive strength value on each curing period decreases than the corresponding control concrete. The reason for the decrement is the lower specific gravity, higher fineness, and porous texture of ferrochrome. Although the compressive strength of different replacements of ferrochrome concrete rises than the reference mix after 365 days (Dash & Patro, 2021). GGBS attained the target strength in a slower manner because of the slow reactivity and retardation characteristics. The early age strength of GGBS concrete is lower than the later age strength. The strength gain is achieved between 28 days and 90 days. Optimum strength was achieved at 20% of GGBS. Steel slag is more active in filling effect compared to GGBS (Palod et al., 2019).

11.2.4.5 Effect on Splitting Tensile Strength

The tensile strength of concrete is normally one-tenth of compressive strength. Generally, concrete is a compression-taking member sometimes the tensile strength of concrete increases in case of adding fibers to the concrete. Splitting tensile strength of concrete incorporating steel slag concrete increases as the replacement increases. The increasing trend is observed in each curing period (Wang & Lin, 2013; Subathra et al. 2014; Singh & Siddique, 2016). The optimum percentage replacement of steel slag is 40%. Blast furnace slag gives optimum tensile strength at a 30% replacement level (Saxena & Tembhurkar, 2018). Splitting tensile strength of ferrochrome slag concrete shows a similar pattern of variation of compressive strength of ferrochrome concrete, that is, the strength increases after 365 days of curing compared to the reference mix but the strength increases within the replacement level as the curing period increases. Ferrochrome of 30% replacement shows 5% lower tensile strength than control concrete. The tensile strength of GGBS incorporated concrete shows a similar path of variation as in the case of compressive strength. Tensile strength at the early curing period is lower and later it attains maximum tensile value at 40% replacement (Ganesh & Murthy, 2019).

11.2.4.6 Leaching Characteristics

Waste from industry consists of toxic components like chromium (Cr), selenium (Se), barium (Ba), and cadmium (Cd). These material causes environmental hazards even incorporated in concrete. Leaching study for such materials is essential to conduct as per EPA method-311 Toxic Characteristics Leaching Procedure (TCLP). Leaching study on steel slag concrete evidenced chromium and vanadium oxide components and are within the allowable limit (Mombelli et al., 2016). Ferrochrome slag contains chromium (Cr), cadmium (Cd), selenium (Se), and other toxic composition that causes environmental pollution. The leaching study was concluded that the leachability of chromium, cadmium, and selenium in the ferrochrome concrete sample is much lower than the raw ferrochrome slag sample (Acharya & Patro, 2016; Dash & Patro, 2021). A leaching study on GGBS revealed that very small amounts of Ba, Cu, Mn, and Sr in the amount of 0.008 ppm, 0.087 ppm, 0.008 ppm, and 0.002 ppm, respectively, were observed. All the observed values are within the allowable limit (Brightson et al., 2014). Saraswathy et al. (2014) studied the leaching of copper slag considering the leaching period. The amount of leaching increases with the leaching period in all copper slag replacements. This study depicts that, at 24 h, the amount of copper leaching is found to be 0.0368 ppm. At the end of 4320 h (180 days), the value of leaching is 0.668 ppm only. The leaching value of copper slag in tap water is well within the allowable limit as per the toxicity characteristics leaching procedure (TCLP) test.

11.2.4.7 Durability Characteristics

A concrete designed for its service life should be checked for durability characteristics. The concrete resistance to corrosion attack, carbonation, permeability, water absorption, alkali-silica reaction, etc. is the different aspects of discussing the durability of concrete. Steel slag in concrete shows less resistance to carbonation and permeability and is significantly resistant to shrinkage (Mo et al., 2017; Wang et al., 2013). Concrete shows a very low water absorption value than control concrete up to 40% of copper slag replacement. Beyond 40%, the segregation and bleeding effect of copper slag concrete increases, thereby increasing the water absorption. The results exhibit low resistance to acid attack when compared to control concrete. RCPT test reveals that after 90 days curing period the total charge passed is found to be very less when compared with the sand mortar. Brindha and Nagan (2011) analyzed the corrosion behavior of copper slag replacement for both sand and cement. The rate of corrosion of slag admixed concrete is maximum for 60% replacement of copper slag to sand. The combination also shows a 10 times greater rate of corrosion than control cylindrical specimens. GGBS explored very good resistance to chloride penetration due to fine pore structure. Alkali in Portland cement and silica in the aggregate exerts pressure on the hardened concrete and cracks. GGBS shows high reluctance to alkali silicate reaction. In ferrochrome concrete, the chloride penetration increases as the replacement increases but it varies between low to a very low category of penetration. The compressive strength losses after chloride attack are found to be 4.02% in ferrochrome concrete, whereas the loss percentage in control concrete is 4.64%. High water absorption, permeability, and presence of voids in ferrochrome concrete expose less resistance to sulfate attack. The presence of MgO in ferrochrome slag causes some expansion in concrete, and thus reduces the shrinkage in concrete (Dash & Patro, 2018).

11.3 Case Studies

The following section deals with some of the case studies regarding the technologies adopted across the country to ensure the development happened is sustainable.

11.3.1 *Case Study 1: CII-Sohrabji Godrej Green Business Center (GBC) (Greenbusinesscentre.com, 2024)*

CII-Sohrabji Godrej Green Business Center (GBC) located in Hyderabad, within a hot and dry climatic zone, the CII-Sohrabji Godrej Green Business Center (GBC) serves as a pioneering model of sustainable office building design in India. The project showcases numerous innovative features that significantly reduce its environmental footprint while optimizing energy efficiency and occupant comfort.

Highlights One of the key highlights of the GBC is its onsite solar generation, which meets nearly 20% of the building's electricity needs. This reliance on renewable energy not only reduces the building's carbon footprint but also exemplifies the potential for integrating solar power into commercial buildings. The circular design of the building plays a crucial role in optimizing solar gain, reducing it by 25% compared to conventional buildings. This design approach, coupled with the use of double-glazed facades, substantially lowers heat gain, thereby minimizing the need for artificial cooling and enhancing the overall energy efficiency.

The GBC is also notable for its use of recycled materials, with 77% of the building components being derived from recycled sources. This commitment to sustainable material use reduces the environmental impact of construction and promotes the principles of a circular economy. The building's design includes two wind towers that enhance natural ventilation, improving indoor air quality and reducing the need for mechanical ventilation systems. Additionally, 90% of the interior spaces receive direct daylight, significantly cutting down on the requirement for artificial lighting and contributing to a healthier, more productive indoor environment.

Achievements
Recognized as the first platinum-rated building in India, the GBC has set a benchmark in sustainable building practices. Its achievements include significant reductions in heat gain and municipal water usage, the latter achieved through the installation of water-saving fixtures that cut water consumption by 30%. These features collectively position the GBC as one of the greenest buildings globally, highlighting the viability of passive design strategies in modern construction.

Long-Term Impacts
The long-term impacts of the GBC's sustainable design are profound. The building maximizes energy savings and minimizes operational costs, providing a cost-effective model for future green buildings. The extensive vegetation around the GBC helps mitigate the Urban Heat Island effect, improving the local microclimate. Moreover, the building's water conservation practices support the sustainable management of local water resources, ensuring their availability for future generations. Overall, the CII-Sohrabji Godrej Green Business Center stands as a testament to the potential of green building practices to create environmentally responsible and economically viable structures.

11.3.2 Case Study 2: Himurja Office Building (Himurja, 2024)

Situated in Shimla, Himachal Pradesh, the Himurja Office Building is a remarkable example of sustainable architecture tailored for cold climates. The building is designed as an office space and leverages innovative passive design strategies to achieve outstanding energy efficiency and thermal comfort without relying on electrical energy for heating, lighting, and hot water during the daytime.

Features

One of the most significant features of the Himurja Office Building is its ability to meet hot water demands entirely through a solar water heater system. This system harnesses solar energy to provide a reliable and sustainable source of hot water, eliminating the need for conventional heating methods that consume electricity. Additionally, the building is equipped with a photovoltaic system that supports its lighting needs, further reducing its dependency on external energy sources and promoting the use of renewable energy.

Achievements

The building's design is highly effective in maintaining a comfortable indoor environment. It boasts a well-insulated structure that minimizes heat loss, which is crucial in the cold climate of Shimla. This insulation not only preserves warmth but also maximizes the benefits of natural lighting, thereby reducing the need for artificial lighting and enhancing energy savings. The strategic use of passive design elements, such as proper insulation and orientation to maximize solar gain, ensures that the building remains warm and well-lit during the day without the need for additional energy input.

Long-Term Impacts

The long-term impacts of the Himurja Office Building are significant and multifaceted. By demonstrating the viability of passive design strategies in cold climates, the building serves as a model for energy-efficient construction in similar regions. It highlights how buildings can be designed to harness natural energy sources effectively, reducing operational costs and environmental impact. Moreover, the success of the Himurja Office Building raises awareness about the importance of energy efficiency among building communities and stakeholders in cold climatic zones. It encourages the adoption of sustainable building practices and technologies, paving the way for a greener and more sustainable future in the construction industry. Overall, the Himurja Office Building stands as a testament to the potential of passive design in achieving high levels of energy efficiency and occupant comfort in cold climates.

11.3.3 Case Study 3: Akshay Urja Bhawan, HAREDA Office

Located in Panchkula, the Akshay Urja Bhawan, also known as the HAREDA Office, is an exemplary model of sustainable office building design in a composite climatic zone. This building embodies the principles of a Net Zero Energy Building (NZEB), primarily through the integration of Building Integrated Photovoltaic (BIPV) systems, which enable it to achieve remarkable energy performance and sustainability.

Highlights: One of the key highlights of Akshay Urja Bhawan is its status as an NZEB, meaning it produces as much energy as it consumes over the course of a year. This is made possible by the extensive use of BIPV systems, which seamlessly integrate solar panels into the building's structure. These systems harness solar energy to meet the building's electricity needs, significantly reducing its dependence on grid power and highlighting the potential of renewable energy in urban settings.

The building's design and construction are fully compliant with the Energy Conservation Building Codes (ECBC), ensuring that it meets stringent energy efficiency standards. Additionally, Akshay Urja Bhawan has achieved a GRIHA 5-star rating, which is a testament to its superior environmental performance and sustainability credentials. This rating recognizes the building's holistic approach to sustainability, including energy efficiency, water conservation, waste management, and occupant comfort.

Achievements

Akshay Urja Bhawan stands as a symbol of energy efficiency and renewable energy in India. Its high energy performance is marked by significant energy savings, which are achieved through the combined effects of its BIPV systems, energy-efficient design, and compliance with ECBC standards. These features not only reduce operational costs but also minimize the building's carbon footprint, making it a benchmark for future office buildings in composite climatic zones.

Long-Term Impacts

The long-term impacts of Akshay Urja Bhawan are far-reaching. By demonstrating the feasibility and benefits of NZEBs, the building promotes the adoption of such practices in both the public and private sectors. It serves as an inspirational example of how buildings can be designed to achieve net-zero energy goals, thereby supporting national and global sustainability targets. Furthermore, Akshay Urja Bhawan encourages the widespread adoption of solar passive design techniques in building construction, showcasing how thoughtful design can harness natural energy sources to create comfortable and efficient living and working spaces. Overall, Akshay Urja Bhawan is a pioneering project that underscores the potential of sustainable building practices to transform the built environment.

11.3.4 Case Study 4: Smart GHAR-III (Green Homes at Affordable Rate) (Greenbusinesscentre.com, 2024)

Located in Rajkot, Smart GHAR-III (Green Homes at Affordable Rate) is a pioneering residential project designed to provide affordable housing with a strong emphasis on energy efficiency and sustainability. Developed under the Pradhan Mantri Awas Yojana (PMAY) scheme, this project exemplifies how green technologies can be integrated into low-cost housing to enhance thermal comfort and reduce energy consumption.

The Smart GHAR-III project comprises 1176 energy-efficient dwelling units, each designed with low-cost measures aimed at achieving significant energy savings. These measures include optimal building orientation, high-performance building envelopes, and the use of energy-efficient appliances and fixtures. The design prioritizes passive cooling techniques to maintain indoor thermal comfort, reducing the reliance on air conditioning and consequently lowering electricity usage.

Achievements

One of the standout achievements of Smart GHAR-III is its significant reduction in electricity consumption, which translates into substantial CO_2 emission reductions. By incorporating energy-efficient design principles and technologies, the project not only lowers operational costs for residents but also contributes to broader environmental sustainability goals. This reduction in energy use underscores the potential of green technologies to make a meaningful impact even in cost-sensitive housing projects.

Long-Term Impacts

In the long term, Smart GHAR-III offers a scalable model for energy-efficient affordable housing that can be replicated in other regions with similar climatic conditions. The project enhances living conditions for economically weaker sections by providing comfortable, healthy, and sustainable living environments. It sets a precedent for future housing developments, demonstrating that affordability and sustainability can go hand in hand. By promoting environmental responsibility and energy efficiency, Smart GHAR-III supports the goal of creating resilient and sustainable urban communities.

11.4 Status of SDG-9

The following section deals about the status of attainment of Sustainable Development Goal 9 as per the report issued by United Nations, Department of Economic and Social Affairs. There has been some progress in reducing the emission of CO_2 in manufacturing sector. But at the same time, the reduction happened was insufficient to meet out the target fixed by 2030. The overall stats of the Sustainable Development Goal 9 are shown in Fig. 11.1.

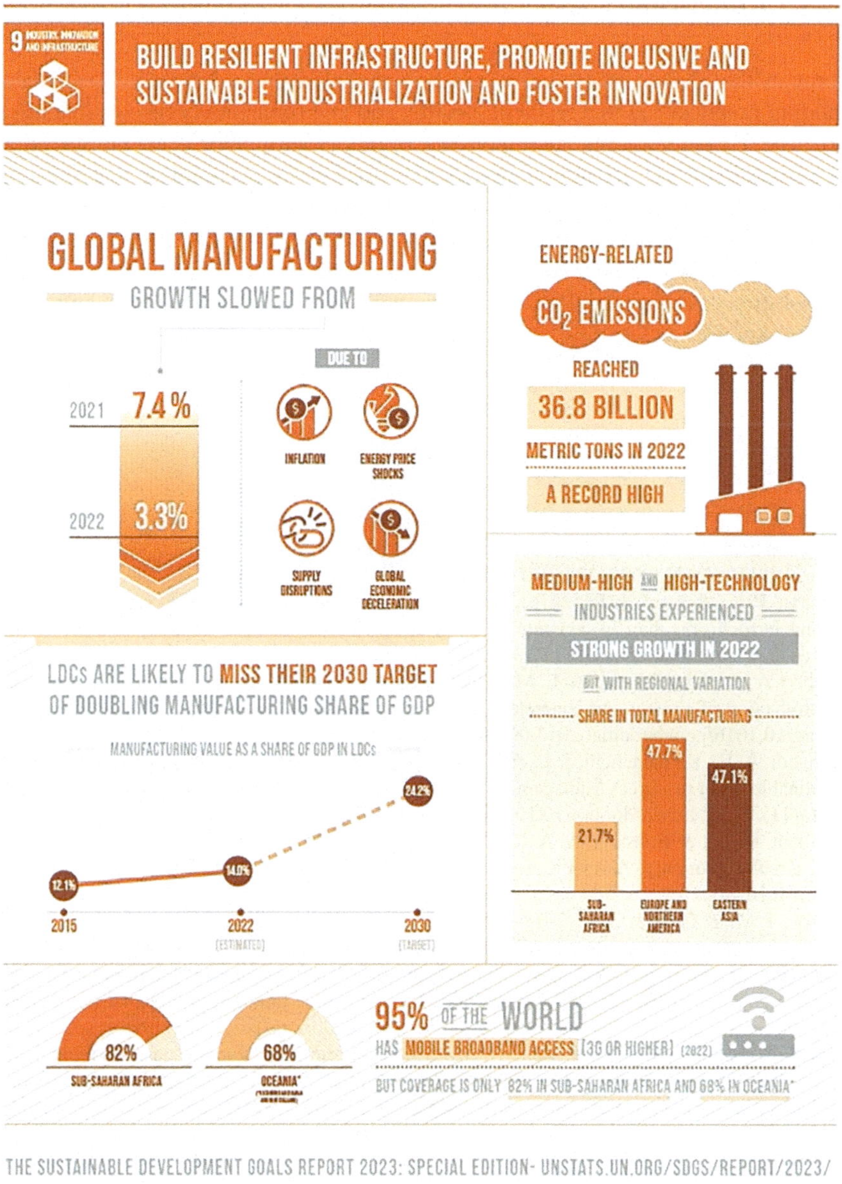

Fig. 11.1 Status of SDG 9. (Source: https://unstats.un.org/sdgs/report/2023/)

11.5 Conclusion

Through the above discussion, it can be concluded that adoption of sustainable building practices led to a resilient, healthy and environmentally responsible infrastructure. By emphasizing the importance of indoor air quality, water conservation, locally available eco-friendly materials/industrial wastes as building materials, the developing countries can make a significant progress toward the attainment of Sustainable Development Goal 9 across the world. As the urbanization continues in all parts of the world, adopting sustainable building practices is one of the promising solutions to mitigate climatic change, conserve valuable resources, and foster the communities.

References

Abera, Y. A. (2024). Sustainable building materials: A comprehensive study on eco-friendly alternatives for construction. *Composites and Advanced Materials, 33*, 1–17. https://doi.org/10.1177/26349833241255957

Acharya, P. K., & Patro, S. K. (2016). Utilization of ferrochrome wastes such as ferrochrome ash and ferrochrome slag in concrete manufacturing. *Waste Management and Research, 34*(8), 764–774. https://doi.org/10.1177/0734242X16654751

Aliabdo, A. A., Abd Elmoaty, A. E. M., & Auda, E. M. (2014). Re-use of waste marble dust in the production of cement and concrete. *Construction and Building Materials, 50*, 28. https://doi.org/10.1016/j.conbuildmat.2013.09.005

Al-Juaidi, A. E., Kaluarachchi, J. J., & Mousa, A. I. (2014). Hydrologic-economic model for sustainable water resources Management in a Coastal Aquifer. *Journal of Hydrologic Engineering, 19*(11), 1–12. https://doi.org/10.1061/(asce)he.1943-5584.0000960

Al-Sinan, M. A., & Bubshait, A. A. (2022). Using plastic sand as a construction material toward a circular economy: A review. *Sustainability (Switzerland), 14*(11). https://doi.org/10.3390/su14116446

Amado, T. M., & Cruz, J. C. D. (2018). Development of machine learning-based predictive models for air quality monitoring and characterization. *Proceedings of TENCON 2018–2018 IEEE Region 10 Conference (Jeju, Korea, 28–31 October 2018) Development* [Preprint], (October).

Arundel, A. V., et al. (1986). Indirect health effects of relative humidity in indoor environments. *Environmental Health Perspectives, 65*(3), 351. https://doi.org/10.2307/3430203

Ashraf, A., Ramamurthy, R., & Rene, E. R. (2021). Wastewater treatment and resource recovery technologies in the brewery industry: Current trends and emerging practices. *Sustainable Energy Technologies and Assessments, 47*(October 2020), 101432. https://doi.org/10.1016/j.seta.2021.101432

ASTM, C. (1995). *ASTM C 618-Standard Specification for Coal Fly Ash and Raw or Calcined Natural Pozzolan for Use in Concrete, 14*(200), 4–5. https://doi.org/10.1520/C0618-17A.10.1520/C0618-19.2

Baawain, M. S., et al. (2020). Public perceptions of reusing treated wastewater for urban and industrial applications: Challenges and opportunities. *Environment, Development and Sustainability, 22*(3), 1859–1871. https://doi.org/10.1007/s10668-018-0266-0

Bhargava, B., et al. (2021). Mitigation of indoor air pollutants using areca palm potted plants in real-life settings. *Environmental Science and Pollution Research, 28*(7), 8898–8906. https://doi.org/10.1007/s11356-020-11177-1

Bostanci, S. C. (2020). Use of waste marble dust and recycled glass for sustainable concrete production. *Journal of Cleaner Production, 251*, 119785. https://doi.org/10.1016/j.jclepro.2019.119785

Brightson, P., Premanand, M., & Ravikumar, M. S. (2014). Flexural behavior of beams incorporating GGBS as partial replacement of fine aggregate in concrete. *Advanced Materials Research, 984–985*, 698–706. https://doi.org/10.4028/www.scientific.net/AMR.984-985.698

Brindha, D., & Nagan, S. (2011). *Durability Studies On Copper Slag Admixed Concrete, 12*(5), 563–578.

Chakrawarthi, V., Darmar, B., & Elangovan, A. (2016). Copper slag concrete admixed with polypropylene fibres. *Gradjevinar, 68*(2), 95–104. https://doi.org/10.14256/jce.1211.2015

Chakrawarthi, V., et al. (2021). Impact resistance of polypropylene fibre-reinforced alkali–activated copper slag concrete. *Materials, 14*(24). https://doi.org/10.3390/ma14247735

Che-Ani, et al. (2014). Rainwater harvesting as an alternative water supply in the future. *European Journal of Scientific Research, 1*(January), 132–140.

Chilamkurthy, K., et al. (2016). *A statistical overview of sand demand in Asia and Europe*. in CTMC-2016.

Chockalingam, T., Vijayaprabha, C., & Leon Raj, J. (2023). Experimental study on size of aggregates, size and shape of specimens on strength characteristics of pervious concrete. *Construction and Building Materials, 385*(March), 131320. https://doi.org/10.1016/j.conbuildmat.2023.131320

Dash, M. K., & Patro, S. K. (2018). Effects of water cooled ferrochrome slag as fine aggregate on the properties of concrete. *Construction and Building Materials, 177*, 457–466. https://doi.org/10.1016/j.conbuildmat.2018.05.079

Dash, M. K., & Patro, S. K. (2021). Performance assessment of ferrochrome slag as partial replacement of fine aggregate in concrete. *European Journal of Environmental and Civil Engineering, 25*(4), 635–654. https://doi.org/10.1080/19648189.2018.1539674

De Capua, C., et al. (2023). An improvement strategy for indoor air quality monitoring systems. *Sensors, 23*(8). https://doi.org/10.3390/s23083999

Finance Corporation, I. (2023) Annual report leadership perspectives results strategy in action critical functions building building Ifc 2023 annual report leadership perspectives results strategy in action critical functions contents

Ganesh, P., & Murthy, A. R. (2019). Tensile behaviour and durability aspects of sustainable ultra-high performance concrete incorporated with GGBS as cementitious material. *Construction and Building Materials, 197*, 667–680. https://doi.org/10.1016/j.conbuildmat.2018.11.240

Gao, J., et al. (2017). Utilization of steel slag as aggregate in asphalt mixtures for microwave deicing. *Journal of Cleaner Production, 152*, 429–442. https://doi.org/10.1016/j.jclepro.2017.03.113

Gleick, P. H. (2003). Water use. *Annual Review of Environment and Resources, 28*(May), 275–314. https://doi.org/10.1146/annurev.energy.28.040202.122849

Global Status Report for Buildings and Construction: Beyond foundations—Mainstreaming sustainable solutions to cut emissions from the buildings sector (2024) *2023 Global Status Report for Buildings and Construction: Beyond foundations—Mainstreaming sustainable solutions to cut emissions from the buildings sector*. https://doi.org/10.59117/20.500.11822/45095.

Gnanadurai, L. T., Renganathan, N. T., & Selvaraj, C. G. (2021). Synthesis and characterization of synthetic sand by geopolymerization of industrial wastes (fly ash and GGBS) replacing the natural river sand. *Environmental Science and Pollution Research, 28*(40), 56294–56304. https://doi.org/10.1007/s11356-021-14223-8

Greenbusinesscentre.com. (2024). CII GBC. [online] Available at: https://greenbusinesscentre.com/. Accessed 13 Aug 2024

Gulandaz, M. A., et al. (2024). Layout of suspension-type small-sized dehumidifiers affects humidity variability and energy consumption in greenhouses. *Horticulturae, 10*(1). https://doi.org/10.3390/horticulturae10010063

Guo, Y., et al. (2019). Utilization of unprocessed steel slag as fine aggregate in normal- and high-strength concrete. *Construction and Building Materials, 204*, 41–49. https://doi.org/10.1016/j.conbuildmat.2019.01.178

Hashim, N. H., Teh, E. J., & Rosli, M. A. (2019). A dynamic botanical air purifier (DBAP) with activated carbon root-bed for reducing indoor carbon dioxide levels. *IOP Conference Series: Earth and Environmental Science, 373*(1), 012022. https://doi.org/10.1088/1755-1315/373/1/012022

Himurja. (2024). *Home—Himurja*. [online] Available at: https://himurja.hp.gov.in/. Accessed 13 Aug 2024

Işil, S. K., & Hasan, B. (2020). Assessment of reusing ferrochrome slag wastes in mortar as SCMs. *Journal Wuhan University of Technology, Materials Science Edition, 35*(6), 1043–1052. https://doi.org/10.1007/s11595-020-2354-x

Játiva, A., Ruales, E., & Etxeberria, M. (2021). Volcanic ash as a sustainable binder material: An extensive review. *Materials, 14*(5), 1–32. https://doi.org/10.3390/ma14051302

Jena, S., & Panigrahi, R. (2019). Performance assessment of geopolymer concrete with partial replacement of ferrochrome slag as coarse aggregate. *Construction and Building Materials, 220*, 525–537. https://doi.org/10.1016/j.conbuildmat.2019.06.045

Kavathekar, A. K., & Bantanur, S. (2022). Study of the influence of indoor plants as an indicator of Biophilic design on CO2 concentrations in a classroom of higher education institute. *Journal of Sustainable Architecture and Civil Engineering, 31*(2), 96–108. https://doi.org/10.5755/j01.sace.31.2.30791

Kurunthachalam, S. K. (2014). Water conservation and sustainability: An utmost importance. *Journal of Waste Water Treatment & Analysis, 05*(02). https://doi.org/10.4172/2157-7587.1000e117

Li, Y., et al. (2023). Design of a high-rate wastewater treatment process for energy and water recovery at biorefineries. *ACS Sustainable Chemistry and Engineering, 11*(9), 3861–3872. https://doi.org/10.1021/acssuschemeng.2c07139

López-Morales, C. A., & Rodríguez-Tapia, L. (2019). On the economic analysis of wastewater treatment and reuse for designing strategies for water sustainability: Lessons from the Mexico Valley basin. *Resources, Conservation and Recycling, 140*(August 2018), 1–12. https://doi.org/10.1016/j.resconrec.2018.09.001

Manhas, S., & Moohmend, A. (2018). A research—Vaguely replacement of fine aggregate with GGBS in concrete. *International Journal of Civil Engineering and Technology, 9*(3), 65–69.

Meisuh, B. K., Kankam, C. K., & Buabin, T. K. (2018). Effect of quarry rock dust on the flexural strength of concrete. *Case Studies in Construction Materials, 8*, 16. https://doi.org/10.1016/j.cscm.2017.12.002

Mo, L., et al. (2017). Accelerated carbonation and performance of concrete made with steel slag as binding materials and aggregates. *Cement and Concrete Composites, 83*, 138–145. https://doi.org/10.1016/j.cemconcomp.2017.07.018

Mombelli, D., et al. (2016). The effect of chemical composition on the leaching behaviour of electric arc furnace (EAF) carbon steel slag during a standard leaching test. *Journal of Environmental Chemical Engineering, 4*(1), 1050–1060. https://doi.org/10.1016/j.jece.2015.09.018

Murari, K., Siddique, R., & Jain, K. K. (2015). Use of waste copper slag, a sustainable material. *Journal of Material Cycles and Waste Management, 17*(1), 13–26. https://doi.org/10.1007/s10163-014-0254-x

Nguyen, T. T. H., et al. (2020). Investigation on compressive characteristics of steel-slag concrete. *Materials, 13*(8). https://doi.org/10.3390/MA13081928

Nisitha, S., Balasubramani, G. & J, P.P. (2023). *Systemic review on indoor plants as an alternative technique for reducing indoor air pollutants*, (November). https://doi.org/10.37896/jxu17.11/092.

Palod, R., Deo, S. V., & Ramtekkar, G. D. (2019). Utilization of waste from steel and iron industry as replacement of cement in mortars. *Journal of Material Cycles and Waste Management, 21*(6), 1361–1375. https://doi.org/10.1007/s10163-019-00889-3

Panda, C. R., et al. (2013). Environmental and technical assessment of ferrochrome slag as concrete aggregate material. *Construction and Building Materials, 49*, 262–271. https://doi.org/10.1016/j.conbuildmat.2013.08.002

Prakash, R., et al. (2021). An investigation of key mechanical and durability properties of coconut shell concrete with partial replacement of fly ash. *Structural Concrete, 22*(S1), E985–E996. https://doi.org/10.1002/suco.201900162

Raju, S., & Dharmar, B. (2020). Studies on flexural behavior of reinforced concrete beams with copper slag and fly ash. *Structural Concrete, 21*(1), 107–116. https://doi.org/10.1002/suco.201800351

Safi, B., et al. (2015). The use of seashells as a fine aggregate (by sand substitution) in self-compacting mortar (SCM). *Construction and Building Materials, 78*, 430–438. https://doi.org/10.1016/j.conbuildmat.2015.01.009

Saranya, P., Nagarajan, P., & Shashikala, A. P. (2018). Eco-friendly GGBS concrete: A state-of-the-art review. *IOP Conference Series: Materials Science and Engineering, 330*(1), 012057. https://doi.org/10.1088/1757-899X/330/1/012057

Saranya, P., Nagarajan, P., & Shashikala, A. P. (2020). Behaviour of GGBS-dolomite geopolymer concrete short column under axial loading. *Journal of Building Engineering, 30*(October 2019), 101232. https://doi.org/10.1016/j.jobe.2020.101232

Saraswathy, V., Karthick, S., & Muralidharan, S. (2014). Corrosion and leaching studies in blended copper slag mortar. *Journal of Engineering and Technology, 4*(1), 29. https://doi.org/10.4103/0976-8580.123801

Saxena, S., & Tembhurkar, A. R. (2018). Impact of use of steel slag as coarse aggregate and wastewater on fresh and hardened properties of concrete. *Construction and Building Materials, 165*, 126–137. https://doi.org/10.1016/j.conbuildmat.2018.01.030

Sezer, G. İ., & Gülderen, M. (2015). Usage of steel slag in concrete as fine and/or coarse aggregate. *Indian Journal of Engineering and Materials Sciences, 22*(3), 339–344.

Shi, C., Meyer, C., & Behnood, A. (2008). Utilization of copper slag in cement and concrete. *Resources, Conservation and Recycling, 52*(10), 1115–1120. https://doi.org/10.1016/j.resconrec.2008.06.008

Singh, G., & Siddique, R. (2016). Strength properties and micro-structural analysis of self-compacting concrete made with iron slag as partial replacement of fine aggregates. *Construction and Building Materials, 127*, 144–152. https://doi.org/10.1016/j.conbuildmat.2016.09.154

Subathra Devi, V., & Gnanavel, B. K. (2014). Properties of concrete manufactured using steel slag. *Procedia Engineering, 97*, 95–104. https://doi.org/10.1016/j.proeng.2014.12.229

Sublett, J. L. (2011). Effectiveness of air filters and air cleaners in allergic respiratory diseases: A review of the recent literature. *Current Allergy and Asthma Reports, 11*(5), 395–402. https://doi.org/10.1007/s11882-011-0208-5

Sung, W. T., & Hsiao, S. J. (2021). Building an indoor air quality monitoring system based on the architecture of the internet of things. *EURASIP Journal on Wireless Communications and Networking, 2021*, 1–41. https://doi.org/10.1186/s13638-021-02030-1

UNDRR. (2023). Annual report 2022, United Nations Office for disaster risk reduction (UNDRR), *United Nations Office for disaster risk reduction* [preprint].

Vijayaprabha, C. (2020). Evaluation of long term compressive strength of concrete made with copper slag and polypropylene fiber. *Journal of Structural Engineering, 441*(6), 441–448.

Vijayaprabha, C., & Brindha, D. (2020). Investigation on cyclic behaviour of FRC beams incorporating copper slag as sustainable waste. *Journal of Materials and Engineering Structures, 7*(January), 293–306.

Vijayaprabha, C., Brindha, D., & Siva, A. (2017). Durability performance of copper slag concrete admixed with polypropylene fiber. *Journal of Structural Engineering (India), 43*(6), 598–606.

Wang, H. Y., & Lin, C. C. (2013). A study of fresh and engineering properties of self-compacting high slag concrete (SCHSC). *Construction and Building Materials, 42*, 132–136. https://doi.org/10.1016/j.conbuildmat.2012.11.020

Wang, Q., et al. (2013). Influence of steel slag on mechanical properties and durability of concrete. *Construction and Building Materials, 47*, 1414–1420. https://doi.org/10.1016/j.conbuildmat.2013.06.044

Yu, X., et al. (2016). Performance of concrete made with steel slag and waste glass. *Construction and Building Materials, 114*, 737–746. https://doi.org/10.1016/j.conbuildmat.2016.03.217

Chapter 12
Autism-Friendly Built Environment Design Criteria for Children

Gökçen Firdevs Yücel Caymaz and Meryem Eren

Abstract The study seeks to analyze the needs of the children diagnosed with autism spectrum disorder (ASD) and identify the requirements for autism-friendly environments. It aims to develop a design guideline that can enhance the quality of spaces for users by organizing an existing building. Therefore, the spatial needs of individuals with ADC were investigated across different criteria including accessibility, safety and security, clarity, predictability and proportionality, spatial organization and flexibility in space arrangement and building design, spatial sequencing, transition spaces and escape areas; as well as surfaces, textures, acoustics, light, color, and sensory integration activities. During the literature review phase, Google Scholar was utilized, and relevant studies were selected by reading the titles, abstracts, and full texts. The criteria outlined in these studies were analyzed in a source and criteria table, identifying common headings. In addition, a checklist that includes these common headings were prepared. Using this checklist, the structure of a kindergarten in Istanbul accommodating children with autism was examined by taking photos and recording videos and suggestions have been presented. The study revealed that Başakşehir Kiraz Çiçeği Kindergarten mostly meets the criteria of an autism-friendly environment for children.

Keywords ASD · Autism-friendly environment · Sensory environment · Spatial design criteria

G. F. Y. Caymaz (✉)
Istanbul Aydin University, Istanbul, Turkey
e-mail: gokcenfyucel@aydin.edu.tr

M. Eren
Architecture Master's Program, Istanbul Aydın University, Istanbul, Turkey
e-mail: meryemeren@stu.aydin.edu.tr

V. Baskaran et al. (eds.), *EcoTech Urbanism*, The Urban Book Series,
https://doi.org/10.1007/978-3-031-83833-0_12

12.1 Introduction

Autism spectrum disorder (ASD) is a neurodevelopmental disorder called "autism." It usually shows itself in the early phases of the life (from infancy to preschool period) and has lasting and far-reaching consequences on an individual's brain function (American Psychological Association; U.S. Centers for Disease Control and Prevention, 2024; World Health Organization, 2023). Leo Kanner, a psychiatrist at John Hopkins University, was the first person to coin the term "autism" (Tohum Autism Foundation, 2017). Kanner, in 1938, examined a five-year-old male patient and subsequently continued to investigate people with similar disorders. As a result, in 1943, he published a case series including 11 individuals. In this case series, it was realized that children exhibited delayed speech and showed stereotypical (repetitive) behaviors. They insisted on maintaining sameness in the environment and they lacked imagination. However, it is realized that they had good memory skills, and they had no physical difference from other children. He called these cases as "autism" (Kanner, 1943). Every individual has idiosyncratic symptoms. Major symptoms include deficits in social functioning, movement, and learning, with abnormalities in sensory processing, and emotional regulation (Public Health Agency of Canada, 2024). The fifth edition of the Diagnostic and Statistical Manual of Mental Disorders (DSM-5) has consolidated all previously distinct subtypes—autistic disorder, Asperger's disorder, childhood disintegrative disorder, and pervasive developmental disorder—under the umbrella term autism spectrum disorder. This significant update emphasizes the heterogeneity and broader dimensions of autism (American Psychological Association, 2024). This revision underscores the extensive heterogeneity inherent in autism, acknowledging the wide range of individual differences while also enhancing diagnostic uniformity (Rosen et al., 2021). While a clear clinical and medical definition has emerged, with the appearance of the neurodiversity movement, the autism community has come to support different views on the understanding of the term, and some researchers agree (Happé & Frith, 2020; Pellicano & Houting, 2022). This viewpoint seeks to destigmatize autism by recognizing it as a form of diversity, thereby challenging the traditionally negative connotations associated with clinical and medical definitions, which often emphasize terms such as "deficits," "disorder," and "dysfunction." Embracing this broader perspective aims to highlight the potential strengths and unique contributions of individuals with autism, thereby fostering more inclusive societal outcomes (Xu, 2023a, b). Consequently, the design and accommodation for individuals with autism should focus on removing barriers while leveraging their potential strengths. Overall, autism encompasses a wide spectrum of characteristics, underscoring the diversity within the autistic population (World Health Organization, 2023). Autism spectrum disorder (ASD) is defined by impairments in verbal and nonverbal communication and social interaction skills and restricted, repetitive behaviors, interests, and activities (World Health Organization, 2019). It is essential to acknowledge that the heterogeneity of autism implies a vast range of sensitivities and abilities, with each individual displaying uniquely different characteristics (Georgiades et al., 2013; Lord et al., 2018, 2022; Ofner et al., 2018).

The word heterogeneity implies the variance and the diversity of symptoms within the autistic population. This heterogeneity is influenced by factors such as chronological age, gender, developmental level, the severity of the syndrome, and environmental factors (Hyman et al., 2020; WHO, 2019). This would explain why autism is commonly referred to as a spectrum disorder. While some children may display mild symptoms and even have advantages in some areas, many may experience more severe symptoms that require individualized support in their daily lives (Lord et al., 2018; Ofner et al., 2018; WHO, 2023). According to a study published in 2023 and conducted in 11 states in 2020 by the U.S. Center for Disease Control and Prevention (CDS), which works under the U.S. Department of Health and Social Services, the prevalence rate of autism in the United States is 1 in every 36 children. The prevalence rate, which was reported to be 1 in 150 in 2000, 1 in 88 in 2008, 1 in 54 in 2016, 1 in 54 in 2016, and 1 in 44 in 2020, has been updated to 1 in 36 children in 2023. Recent studies show that the prevalence rate of autism is increasing (Centers for Disease Control and Prevention, 2022). Impairments in the processing, integration, and organization of sensory information are common characteristics of individuals with autism (Williams, 2005). Balance, proprioception (the ability to be consciously and unconsciously aware of the position of body parts), smell, hearing, sight, taste, and touch are important for individuals with autism and these systems are independent of each other. It is therefore possible for a person to be hypersensitive to both tactile and auditory sensation at the same time (Saitelbach, 2016). Another behavioral pattern that characterizes the interaction between individuals with ASD and the environment is described as "insistence on sameness, inflexible adherence to routines, or ritualized patterns of verbal or nonverbal behavior (e.g., extreme distress at small changes, difficulties with transitions, need to take the same route every day)" (American Psychological Association, 2024).

A sustainable city is one that can address human needs while ensuring the continuity and resilience of urban systems for future generations (Ertürk, 1996). To achieve sustainability, a city must encompass three critical components: economic, social, and environmental. According to literature, these components are pivotal in fulfilling the sustainability criteria of a city (Eurostat, 2011). An analysis of the subcategories of these criteria reveals that cities which ensure societal comfort and adequately respond to social needs are more likely to be sustainable. Given the increasing prevalence of autism spectrum disorder (ASD) each year, it is imperative that urban design and smart city initiatives incorporate considerations for individuals with ASD. This inclusion is essential to creating cities that are not only sustainable but also inclusive and responsive to the diverse needs of their populations.

In today's context, there is a renewed focus on redesigning-built environments with the framework of neurodiversity. Thus, regarding the importance of the built environment for individuals with ASD, design criteria have been proposed to enhance their comfort and increase their autonomy. This study aims to identify the requirements for autism-friendly spaces by analyzing the needs of children with

Table 12.1 Selected literatures about designing about autistic children

	Author	Title	Focused area
1	Hebert (2003)	Design Guidelines of a Therapeutic Garden for Autistic Children	Built environment design
2	Wilson (2006)	Sensory Gardens for Children with Autism Spectrum Disorder	Built environment design
3	Mostafa (2008)	An Architecture for Autism: Concepts of Design Intervention for the Autistic User	Interior design
4	Linehan (2008)	Guidelines and Design of Outdoor Spaces for Children with Autism Spectrum Disorder	Built environment design
5	Sachs & Vincenta (2011)	Outdoor Environments for Children with Autism and Special Needs	Built environment design
6	Gaines et al. (2014)	The Perceived Effects of Visual Design and Features on Students with Autism Spectrum Disorder	Interior design
7	Mostafa (2008)	Designing For Autism: An ASPECTS Post-Occupancy Evaluation of Learning Environments	Interior design
8	Barakat et al. (2018)	Nature as a Healer for Autistic Children	Built environment design
9	Wagenfeld et al. (2019)	Designing an Impactful Sensory Garden for Children and Youth with Autism Spectrum Disorder	Built environment design
10	Mostafa (2021)	The Autism Friendly University Design Guide	Built environment design
11	Xu (2023a, b)	An Application of Autism-Responsive Landscape Design in Commercial Plaza Space	Built environment design

autism spectrum disorder (ASD) and to create a design guideline to improve the quality of spaces for individuals by organizing an existing structure.

The identified sources are listed in chronological order and their titles and focus areas are shown in Table 12.1.

The frequency of mentions of common criteria in the selected sources is given in Table 12.2.

12.2 Method

12.2.1 Research Area

Within the framework of the determined criteria, a kindergarten building in Başakşehir, Istanbul, which includes children with autism, was examined on May 17, 2024. The building consists of 900 m^2 interior space and 1600 m^2 garden. There is a sand pool, orchard, playground, and playground tracks in the garden. The identity details of the structure are provided in Table 12.3.

Table 12.2 The frequency of mentions of common criteria in the selected sources

	Criteria	1	2	3	4	5	6	7	8	9	10	11
Space fiction related criteria	Accessibility	✓	✓	✓	✓	✓	×	✓	✓	✓	✓	✓
	Safety and security	✓	✓	✓	✓	✓	×	✓	✓	✓	✓	✓
	Building design clarity, predictability, and proportion	✓	✓	✓	✓	✓	✓	✓	✓	✓	✓	✓
	Flexibility in space organization and building design	✓	×	✓	×	✓	×	✓	×	×	✓	✓
	Spatial ranking	✓	✓	✓	✓	✓	✓	✓	✓	✓	✓	✓
	Transition zones	✓	✓	✓	✓	✓	✓	✓	✓	✓	✓	✓
	Escape areas	✓	✓	✓	✓	✓	✓	✓	✓	✓	✓	✓
Sensory input related Criteria	Surfaces and textures	✓	✓	✓	✓	✓	×	✓	✓	✓	✓	✓
	Acoustics	✓	×	✓	✓	✓	×	✓	×	✓	✓	✓
	Light	✓	✓	✓	✓	✓	✓	✓	✓	✓	✓	✓
	Color	✓	✓	✓	✓	✓	✓	✓	✓	✓	✓	✓
	Sensory integration activities	✓	✓	✓	✓	✓	✓	✓	✓	✓	✓	✓

Table 12.3 Research area

	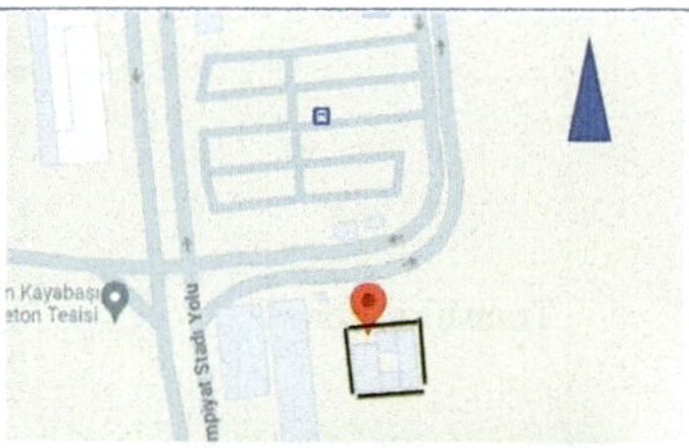 
Name of the area	Başakşehir Kiraz Çiçeği Kindergarten
Production year	2020
Location	Istanbul, Turkey
Coordinates	41°05′57.0″N, 28°46′36.8″E
Project area	2500 m^2

Google Maps (2024)

Table 12.4 Photographs taken in the sample area

Criteria-related to spatial setup	Safety and security	 
	Structure design clarity and proportion	 
	Transition zones	 

(continued)

Table 12.4 (continued)

Sensory input-related criteria	Surfaces and textures	 
	Color	 
	Sensory integration activities	 

Photos were taken by authors

Table 12.5 Evaluations of research area

Criteria-related to spatial setup	Accessibility	The building is suitable for the use of disadvantaged children	✓	
	Safety and security	Enclosure of the building to prevent escape from inside and controlled access from outside	✓	
		Presence of an alarm system	✓	
		Avoid sharp edges	✓	
		Cable equipment is switched off	✓	
		Use of break-resistant materials	✓	
		Not using plants that can cause injury or poisoning	✓	
		Use of climb-proof walls in fences	✓	
	Structure design clarity and proportion	It has a simple plan	✓	
		The spaces are not very large	✓	
		Spaces are not too long	X	
		Grouping and segmenting similar activity areas (areas with the same function and sensory message) rather than having many activities together	✓	
	Spatial ranking	Ensuring that the roads providing circulation between activity areas are simple, recognizable, and actively used	✓	
		Use of one-way routes for transitions between areas	✓	
	Transition zones	Indicating the transition to a different area by using texture or color elements	✓	
		Creating spaces for emotion control in space exchange	X	
		Providing a smooth transition between the transition zone and the next activity area by using structurally or vegetatively similar elements	✓	
	Escape zones	Bamboo tunnels, viewing holes, stunted tree branches, natural hidden areas in the area, tunnels, tents, and gazebos covered with plants such as ivy, separated area using elements such as curtains and tulle	X	

(continued)

Table 12.5 (continued)

Sensory input-related criteria	Surfaces and textures	Use of soft- and smooth-textured materials	X	
		Use of nonreflective materials	✓	
	Acoustics	The building is not located in a noisy environment	✓	
		Use techniques to reduce environmental noise (sound insulation or green walls)	✓	
		Use of anti-traumatic and sound absorbing materials for flooring (carpet, wood, grass, etc.)	✓	
	Light	Use of natural lighting	✓	
		Creating shade areas	X	
	Color	Use of matte and pastel colors	✓	
		Avoiding the use of contrasting colors	✓	
	Sensory integration activities	Use of game elements to reinforce balance	✓	
		No overstimulating visuals	✓	
		Not using plants that emit intense odor	✓	
		No noisy materials	✓	
		No poisonous and allergenic plants	✓	
		No thorny plants	✓	

12.2.2 Methodology

In the study, studies in which the design criteria were determined by reading the title, abstract, and the last full text in the scans made with the relevant topics on Google Scholar were selected. The criteria specified in the selected studies are grouped under two main headings: those related to space organization and those related to sensory input. Sources and criteria were examined in the table and common headings were identified. A checklist of common headings was prepared. Based on the checklist created, the structure of a kindergarten in Istanbul that accommodates inclusive students with autism was examined and photographs and videos were taken. The existing structure was evaluated according to the items in the checklist and the criteria that the structure meets ✓. The criteria that the structure does not meet are expressed using X symbols and the result was analyzed.

12.3 Research Results

The structure meets all criteria under the heading of safety and security. Some areas of the structure are disproportionate due to being too tall. All criteria for spatial ranking are met in the structure. It was found that there were no emotional control areas during the change of space and no escape areas to provide sensory control within the building. It was determined that not all of the materials used on the surface meet the soft ground criteria. All the criteria determined to ensure acoustic comfort in the space are met. It has been observed that shadow areas that can provide escape from natural light have not been created in the building garden. It was noted that all criteria for color and sensory integration activities were met. The results of the examination carried out in the sample area within the scope of the determined criteria and related area photos with criteria are given in Tables 12.4 and 12.5.

In Table IV, the photographs of the parts that the main headings identified within the scope of the research represent in the existing structure are expressed in the table.

12.4 Discussion

The evaluation was made on this area by selecting a sample school in Istanbul. Accessibility, safety and security, structure design clarity and proportion, spatial ranking, transition zones, and escape zones—are considered to be important in a successful autism open space—were evaluated under the title of spatial setup. Surfaces and textures, acoustics, light, color, and sensory integration activities were evaluated using a checklist including sensory input titles. As a result of the evaluation of the checklist, several issues were observed in the clarity and proportionality of the building design, related to specific criteria. These issues include the need to avoid excessively long spaces, the creation of spaces for emotional regulation during spatial transitions, the lack of escape spaces, the use of smooth-textured materials in terms of surfaces and textures, and the creation of shaded areas in terms of lighting.

As a result of the evaluation of the checklist in the clarity and proportionality of the building design, several issues were observed concerning specific criteria. These issues include the necessity to avoid excessively long spaces, the creation of areas for emotional regulation during spatial transitions, the lack of escape areas, the use of smooth-textured materials in terms of surfaces and textures, and the creation of shadowed areas with respect to lighting.

Routine adherence and predictability make transitions troublesome for individuals with autism (Lequia et al., 2015). In case of a change in the environmental conditions of children with autism, symptoms such as anxiety, panic, and irritability may occur (Steingard et al., 1997). Transition zones, created to facilitate both spatial separation and emotional segmentation, help the user to recalibrate their senses when moving from one stimulus zone to another (Mostafa, 2008). Transition zones should not create sudden emotional changes. They should be planned to allow for a smooth transition (Clouse et al., 2020).

Transitions between spaces should be as distraction-free, smooth, minimal disruption, and one-way as possible, using transition zones (Mostafa, 2021). Spatial sequencing creates orderliness in the spaces and prevents the child from being taken by surprise or becoming restless. Regularity will help to clearly define the function and meaning of the environment. The regularity of processes will contribute to the coherence of activities. Organized environments will contribute to individuals with autism to better understand their environment and work effectively (Schopler et al., 1995).

When personal space is violated, the person retreats to a safe zone in order to restore the safety distance and experience a sense of personal space (Gessaroli et al., 2013). Children with autism also lack the ability to cope with crowds, and crowded spaces trigger their sensitivities. Therefore, crowded environments cause stress and restlessness in individuals with autism. It is important to create special escape spaces for individuals with autism to get rid of the stress and restlessness caused by crowded environments (Hilton et al., 2010). According to Steele and Ahrentzen (2016), privacy spaces for individuals with autism contribute to controlling unwanted social interactions. The clear demarcation of public and private spaces with clearly defined boundaries contributes to identifying which are openly shared spaces and which are private spaces and helps to avoid ambiguity. In transitions between sensory compartmentalized areas, smooth transitions should be created, as in the transition zones criterion. In addition, in cases of sensory overload, escape areas should be created for the individual to recalibrate themselves (Ghazali et al., 2018).

According to Leestma (2015), to prevent injuries to children with autism, soft surfaces such as rubber and carpets should be used and hard floors such as concrete and bricks should be covered.

Children diagnosed with autism are sensitive to light. Therefore, stress levels should be kept under control by creating shade areas. According to McAllister and Maguire (2012), the design for children with autism should be simple in layout and use nonglare and indirect lighting methods. Therefore, excessive brightness and excessive darkness should be avoided in lighting. Soft and diffused lighting should be created in the space. By utilizing light and shadow, different effects can be created and shadows and light can be transformed into activity areas and turned into games (Bagdett 2020).

The building satisfies all criteria under the categories of safety and security, spatial sequencing, acoustics, and color and sensory integration activities.

In areas where children with autism feel safe, their aggressive behaviors decrease. It is important that the outdoor space created for children with autism is adequately enclosed. Enclosed outdoor spaces not only prevent uncontrolled wandering but also prevent third parties from entering and disturbing children. Not only the whole area be enclosed, but also the different activity and recreational areas. For example, a child who has to go through the swinging area cannot think that the swings can harm them. For this reason, activity areas should be enclosed or large spaces should be provided for circulation (Herbert, 2003).

Routine adherence and predictability make transitions troublesome for individuals with autism (Lequia et al., 2015). In case of a change in the environmental conditions of children with autism, symptoms such as anxiety, panic, and irritability may occur (Steingard et al., 1997). Transition zones, created to facilitate both spatial separation and emotional segmentation, help the user to recalibrate their senses when moving from one stimulus zone to another (Mostafa, 2008). Transition zones should not create sudden emotional changes. They should be planned to allow for a smooth transition (Clouse et al., 2020).

Noise to which a normal person is exposed has both physiological and psychological negative consequences (Kanakri et al., 2017). Noise can cause undesirable consequences such as depression, sadness, low self-esteem, anxiety, anger, restlessness, distraction, difficulty concentrating, and hearing loss (Goines & Hagler, 2007). Individuals with autism are often hypersensitive to sounds (Bhatara et al., 2013; Stiegler & Davis, 2010). According to Saitelbach (2016), approximately 70% of individuals with autism have sensitivity to noise. Autism was discovered, in part, due to the discovery of individual's sensitivity to noise. Acoustics is therefore the most influential feature of the sensory environment in autism. The acoustic criterion recommends that sound and noise should be minimized and controllable in spaces used by individuals with autism (Mostafa, 2008).

According to Campelo (2017), all senses are stimulated by plants, including sight, taste, touch, and hearing. Touching plants have a calming effect on individuals with autism (Etherington, 2012). Ramps, balance bridges, and climbing areas that support gross motor skills should be designed according to different levels of difficulty. Since swings that oscillate in a linear direction will negatively affect children with autism, multidirectional swings will help to improve the sense of balance (Raising an Extraordinary Person, 2024).

12.5 Conclusion

Considering the design requirements is important in ensuring the comfort of physical space environments regardless of the user group. Users with autism are individuals with special requirements in terms of universal design criteria. In the studies to be carried out, it should be ensured that individuals in this special group are not separated from the society. In this study, which design components should be considered for the physical spaces of children with ASD, space organization and sensory input criteria were examined. As a result of the evaluation of the built environment design criteria for children with autism in Başakşehir Kiraz Çiçeği Kindergarten, it was determined that all of the criteria for accessibility, safety and security, spatial ordering, acoustics, and sensory integration activities were met, while the criteria of not making the spaces too long, creating areas for emotion control during space change, creating escape areas, using soft- and smooth-textured materials, and creating shade areas were not met. This study is important in terms of determining the physical, physiological, psychological, and social needs of individuals with autism.

References

American Psychological Association (2024). *Autism spectrum disorder*. https://www.apa.org/topics/autism-spectrum-disorder

Badgett, A. (2020). *Playing on the spectrum: Exploring how to create playgrounds more accessible for children with autism spectrum disorder*. Master dissertation, University of Washington. https://www.proquest.com/docview/2445948729?pq-origsite=gscholar&fromopenview=true&sourcetype=Dissertations%20&%20Theses

Barakat, H., Bakr, A. F., & El-sayad, Z. (2018). Nature as a healer for autistic children. *Environmental Science & Sustainable Development, 3*(1), 42–62. https://doi.org/10.21625/essd.v3iss1.277

Bhatara, A., Quintin, E. M., Fombonne, E., & Levitin, D. J. (2013). Early sensitivity to sound and musical preferences and enjoyment in adolescents with autism spectrum disorders. *Psychomusicology: Music, Mind, and Brain, 23*(2), 100. https://psycnet.apa.org/buy/2013-26924-002

Campelo, A. (2017). Smell it, taste it, hear it, touch it and see it to make sense of this place. In *Handbook on place branding and marketing* (pp. 124–144). Edward Elgar Publishing. https://doi.org/10.4337/9781784718602.00018

Centers for Disease Control and Prevention. (2022). *Data & statistics on autism spectrum disorder*. Centers for Disease Control and Prevention.

Clouse, J. R., Wood-Nartker, J., & Rice, F. A. (2020). Designing beyond the Americans With Disabilities Act (ADA): Creating an autism-friendly vocational center. *HERD, 13*(3), 215–229. https://doi.org/10.1177/1937586719888502

Ertürk, H. (1996). *Sürdürülebilir Kentler*. Yeni Türkiye Habitat II Özel Sayısı. Yeni Türkiye Medya Hizmetleri, Ankara.

Etherington, N. (2012). Gardening for children with autism spectrum disorders and special educational needs: Engaging with nature to combat anxiety, promote sensory integration and build social skills. *Jessica Kingsley Publishers*. https://uk.jkp.com/products/gardening-for-children-with-autism-spectrum-disorders-and-special-educational-needs?afid=googleprint&variant=36608430669989

EUROSTAT. (2011). *Sustainable development in the European Union: 2011 monitoring report of the EU sustainable development strategy*. Publications office of the European Union.

Gaines, K. S., Curry, Z., Shroyer, J., Amor, C., & Lock, R. H. (2014). The perceived effects of visual design and features on students with autism spectrum disorder. *Journal of Architectural and Planning Research, 31*(4), 282–298. http://www.jstor.org/stable/44113088

Georgiades, S., Szatmari, P., & Boyle, M. (2013). Importance of studying heterogeneity in autism. *Neuropsychiatry, 3*(2), 123. https://www.jneuropsychiatry.org/peer-review/importance-of-studying-heterogeneity-in-autism-neuropsychiatry.pdf

Gessaroli, E., Santelli, E., di Pellegrino, G., & Frassinetti, F. (2013). Personal space regulation in childhood autism spectrum disorders. *PLoS One, 8*(9), e74959. https://doi.org/10.1371/journal.pone.0074959

Ghazali, R. B., Sakip, S. R., & Samsuddin, I. (2018). *A Review of Sensory Design Physical Learning Environment for Autism Centre in Malaysia*. https://doi.org/10.21834/e-bpj.v3i7.1262

Goines, L., & Hagler, L. (2007). Noise pollution: A modem plague. *Southern Medical Journal, 100*(3), 287–294. https://docs.wind-watch.org/Goines-Hagler-2007-Noise_pollution__a_modern_plague.pdf

Google Maps. (2024). https://goo.gl/maps/ZX1xcXmr9EpTLNgCA

Happé, F., & Frith, U. (2020). Annual Research Review: Looking back to look forward—Changes in the concept of autism and implications for future research. *Journal of Child Psychology and Psychiatry, and Allied Disciplines, 61*(3), 218–232. https://doi.org/10.1111/jcpp.13176

Hebert, B. B. (2003). *Design guidelines of a therapeutic garden for autistic children*. Louisiana State University and Agricultural & Mechanical College. Master dissertation, Louisiana State

University. https://www.proquest.com/docview/2675222151?pq-origsite=gscholar&fromopenview=true&sourcetype=Dissertations%20&%20Theses

Hilton, C. L., Harper, J. D., Kueker, R. H., Lang, A. R., Abbacchi, A. M., Todorov, A., & LaVesser, P. D. (2010). Sensory responsiveness as a predictor of social severity in children with high functioning autism spectrum disorders. *Journal of Autism and Developmental Disorders, 40*(8), 937–945. https://doi.org/10.1007/s10803-010-0944-8

Hyman, S. L., Levy, S. E., Myers, S. M., Kuo, D. Z., Apkon, S., Davidson, L. F., et al. (2020). Identification, evaluation, and management of children with autism spectrum disorder. *Pediatrics, 145*(1) https://publications.aap.org/pediatrics/article/145/1/e20193447/36917/Identification-Evaluation-and-Management-of

Kanakri, S. M., Shepley, M., Varni, J. W., & Tassinary, L. G. (2017). Noise and autism spectrum disorder in children: An exploratory survey. *Research in Developmental Disabilities, 63*, 85–94. https://doi.org/10.1016/j.ridd.2017.02.004

Kanner, L. (1943). Autistic disturbances of affective contact. *Nervous child, 2*(3), 217–250. http://www.th-hoffmann.eu/archiv/kanner/kanner.1943.pdf

Leestma, D. (2015). *Designing For The Spectrum: An Educational Model for the Autistic User (Master dissertation)*. https://doi.org/10.13016/M2Z04V

Lequia, J., Wilkerson, K. L., Kim, S., & Lyons, G. L. (2015). Improving transition behaviors in students with autism spectrum disorders: a comprehensive evaluation of interventions in educational settings. *Journal of Positive Behavior Interventions, 17*(3), 146–158. https://doi.org/10.1177/1098300714548799

Linehan, J. (2008). *Landscape for autism: Guideline and design of outdoor space for children with autism spectrum disorder*. University of California, Davis: A Senior Project in the Landscape Architecture Program.

Lord, C., Elsabbagh, M., Baird, G., & Veenstra-Vanderweele, J. (2018). Autism spectrum disorder. *Lancet (London, England), 392*(10146), 508–520. https://doi.org/10.1016/S0140-6736(18)31129-2

Lord, C., Charman, T., Havdahl, A., Carbone, P., Anagnostou, E., Boyd, B., Carr, T., de Vries, P. J., Dissanayake, C., Divan, G., Freitag, C. M., Gotelli, M. M., Kasari, C., Knapp, M., Mundy, P., Plank, A., Scahill, L., Servili, C., Shattuck, P., Simonoff, E., et al. (2022). The Lancet Commission on the future of care and clinical research in autism. *Lancet (London, England), 399*(10321), 271–334. https://doi.org/10.1016/S0140-6736(21)01541-5

McAllister, K., & Maguire, B. (2012). A design model: The Autism Spectrum Disorder Classroom Design Kit. *British Journal of Special Education, 39*(4), 201–208.

Mostafa, M. (2008). An architecture for autism: Concepts of design intervention for the autistic user. *International Journal of Architectural Research, 2*(1), 189–211. https://doi.org/10.1111/1467-8578.12006

Mostafa, M. (2021). *The autism friendly university design guide*. Dublin City University.

Ofner, M., Coles, A., Decou, M. L., Do, M., Bienek, A., Snider, J., & Ugnat, A. (2018). *Autism spectrum disorder among children and youth in Canada 2018*. Public Health Agency of Canada. https://www.canada.ca/en/public-health/services/publications/diseases-conditions/autism-spectrum-disorder-children-youth-canada-2018.html

Pellicano, E., & den Houting, J. (2022). Annual research review: Shifting from 'normal science' to neurodiversity in autism science. *Journal of Child Psychology and Psychiatry, 63*, 381–396. https://doi.org/10.1111/jcpp.13534

Public Health Agency of Canada. (2024). Autism: About, causes and co-occurring conditions. https://www.canada.ca/en/public-health/services/diseases/autism-spectrum-disorder-asd.html

Raising an Extraordinary Person (2024). *Beyond play: The benefits of swinging for children with autism*. https://hes-extraordinary.com/autism-swinging-benefits

Rosen, N. E., Lord, C., & Volkmar, F. R. (2021). The diagnosis of autism: From Kanner to DSM-III to DSM-5 and beyond. *Journal of Autism and Developmental Disorders, 51*(12), 4253–4270. https://doi.org/10.1007/s10803-021-04904-1

Sachs, N., & Vincenta, T. (2011). Outdoor environments for children with autism and special needs. *Implications, 9*(1), 1–8.
Saitelbach, J. (2016). *Therapeutic streetscapes: A new bell street designed to accommodate those on the autism spectrum*. Doctoral dissertation, University of Washington. https://digital.lib.washington.edu/researchworks/items/ba5527aa-660c-4163-b1ea-5135d89a6cbf
Schopler, E., Mesibov, G. B., & Hearsey, K. (1995). Structured teaching in the TEACCH system. *Learning and Cognition in Autism*, 243–268.
Steele, K., & Ahrentzen, S. (2016). *At home with autism: Designing housing for the spectrum* (1st ed.). Bristol University Press. https://doi.org/10.2307/j.ctt1t88z8x
Steingard, R. J., Zimnitzky, B., DeMASO, D. R., Bauman, M. L., & Bucci, J. P. (1997). Sertraline treatment of transition-associated anxiety and agitation in children with autistic disorder. *Journal of Child and Adolescent Psychopharmacology, 7*(1), 9–15. https://doi.org/10.1089/cap.1997.7.9
Stiegler, L. N., & Davis, R. (2010). Understanding sound sensitivity in individuals with autism spectrum disorders. *Focus on Autism and Other Developmental Disabilities, 25*(2), 67–75. https://doi.org/10.1177/1088357610364530
Tohum Autism Foundation (2017). https://tohumotizm.org.tr/otizm/otizm-spektrum-bozuklugu/
U.S. Centers for Disease Control and Prevention. (2024, May 14). *About autism spectrum disorder*. https://www.cdc.gov/autism/about/index.html
Wagenfeld, A., Sotelo, M., & Kamp, D. (2019). Designing an impactful sensory garden for children and youth with autism spectrum disorder. *Children, Youth and Environments, 29*(1), 137–152.
Williams, D. L., Goldstein, G., Carpenter, P. A., & Minshew, N. J. (2005). Verbal and spatial working memory in autism. *Journal of Autism and Developmental Disorders, 35*, 747–756. https://link.springer.com/article/10.1007/s10803-005-0021-x
Wilson, B. J. (2006). *Sensory gardens for children with autism spectrum disorders. The University of Arizona*. Master dissertation, University of Arizona. https://www.proquest.com/docview/305355073?pq-origsite=gscholar&fromopenview=true&sourcetype=Dissertations%20&%20Theses
World Health Organization. (2019). 6A02 Autism spectrum disorder. *International statistical classification of diseases and related health problems*.
World Health Organization (2023, November 15). *Autism*. https://www.who.int/news-room/fact-sheets/detail/autism-spectrum-disorders
Xu, Z. (2023a). *An application of autism-responsive landscape design in commercial plaza space*. Doctoral dissertation, University of Guelph. https://hdl.handle.net/10214/27626
Xu, Z. (2023b). *An Application of Autism-responsive Landscape Design in Commercial Plaza Space*. Doctoral dissertation, University of Guelph. https://atrium.lib.uoguelph.ca/items/123d33fb-0382-4150-abcb-9d70a9d00fc4

Chapter 13
Approaches to Assessing IEQ in Urban School Classrooms: Focusing on Primary and Secondary Education

P. Khizar Aadil and M. Naveenkumar

Abstract The indoor environmental quality (IEQ) is described as "the quality of a building's environment concerning the health and well-being of its occupants; it encompasses design, analysis, and operation aspects that result in buildings that are energy-efficient, healthy, and welcoming," according to the study. One-fifth of all people spend over 30% of their time in educational facilities, such as schools, universities, and colleges. School buildings are among the most often occupied building types by children and it is imperative to support building occupants' health and well-being regarding indoor environmental quality, particularly thermal comfort, and visual comfort in primary and secondary school classrooms around the globe. This chapter looks into the donation of previous literary journals on the impact of IEQ and its approaches to assessing the general well-being, daylighting, thermal comfort, and visual comfort of pupils in urban primary and secondary school classrooms that were published during the last 10 years (2014–2024). By using PRISMA methodology, this chapter conducts a systematic literature review and examines the ways to study and recognize the appropriate methodology for the evaluation of IEQ (indoor environmental quality) in the primary and secondary school classroom premises worldwide at various climatic zones.

Keywords Thermal comfort · Visual comfort · Assessment models · Daylight · Glare · Well-being · Climatic responsive · Sustainable indoor environmental quality

---

P. Khizar Aadil (✉)
School of Environment Architecture and Design, SRM Institute of Science and Technology-Ramapuram, Chennai, Tamil Nadu, India

M. Naveenkumar (✉)
Department of Civil Engineering, Easwari Engineering College, Chennai, Tamil Nadu, India

V. Baskaran et al. (eds.), *EcoTech Urbanism*, The Urban Book Series,
https://doi.org/10.1007/978-3-031-83833-0_13

13.1 Introduction

EcoTech urbanism is an approach to city planning and design that integrates ecological principles with advanced technology to create sustainable, efficient, and livable urban environments. A key component of EcoTech urbanism is enhancing indoor environmental quality (IEQ), which is crucial for the well-being and productivity of residents. A building's indoor environmental quality (IEQ) plays a significant role in determining how well-off and comfortable its residents are perceived to be (Jamaludin et al., 2016). Nonetheless, every building uniquely impacts the internal environment and the people who live there (Jamaludin et al., 2016). IEQ is regarded as a measure of occupant's comfort since it highlights areas that need improvement in the variables that impact the interior environment (Mihai & Iordache, 2016). Consequently, IEQ is viewed by the majority of worldwide sustainability rating systems as a crucial component for creating sustainable buildings (Al-Sulaihi et al., 2015). Since educational buildings house large populations, it is even more important to provide adequate IEQ (Al-Sulaihi et al., 2015). The learning environment is one of the spaces that need attention as it is related to student's well-being as well as their learning performance (Jamaludin et al., 2016). IEQ components have significant effects on occupants (Samad et al., 2017). Included in the IEQ components that have a major impact on occupants is thermal comfort, which is influenced by air temperature, humidity, radiation, interior illumination, air movement, activities, clothing, and climate change (Samad et al., 2017). In addition to improving health and happiness, visual comfort in schools also improves learning and visual performance (Korsavi et al., 2016). IEQ in terms of classrooms, there is a great demand for a credible and trustworthy approach to research students' adaptable behavior (Korsavi & Montazami, 2019). IEQ in classrooms can affect school children's comfort, health, and performance (Zhang & Bluyssen, 2021). Children in primary school have different perceptions of IEQ than adults and have less control over their surroundings (Korsavi et al., 2022). Additionally, kids typically do not realize how their environmental adaptation behaviors affect their energy, well-being, and comfort with poor IEQ (Korsavi et al., 2022). According to the study, inadequate IEQ conditions in school classrooms can lead to not only discomfort but also to negative and unhealthy outcomes, such as inattention and lack of concentration (Demozzi et al., 2023). An unsatisfactory IEQ condition could hamper pupils' learning performance. (Demozzi et al., 2023). Therefore, it is crucial to establish IEQ circumstances that satisfy the needs of the students (Carton et al., 2023), as IEQ is a multi-criteria indicator (Diaz Cisternas, 2023). Here in this chapter, a systematic meta-analysis is conducted with the IEQ indicators such as thermal, and visual comfort exclusively to understand its impact on the well-being of primary and secondary school children inside the classroom environment in urban zones along with the recognition of the types of approaches adopted for the evaluation of these two components and discussing its level of applicability, factors for consideration, accuracy, and limitations. The analysis is done based on a systematic review with the donation of previously published literary journals using a meta-analysis method called

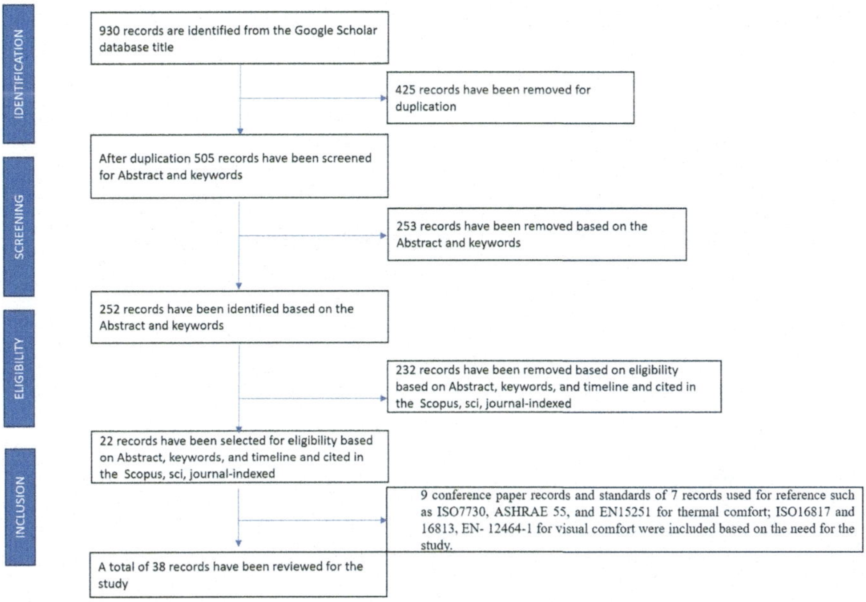

Fig. 13.1 Preferred reporting items for systematic review and meta-analysis

PRISMA (Preferred Reporting Items for Systematic Review and Meta-Analysis) for the period of (2014–2024), as shown in Figs. 13.1 and 13.2.

13.2 Methodology

Preferred Reporting Items for Systematic Review and Meta-analysis were used for this study. Area total of 930 records identified from the Google Scholar database based on the title of IEQ classroom, excluding the universities and higher secondary level and duplication of 425 records have been removed for screening. After screening 505 records, 249 records have been removed for eligibility based on the timeline, abstract, and keywords.

The journals were reduced to 22 entries by choosing those that were indexed by SCI and ranked by Scopus for the previous ten years (2014–2024). Based on the requirements of the study, nine conference paper records and reference standards, including ISO7730, ASHRAE 55, and EN15251 for thermal comfort and ISO16817 and 16813, and EN-12464-1 for visual comfort, were added in addition to the journals. Overall, 38 journal records have been examined for the study, as shown in Figs. 13.1 and 13.2.

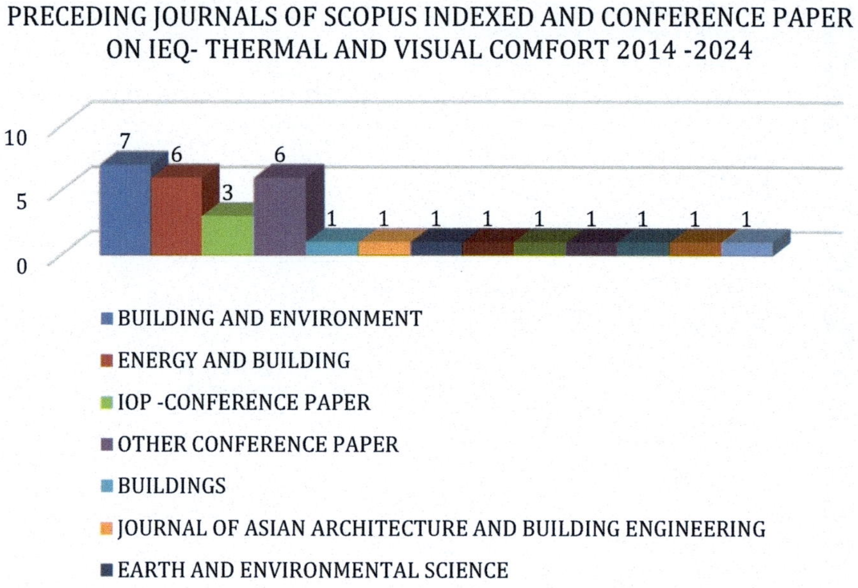

Fig. 13.2 The number of records of preceding Scopus-indexed, SCI journals and conferences on IEQ-thermal and visual comfort in primary and secondary classroom environments from 2014 to 2024

13.3 Thermal Comfort as IEQ Indicator in the Classroom

The state of mind that conveys happiness with the thermal environment and is measured by subjective assessment is known as thermal comfort, according to ANSI/ASHRAE standard 55. The thermal comfort criteria were deemed inappropriate for evaluating children's thermal environments, as evidenced by recent thermal comfort studies conducted in a lightweight junior school building (Teli et al., 2014). Children were more susceptible to elevated temperatures than adults (Teli et al., 2014). This caused the summertime thermal perception of the students to be under estimated (Teli et al., 2014). As the children's metabolic rate plays a significant role, it must be considered in determining the perceived thermal comfort for primary and secondary school children (Almeida et al., 2016). Apart from metabolic rate, students rely on clothing insulation, especially in schools without air conditioning systems (Kim & de Dear, 2018). Therefore, it is critical to raise knowledge among educators, students, and school building designers about how to manage the classroom atmosphere without the need for air conditioning (Mishan & Bahadur, 2019). Ratajczak et al. (2023) emphasized the benefits of being in nature for creativity and problem-solving abilities.

13.3.1 Thermal Comfort Assessment Models

The indoor environment is evaluated for human thermal comfort-discomfort using the Predicted Mean Vote (PMV) and Predicted Percentage of Dissatisfaction (PPD) indices; however, the objective approach also included data processing using ISO 7730 and instrumentation measurements (Papazoglou et al., 2019). The objective approach measurements were made of the globe temperature, airflow velocity, relative humidity, and dry-bulb temperature (Yun et al., 2014). Concurrently, the opinions of the students regarding thermal comfort using age-appropriate questionnaires are to be recorded according to the EN 15251-adaptive model method (Almeida et al., 2016). In addition, the votes for thermal sensation and thermal preference effects are to be recorded in terms of the questionnaire that shows the impact of children's clothing insulation and children's metabolic rate that has to be taken into consideration in the thermal comfort numerical analysis. This is because the PMV model shows some restrictions in the assessment of thermal comfort for naturally ventilated school classrooms (Almeida et al., 2016). Therefore, naturally ventilated classrooms need to be addressed with appropriate thermal comfort standards because there are currently no thermal comfort requirements in any of the climatic zones for classrooms in schools (Jindal, 2018). Hence, in defining and choosing future IEQ mitigation methods, assessing the ventilation-overheating interactions of various places, climatic zones, and regions may be crucial (Grassie et al., 2022). Apart from the analytical model, other simulation models were incorporated into the assessment process to develop modeling different classroom typologies (Grassie et al., 2022) (Table 13.1).

13.3.2 PMV Model

Thermal comfort can be measured using the Predicted Mean Vote (PMV) model, which P.O. Fanger created. It is a commonly used technique. It makes predictions about a broad population's average thermal perception based on a variety of environmental and individual characteristics. The PMV model, which is based on the heat balance of the human body, takes into account the following factors: air temperature, mean radiant temperature, air velocity, garment insulation, metabolic rate, and humidity (Fig. 13.3).

13.3.3 Adaptive Model

People adjust to their thermal surroundings through behavioral, physiological, and psychological mechanisms, according to the adaptive model of thermal comfort. This concept works well in naturally ventilated buildings where residents have

Table 13.1 Methodologies applied for the assessment of thermal comfort in primary and secondary schools at different climatic zones worldwide

Author and year	The methodology adopted for the assessment of thermal comfort	Country	Climatic zone	Findings
Yun et al. (2014)	PMV, ISO 7730 suggested using anticipated mean vote and PPD, or predicted percentage dissatisfied. Standards EN-15251: Instruments are used to measure humidity, air temperature, air velocity, and radiant temperature	Korea	Temperate climate	Both the PMV and EN-15251 model shows limitations detected in the assessment
Kim and de Dear (2018)	PMV model	Australia	Temperate and subtropical	The predicted mean vote model (PMV) is more applicable for school classrooms with air-conditioning systems compared to classrooms without air-conditioning system
Hamzah et al. (2020)	Humphrey's adaptive model	Indonesia	Tropical climate	Humphrey's adaptive model has a high impact on the thermal perception
Grassie et al. (2022)	Simulation demonstration modeling	UK	Temperate	Simulations have shown an impact on hotter climates where ventilation is more crucial to regulating temperature rise and maintaining IEQ
Diaz Cisternas (2023)	Post occupancy IEQ score = w1 × thermal comfort score + w2 × IAQ score + w3 × lighting quality score + w4 × acoustic quality score	Chile	Mediterranean climate	Has impact on the post-evaluation occupants' thermal comfort perception

greater influence over their surroundings (e.g., through opening windows or adjusting clothing). Unlike the PMV model, which is more rigid and relies on a heat balance approach, the adaptive model is based on empirical data from field studies and relates indoor comfort temperatures to outdoor conditions, as shown in Fig. 13.4.

13.4 Summary and Differences

- *Predictive basis*: The PMV model relies on a static heat balance approach, while the adaptive model is based on dynamic adaptation to thermal conditions.

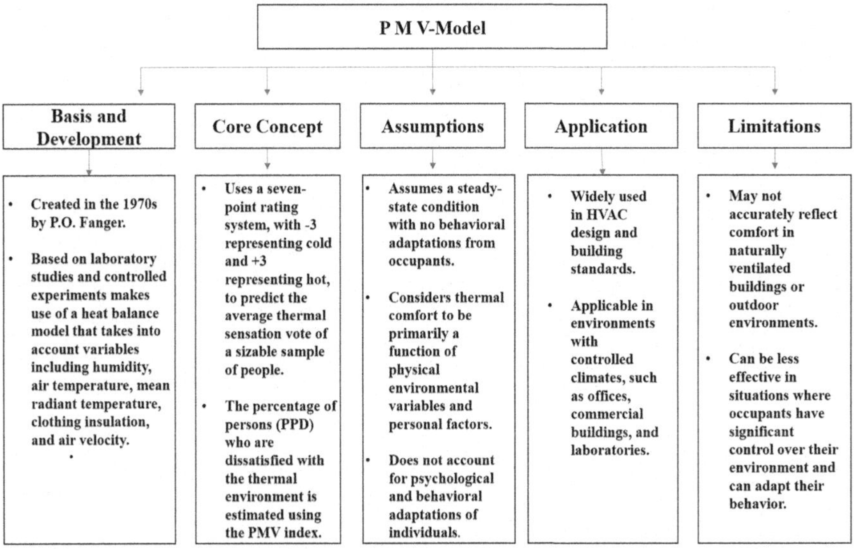

Fig. 13.3 The PMV model application and limitation

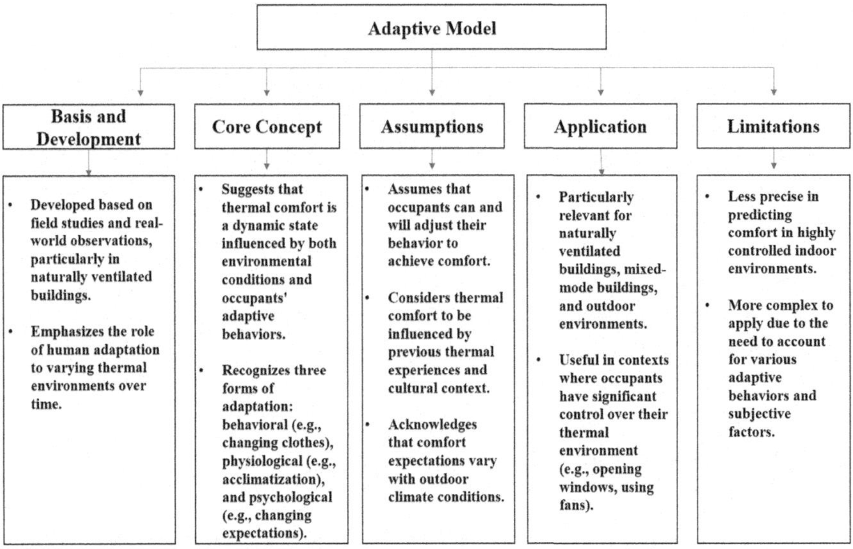

Fig. 13.4 The adaptive model application and limitation

- *Assumptions about occupants*: The PMV model assumes no behavioral adaptation, whereas the adaptive model assumes that occupants will adapt their behavior to achieve comfort.

- *Applicability*: The PMV model is best suited for controlled environments with stable indoor conditions, while the adaptive model is more applicable to naturally ventilated buildings and environments where occupants have more control.
- *Factors considered*: The PMV model focuses on physical environmental variables and personal factors, while the adaptive model includes psychological and behavioral factors and takes into account varying expectations based on outdoor climate conditions.

In summary, the PMV model offers a more rigid and controlled approach to predicting thermal comfort, whereas the adaptive model provides a flexible framework that better captures the variability of human responses to different thermal environments.

13.5 Visual Comfort as IEQ Indicator in the Classroom

Visual comfort in schools enhances health and well-being and provides satisfaction in learning and performing tasks (Korsavi et al., 2016). Children's performance in the classroom is impacted by the fact that they spend a lot of time in the classroom and participate in activities that demand not only mental effort but also physical labor, such as taking notes and looking at the board or smart board. These activities may cause children to become less comfortable with different viewing distances and levels of focus (Noda et al., 2020). Natural lighting is crucial to the design of educational facilities because it assures energy conservation, fosters a pleasant atmosphere, and promotes healthier conditions (Michael & Heracleous, 2017). One of the most crucial aspects of sustainable building is the use of daylighting, which is regrettable that designers overlook this (Moazzeni & Ghiabaklou, 2016). Therefore, architects and designers typically play an important role in designing a building's envelope design early in the design process considering key parameters influencing visual comfort, such as lighting levels, glare, color, temperature, and spatial arrangement (Alkhatatbeh et al., 2023).

13.5.1 Visual Comfort Assessment Model

A visual comfort assessment model for primary and secondary school classrooms can be structured to integrate both subjective perceptions and objective measurements. The pupils' judgment of their visual comfort is systematically explained by a structural equation modelling (SEM) method (Fakhari et al., 2021). Classroom visual comfort is influenced by several elements, both directly and indirectly, and its impacts are quantified (Fakhari et al., 2021). When children are reading from different viewing distances, their subjective impression may change based on how well they do tasks in the classroom, such as looking at the blackboard and taking notes

(Yao et al., 2024). Perceived blackboard glare was shown to be correlated with the percentage of space where disability glare probability (DGP) using dynamic daylighting metrics (Liu et al., 2023).

The objective measurements of visual comfort in space are quantitative metrics that can be accurately assessed using various tools and techniques.

These measurements focus on lighting conditions, spatial distribution of light, and potential sources of visual discomfort.

The tools for objective measurements

- Lux meter: For measuring horizontal and vertical illuminance
- Luminance meter: For measuring surface luminance and assessing glare
- Colorimeter/spectrophotometer: For measuring CCT and CRI
- Daylight meters: For measuring daylight factors
- Simulation software: Tools like Radiance, DIVA for Rhino, and Honeybee +
- Ladybug for advanced daylight simulations and analyses

By systematically measuring these objective parameters, it is possible to evaluate visual comfort in educational environments comprehensively. This data can then be used to optimize lighting design, enhance daylighting strategies, and ensure a comfortable visual environment conducive to learning and productivity. Design elements including the building's orientation, classroom size, light window size, and classroom arrangement have an impact on visual comfort of the children in the form of physiology, psychology, and behavior of elementary and secondary school pupils (Yao et al., 2024) (Table 13.2).

13.5.2 Subjective Assessments

Subjective assessments gather qualitative data on users' perceptions and experiences related to visual comfort. It is divided into two types, as shown in Fig. 13.5.

13.5.3 Objective Measurements

Objective measurements provide quantifiable data on various environmental parameters affecting visual comfort. It is carried out under different levels, as shown in the Fig. 13.6.

- *Basis of measurement*: Objective measurements are based on quantifiable data and physical observations, while subjective measurements are based on personal judgment and perception.
- *Reliability*: Objective measurements are generally considered more reliable and replicable because they are based on standard methods and instruments, whereas subjective measurements can vary significantly between individuals.

Table 13.2 Methodologies applied for the assessment of visual comfort in primary and secondary schools at different climatic zones worldwide

Author and year	The methodology adopted for the assessment of visual comfort	Country	Climatic zone	Findings
Korsavi et al. (2016)	LEED v4 subjective questionnaire and simulation-based dynamic daylight measures, such as spatial daylight autonomy (sDA) and annual sunlight exposure (ASE)	Iran	Hot desert	Variation of the result between the subjective survey and simulation was found
Ashrafian and Moazzen (2019)	Simulation of a virtual model based on a case study considering the climatic characteristics	Turkey	Mediterranean	The effects observed on occupant comfort and energy consumption of various glazing ratios and window designs
Natalia and Suharjanto (2022)	Simulation comparison with the book Standard SNI 6197: 2011 of Energy Conservation in Lighting Systems	Indonesia	Tropical climate	The need to record perception impact for more accuracy
Yao et al. (2024)	Using the ergo LAB human-factor platform, the students' electrodermal indicators and simultaneous subjective scoring were carried out	China	Temperate continental monsoon climate zone	The building's orientation, classroom size, light window size, and classroom arrangement have an impact on the visual comfort of the children

- *Bias*: Subjective measurements are susceptible to bias introduced by individual differences in perception, interpretation, or preferences, whereas objective measurements aim to minimize such biases through standardization.

In many fields, both types of measurements can complement each other. Objective measurements provide a quantitative basis for analysis and decision-making, while subjective measurements offer insights into personal experiences, preferences, and perceptions that quantitative data alone may not capture.

13.5.4 Lighting Quality

Lighting quality encompasses various factors that contribute to a visually comfortable and effective lighting environment. It involves not only the quantity of light but also its distribution, color, and how it interacts with the surrounding space.

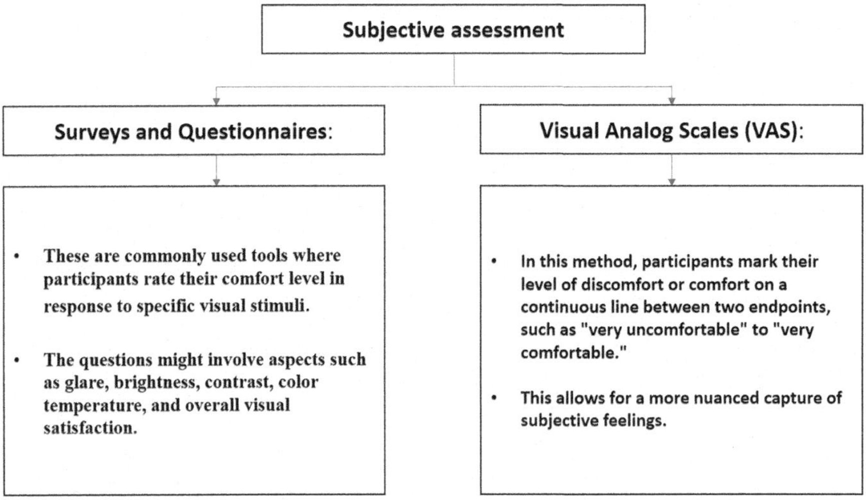

Fig. 13.5 The subjective assessment model types

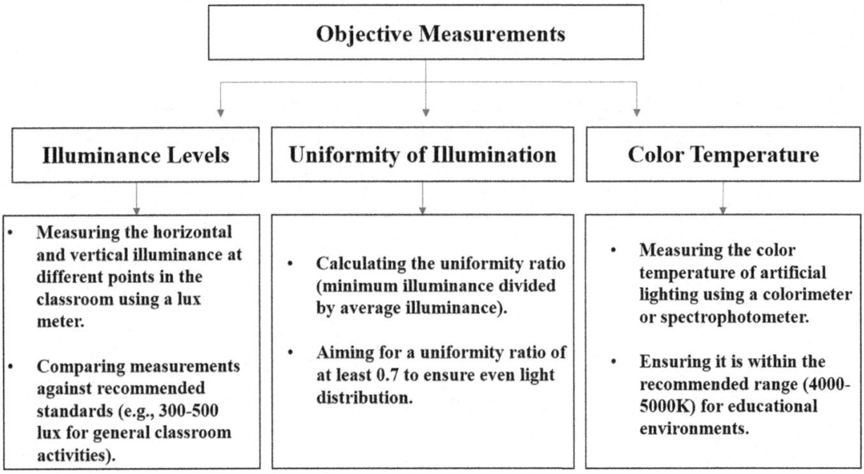

Fig. 13.6 The objective model for thermal comfort assessment

13.6 Glare Assessment

Glare assessment is a crucial component of lighting design, aimed at ensuring visual comfort by minimizing the discomfort and visibility issues caused by excessive or poorly directed light. There are several methods and metrics used to assess and quantify glare, particularly in relation to daylighting and artificial lighting.

13.6.1 Direct Glare

- Using a glare meter to identify potential sources of direct glare from lighting fixtures and windows
- Ensuring compliance with glare control recommendations (e.g., using diffusers, shades, or antiglare coatings)

13.6.2 Reflected Glare

- Evaluating reflective surfaces such as whiteboards, desks, and screens
- Using the luminance meters to measure reflected light and identify problematic areas

13.7 Daylight Evaluation

Daylight evaluation is a critical aspect of building design that focuses on maximizing the benefits of natural light while minimizing its potential drawbacks, such as glare and heat gain. Effective daylighting can enhance visual comfort, reduce energy consumption, and improve the overall well-being of occupants.

13.7.1 Daylight Factor

- Calculating the daylight factor (DF) using measurements of indoor and outdoor illuminance
- Aiming for a DF of 2–5% to balance natural and artificial lighting

13.7.2 Daylight Distribution

- Mapping daylight penetration using sensors or simulation software to ensure even daylight distribution across the classroom.

13.8 Surface Reflectance

Albedo, another name for surface reflectance, is the measurement of how much radiation or light a surface reflects. A larger percentage denotes a higher reflectance, whereas a lower percentage denotes more absorption. It is given as a percentage. The distribution and intensity of natural light in a place are influenced by surface reflectance, which is a key factor in daylighting design. Here is a detailed look at the importance of surface reflectance and how it can be evaluated and optimized in building design.

13.8.1 Reflectance Measurements

- Measuring the reflectance of walls, ceilings, floors, and furniture using a reflectance meter
- Ensuring surfaces have appropriate reflectance values (e.g., ceilings >70%, walls 50–70%, floors 20–40%)

Creating a visually comfortable classroom involves careful consideration of lighting, color schemes, and ergonomic design. These factors not only enhance student well-being but also improve academic performance. Schools should prioritize visual comfort in classroom design to foster a better learning environment. Therefore assessments of thermal and visual comfort were required in order to assess the IEQ level in urban primary and secondary school classrooms.

13.9 Conclusion

The methodologies adopted in the past 10 years between 2014 and2024 were found to be categorized into three types of major:

- Subjective
- Objective
- Simulation

Which were used to assess the thermal and visual comfort of IEQ in urban primary and secondary school classroom environments worldwide. Its impacts were determined under three major criteria such as

- Personal
- Architectural
- Climatic factors

Personal criteria deal with the individual perception and behavior based on metabolic rate, clothing insulation, age, and gender, which records a high level of impact on thermal comfort perception. Personal parameters that address individual behaviour and perception based on age, gender, garment insulation, and metabolic rate have a significant impact on how thermal comfort is perceived. In the meantime, architectural factors like window placements, wall-to-window ratios, and the use of antireflective materials and shading devices are all part of visual comfort.

The EcoTech urbanism combines technology and ecological principles to create environments that are not only sustainable but also improve the quality of life for inhabitants. By focusing on energy efficiency, air quality, lighting, and thermal comfort, this approach ensures that indoor spaces are healthy, comfortable, and conducive to well-being.

References

Alkhatatbeh, B. J., Kurdi, Y., & Asadi, S. (2023). Enhancing learning environments: Exploring optimal classroom design connected to double-loaded corridors across the US climate zones. *Energy and Buildings, 298*, 113562.

Almeida, R. M., Ramos, N. M., & De Freitas, V. P. (2016). Thermal comfort models and pupils' perception in free-running school buildings of a mild climate country. *Energy and Buildings, 111*, 64–75.

Al-Sulaihi, I., Al-Gahtani, K., Alsugair, A., & Tijani, I. (2015). Assessing indoor environmental quality of educational buildings using BIM. *Journal of Environmental Science and Engineering B, 4*, 451–458.

Ashrafian, T., & Moazzen, N. (2019). The impact of glazing ratio and window configuration on occupants' comfort and energy demand: The case study of a school building in Eskisehir, Turkey. *Sustainable Cities and Society, 47*, 101483.

Carton, Q., Mennes, F., Broeck, S. V., Van Roy, V., Kolarik, J., & Breesch, H. (2023). Evaluation of indoor environmental quality and pupils' satisfaction in Flemish primary schools. In *18th Healthy Buildings Europe Conference*.

Demozzi, G., Zaniboni, L., Pernigotto, G., & Gasparella, A. (2023). Impact of visual, thermal and indoor air quality conditions on students' wellbeing and learning performance in a primary school of Bolzano, Italy. *Building Simulation Applications BSA, 2022*, 489–497.

Diaz Cisternas, M. A. (2023). Methodology for the evaluation of indoor environmental quality and comfort in school classrooms through a multicriteria index.

Fakhari, M., Vahabi, V., & Fayaz, R. (2021). A study on the factors simultaneously affecting visual comfort in classrooms: A structural equation modeling approach. *Energy and Buildings, 249*, 111232.

Grassie, D., Karakas, F., Schwartz, Y., Dong, J., Milner, J., Chalabi, Z., Mavrogianni, A., & Mumovic, D. (2022, July). Modeling UK school performance by coupling building simulation and multi-criteria decision analysis. In: *17th international conference of the international society of indoor air quality & climate*. International Society of Indoor Air Quality and Climate (ISIAQ).

Hamzah, B., Mulyadi, R., Amin, S., & Kusno, A. (2020). Adaptive thermal comfort of naturally ventilated classrooms of elementary schools in the tropics. In *IOP conference series: Earth and environmental science* (Vol. 402, No. 1, p. 012021). IOP Publishing.

Jamaludin, N. M., Mahyuddin, N., & Akashah, F. W. (2016). Assessment of indoor environmental quality (IEQ): Students' well-being in university classroom with the application of landscaping. In *MATEC web of conferences* (Vol. 66, p. 00061). EDP Sciences.

Jindal, A. (2018). Thermal comfort study in naturally ventilated school classrooms in composite climate of India. *Building and Environment, 142*, 34–46.

Kim, J., & de Dear, R. (2018). Thermal comfort expectations and adaptive behavioural characteristics of primary and secondary school students. *Building and Environment, 127*, 13–22.

Korsavi, S. S., & Montazami, A. (2019). Developing a valid method to study adaptive behaviours with regard to IEQ in primary schools. *Building and Environment, 153*, 1–16.

Korsavi, S. S., Zomorodian, Z. S., & Tahsildoost, M. (2016). Visual comfort assessment of daylit and sunlit areas: A longitudinal field survey in classrooms in Kashan, Iran. *Energy and Buildings, 128*, 305–318.

Korsavi, S. S., Jones, R. V., & Fuertes, A. (2022). Operations on windows and external doors in UK primary schools and their effects on indoor environmental quality. *Building and Environment, 207*, 108416.

Liu, B., Liu, Y., Deng, Q., & Hu, K. (2023). A study on daylighting metrics related to the subjective evaluation of daylight and visual comfort of students in China. *Energy and Buildings, 287*, 113001.

Michael, A., & Heracleous, C. (2017). Assessment of natural lighting performance and visual comfort of educational architecture in Southern Europe: The case of typical educational school premises in Cyprus. *Energy and Buildings, 140*, 443–457.

Mihai, T., & Iordache, V. (2016). Determining the indoor environment quality for an educational building. *Energy Procedia, 85*, 566–574.

Mishan, S., & Bahadur, R. H. (2019, July). Study on adaptive thermal comfort in naturally ventilated secondary school buildings in Nepal. In *IOP conference series: Earth and environmental science* (Vol. 294, No. 1, p. 012062). IOP Publishing.

Moazzeni, M. H., & Ghiabaklou, Z. (2016). Investigating the influence of light shelf geometry parameters on daylight performance and visual comfort, a case study of educational space in Tehran, Iran. *Buildings, 6*(3), 26.

Natalia, S., & Suharjanto, G. (2022, February). The openings and lighting design strategies of primary school in Jakarta. In *IOP conference series: Earth and environmental science* (Vol. 998, No. 1, p. 012036). IOP Publishing.

Noda, L., Lima, A. V., Souza, J. F., Leder, S., & Quirino, L. M. (2020). Thermal and visual comfort of schoolchildren in air-conditioned classrooms in hot and humid climates. *Building and Environment, 182*, 107156.

Papazoglou, E., Moustris, K. P., Nikas, K. S. P., Nastos, P. T., & Statharas, J. C. (2019). Assessment of human thermal comfort perception in a non-air-conditioned school building in Athens, Greece. *Energy Procedia, 157*, 1343–1352.

Ratajczak, K., Amanowicz, Ł., Pałaszyńska, K., Pawlak, F., & Sinacka, J. (2023). Recent achievements in research on thermal comfort and ventilation in the aspect of providing people with appropriate conditions in different types of buildings—Semi-systematic review. *Energies, 16*(17), 6254.

Samad, M. H. A., Aziz, Z. A., & Isa, M. H. M. (2017, October). Indoor environmental quality (IEQ) of school classrooms: Case study in Malaysia. In *AIP conference proceedings* (Vol. 1892, No. 1). AIP Publishing.

Teli, D., Jentsch, M. F., & James, P. A. (2014). The role of a building's thermal properties on pupils' thermal comfort in junior school classrooms as determined in field studies. *Building and Environment, 82*, 640–654.

Yao, F., Bo, G., & Zitong, W. (2024). Association between visual distance and students' visual perception comfort in primary and secondary classrooms. *Journal of Asian Architecture and Building Engineering, 23*(2), 569–581.

Yun, H., Nam, I., Kim, J., Yang, J., Lee, K., & Sohn, J. (2014). A field study of thermal comfort for kindergarten children in Korea: An assessment of existing models and preferences of children. *Building and Environment, 75*, 182–189.

Zhang, D., & Bluyssen, P. M. (2021). Actions of primary school teachers to improve the indoor environmental quality of classrooms in The Netherlands. *Intelligent Buildings International, 13*(2), 103–115.

Chapter 14
Development of a Risk Response Strategy for Highway Construction Projects

Anshul Gupta, T. Shanmuga Priya, R. Prasanna Venkatesan, and Rukku B. Thoppil

Abstract Construction projects on highways contain complications, including many stakeholders, environmental considerations, and technological difficulties. Minimizing adverse effects and guaranteeing project success depends on effective risk management. The current practices in the construction sector and the science cannot frame the affiliation between the project work and its features, the project aspects of uncertainty and complexity of the risk profile, and its requisite response strategy. A questionnaire survey is conducted to assess the risk factors that have a significant influence on highway construction projects in India are analyzed, and these factors are analyzed using the statistical method to develop a risk response strategy based on the Relative Importance Index and to provide suggestive measures that can improve the overall highway construction works. Two case studies are provided to investigate the risk management techniques applied in actual situations. Descriptive statistics is utilized to determine prevalent risks and their potential outcomes, effects, and mitigation techniques. In addition, an Artificial Neural Network (ANN) model is created to evaluate and forecast possible risks throughout the project. The results underscore the need to employ proactive risk identification and mitigation strategies to improve highway construction projects' overall resilience and performance.

Keywords Artificial Neural Network · Highway construction projects · Relative Importance Index · Risk management

---

A. Gupta · T. Shanmuga Priya (✉) · R. Prasanna Venkatesan · R. B. Thoppil
School of Civil engineering, VIT University, Vellore, India
e-mail: shanmugapriya.t@vit.ac.in

V. Baskaran et al. (eds.), *EcoTech Urbanism*, The Urban Book Series,
https://doi.org/10.1007/978-3-031-83833-0_14

14.1 Introduction

India has one of the most extensive road networks in the world. Over 64.5% of goods and 90% of passenger traffic in India are transported through its road network system (Vishwakarma et al., 2016). Studies demonstrated that highway projects have higher dangers compared to other construction works. Highway projects are spread over a more extensive geographic region, facing risks from underground conditions (Zhao et al., 2016). Another major issue with highway construction projects is the need for a documented inventory of significant data for the completed projects. This increases uncertainty for risk events, which may cause more conservative choices for all parties, such as the client, contractors, and employers. Therefore, risk management is vital to avoid critical risk events and take proper mitigation measures to alleviate these risks over the construction period. The construction segment is confronted with unique projects in which complexity and uncertainty are significant. The levels of complexity and uncertainty are firmly linked to the project, and these influence the project's risk profile. These factors threaten the likelihood of success (Dziadosz & Rejment, 2015).

The Guideline for Project Management mentioned that the risk identification process determines what could happen, how it affects the objectives of a project, and the way these things may happen. It is vital to ensure that the broadest array of risks has been identified (Institute and PMBOK, 2013). The project management body of knowledge (PMBOK) defines the standard procedure to identify the risk factors based on an iterative process since new risks might evolve or be known as the project develops throughout. Delays affect each party involved in the highway construction project in different ways. As for the contractor or owner, it affects contrarily but not identical to the consultant, who is considered the minimum party affected (Aziz et al., 2016). Risk management is crucial, especially in technology-based projects. It aims to reduce risk. Much research has been done on risk management standards. The standards influence managing various risks and are generally described in an intangible way as they have to apply to all projects. The risk management process should be unified with project management (Ahmadi et al., 2017). It is not merely a management tool but must be considered part of project management. Risk management must be considered a continuing review process in the Project Life Cycle. The risk management assessment helps identify the weaknesses, strengths, and opportunities to avoid the risks (De Marco & Thaheem, 2014). Identification and assessment of risks are critical factors for any risk management process. Various risk factors are involved from the planning stage till its completion. One of the concepts that has been widely used in the field of risk management is the risk management procedure. The process involves four steps: risk identification, risk assessment, risk mitigation, and monitoring (Gupta & Thakkar, 2018). There can be several methods and procedures that can be used to handle the risks at each stage.

Analysis of risk and its management is an essential part of decision-making practice in any construction industry. Due to more significant capital investment, complexity, and reliance on social, economic, and political challenges, India's highway projects are more likely to have more substantial uncertainties and risks than further construction projects. The construction industry, its clients, and its contractors are widely linked with higher risks due to the nature of micro, meso, and environments specific to construction, affecting each participant. However, the construction industry has an abysmal reputation for coping with these risks, as many projects fail to meet the cost targets and deadlines. Effective management and analysis of construction risks remain an immense challenge to construction industry experts (Rezakhani, 2012). Factors such as the importance of risks, management tools, existing status, and barriers to risk management tools must be considered. How people in the industry perceive each risk is also essential for proper and effective risk management. It is often clear that the risk assessment and its strategy are the most settled aspects linked to well-made systems and risk management processes (Cheraghi et al., 2017). This research focuses on identifying risk associated with development of transportation road infrastructure. This study classifies risks in to the following major classes: social risks, technical documentation risks, construction risks, traffic risks, time risks, design risks, market risks, management risks, legal risks, financial risks, and safety and environmental risks to develop a risk response strategy for the whole project life cycle to enhance overall life quality. Also, providing sustainable solutions and guidelines to mitigate risks evolved at any phase of construction.

Various methods, such as RII and ANN, are incorporated for risk evaluation and analysis. Project management endeavors to enhance the management processes by inspecting project structure, external environment, processes, organizational environment, and procedures. The process increments the existing business practices, provides case studies, and brings out risk mitigation strategies to evacuate the risk events. The risk analysis determines the influence of various risk factors on the construction project. These risk events and factors form a cumulative impact on one or more project characteristics. It is often easy to mitigate the risk factors if they are bunched in sets and dealt with at a higher level than centring on any specific risk event; in such a case, the project work is expected to be micro-managed (Yu et al., 2015).

Artificial Neural Networks (ANNs) are like artificial intelligence systems with wide risk management applications. The neural network system may identify, evaluate, and assess risks. ANN is a data processing tool that fakes the human nervous system (Waziri et al., 2017). It can study from the experiences, generalize from previous data to new data, and abstract essential data from the input. ANN has many advantages over other conventional systems. At the same time, most statistical methods are parametric in nature, requiring more statistical backgrounds; ANN is a nonparametric model with backpropagation. This algorithm is broadly used for forecasting problems and solving classification.

Even though the backpropagation convergence is slower in its process, it is more guaranteed. The techniques mainly focus on pattern recognition and completion. The study proposes a framework that could assess project risks using Artificial Neural Networking by incorporating the RII values as the input data. The framework has three phases: risk management, ANN training, and framework. Various risk factors linked with highway construction works were identified and analyzed to determine the severity of each risk factor. The main objectives of the study are to identify the risk factors that have a significant influence on highway construction projects of Indian industry, to analyze and rank these factors associated using the statistical method to develop a risk response strategy, and to provide suggestive measures that can improve the overall highway construction works. The research aimed first to identify various risk factors and risk management tools practiced for managing risks in highway construction projects in India. Secondly, the questionnaire survey will be evaluated, and the risks associated with highway construction projects will be ranked. Third, the efficiency of the risk management tool chosen to manage this risk must be analyzed. Lastly, strategies should be developed, and mitigation measures should be suggested to reduce the risk involved in the highway project. The study recommends creating a risk response strategy based on the factors associated with highway construction projects. The study focuses on risk management and its approach to be adopted in the different phases of India's highway construction projects. The factors that can influence the performance of risk management are identified.

14.2 Research Process

In this study, developing a risk response strategy for highway construction projects to establish specific facts and reach a new conclusion. The collection and application of an appropriate methodology is vital to attaining this systematic study. It provides a systematic path with a preliminary set of objectives and conclusions. It incorporates several steps, such as the collection and analysis of data within it, as mentioned in Fig. 14.1.

Initiating the identification of the project, a thorough literature survey is conducted. Studies have shown that most highway construction projects involve some significant risk factors. The factors linked with these risks for highway projects were identified and studied. The data for the risk factors were identified and obtained through two sections. A literature study was conducted first for risk identification and then through expert interaction. These risks were sorted into different groups based on the field of each risk factor. A questionnaire survey was then created using Google Forms and circulated among various agencies, such as contractors, clients, supervisors, and engineers in this field. The tools that are to be used for the risks were also identified. The data obtained from the questionnaire survey was then considered for further risk analysis. The risk value of each factor was first obtained using the Relative Importance Index method. With the data obtained from RII, the study was done using Artificial Neural Networking, and each factor's severity was identified. Further, mitigation measures have been suggested for the most severe risks.

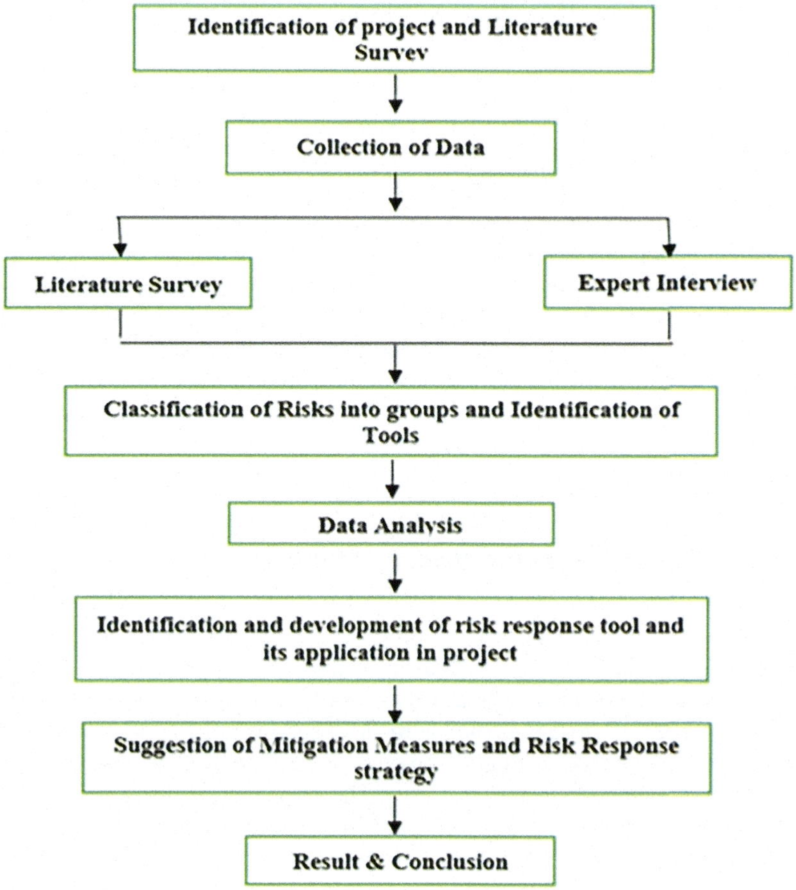

Fig. 14.1 Research process

14.2.1 Case Study

Case studies were done on different ongoing projects to identify the severity of risk factors and their influence on the project. Interviews were conducted with project coordinators for both case studies. In addition to the risk manager, the project manager and project director were also interviewed. Based on the project documentation and review, the interviews were directed to analyze and improve the preliminary case findings. The thesis only comprises the summary and conclusions of case studies. The projects are studied individually and contain a case study with the general project characteristics, information, financial aspects, risks, and further findings. A case study of economic and time risk has been conducted here.

14.2.2 Case I: Kollam Bypass Project

A study was done on the "Kollam Bypass Project," which was the construction of two lanes with paved shoulders (486.500 km to 499.500 km) on NH-47 in the state of Kerala. The work was in the engineering, procurement, and construction mode of agreement. The overall length of the project was 13 km, which included three major bridges, one vehicular underpass, one toll plaza, and 19 culverts in total. The overall completion period of the project was 30 months, with a defects liability period of 48 months. The overall contract price was ₹2,772,435,655. With the allowable cost at the completion period being ₹3,473,784,900, there will be a total variation of ₹100,604,878 at the time of project completion. The variation in the price may be due to various risk factors incorporated within the project. The case study given here is a pure example of financial risks linked with construction projects and, hence, need to be eradicated at the initial stage of the project itself.

14.2.3 Case II: Proposed Bridge at Kollam

A case study was done on the project "Proposed Bridge at Kollam for the Extension of Asramam Link Road to Thoppilkadavu," which was the proposal of a bridge for the extended road at Kollam. The mode of implementation used here was the engineering, procurement, and construction mode of agreement. The study identifies the significance of time risks within the projects. For the analysis, c, instruction computer software, namely "Candy," was used. The graph delivers the actual plot of the work done and its estimated values as per the date and schedule. An amount of work worth ₹54,68,86,933 was to be done in actuality as per the initial schedule, but the actual work was done for ₹25,76,34,674 only. Hence, due to the risk factors associated, there was a time delay. There has been an overall delay in the project work worth ₹289,252,259 due to the associated risk factors. The study signifies the importance of time risk and its significance on the project. The case study showed differences in the overall project due to the impact of various risks. The first study displays the financial figures for the initial budget, working budget, actual costs, and the result of this variation on the project.

14.3 Technical Specification

The research takes a historical look and examines how these factors are identified. The research question of the study mainly concentrates on inspecting which factors influence the risks linked with the construction of highway projects and, after that, measuring the impact of these factors on the project's success. The data collection assignment is initialized after the research methodology has been identified.

14.3.1 Data Collection

The data collected from all the papers identified in the literature review are used to formulate the questionnaire survey. The factors shortlisted for the questionnaire are based on the repetition of the factors in various departments causing risks in highway construction works and through discussion with experts.

14.3.2 Identification of Risk Factors

Two criteria were chosen to identify risk factors: literature surveys and expert interviewing. The most predominant risk factors are sorted out from these risks based on their severity. The severity of these risks is identified through the questionnaire survey. The risk factors identified are categorized under different subheads. This risk identification phase casts an extensive net to collect all the sound risks that can be identified. Based on the degree of analysis, these risks could be related to the programs, projects, activities, or enterprise (Table 14.1).

A total of 37 risks are identified under 11 sections. Right of way, approval and clearances, construction risks, traffic risks, time risks, design risks, market risks, management risks, legal risks, financial risks, and safety and environmental risks are the different subheads identified. Following are the crucial risk factors that lead to the overall time, cost, and other delays of highway construction projects in India:

14.3.3 Questionnaire Survey

The respondents were offered 37 risk factors gathered through the literature in the prepared questionnaire. The questionnaire comprises questions based on these factors and is formed on 5-point Likert scale. Frequency is the sum of the times a score or rating is allotted for each factor.

$$1 \text{—Very low}; 2 \text{—Low}; 3 \text{—Medium}; 4 \text{—High}; 5 \text{—Very high}$$

The mean target respondents for the survey are professionals in highway construction. The respondents are chosen based on their qualifications and field experiences in highway projects. It includes contractors, clients, project managers, engineers, supervisors, design managers, bid managers, and graduate engineers working in different companies with different experience levels.

Table 14.1 List of risk factors

I	Right of way
	Lack of availability of continuous ROW/desired width
	Property demolition/land acquisition, including private, railway, and defence property
	Stoppage of work/rework due to lack of clarity on alignment
	Rock and soil drill ability/suitability
	Improper soil nailing/filling and retaining structures
II	Approval and clearances
	Risks associated with sharp curves/gradient provision
	Underground utilities
	Proactive governmental approach
III	Construction risks
	Rework due to poor quality of materials
	Accidents during construction
	Inadequate data collection/poor definition of project scope
	Labor disputes/availability/shortage
	Lack of efficiency/shortage of equipment
	Less productivity of workers
IV	Traffic risks
	Traffic intensity/network optimization
	Permissions including approvals and local traffic police
	Inadequate underpasses, service roads, and at-grade junctions
V	Time risks
	Improper time allocation—Inability to execute the work within the specified period
	Delay in decision-making and mobilization
VI	Design risks
	Lack of proper site investigation
	Risks associated with cross-drainage works
	Incomplete/lack of coordination in design
	Alteration in design
VII	Market risks
	Variations in material types, prices, and specifications during construction
VIII	Management risks
	Change of ruling parties/top management
	Poor site management/inefficiency of subcontractor
	The poor leadership quality of the associated project manager
IX	Legal risks
	Unfairness/inexperience in tendering
	Lack of consistency between drawings, bills of quantities, and specifications
	Improper verification of contract documents/undocumented change orders
	Approval procedures linked with administrative government departments

(continued)

Table 14.1 (continued)

X	Financial risks
	Payment-related problems from the contractor/owner side
	Change in bank formalities and lenders
	Fluctuation of material price/interest rate
XI	Safety and environmental risks
	Theft of site materials
	Adverse weather conditions/terrain conditions of the site
	Problems with local people
	Natural disaster (flood, earthquake, etc.)

14.3.3.1 Risk Analysis

Data collected through the questionnaires are analyzed by evaluating primary and secondary data for attaining the desired objective. Weights are given for each risk factor using RII analysis, which then analyzes these based on their importance. The study is done for the data collected and the respondents. Based on this, the respondents and data were analyzed.

A total of 161 responses were collected through the survey, and the data was to be analyzed. Almost 6.20% of the respondents were under government departments, and the other 5.6% engaged with other kinds, such as semi-government projects. About 87.6% of the respondents are professionals with an experience of 0–10 years. About 9.3% of the respondents have experience of 10–20 years, and the rest have a higher level of experience of more than 20 years.

14.3.4 RII Method of Analysis

The questionnaire comprises 37 risk-attributing factors under 11 sections. The influence of each factor on the overall project is examined in terms of its criticality using RII, which is calculated using the equation:

$$\mathrm{RII} = \Sigma W / \left(A * N\right) \tag{14.1}$$

where,

W—Weightage for each factor given by the respondent (here from 1–5).
N—Overall figure of respondent.
A—Highest weightage is given (here 5).

After calculating RII, the data is now fed as the input for further analysis; the severity of the factor can be identified by the value given. Based on the RII analysis with a value of 0.641, improper time allocation—inability to execute the work within a specified period is the highest and most critical risk factor, while with a value of 0.501, accidents during construction are regarded as the least critical risk factor, with the most negligible value for RII.

14.3.5 *Analysis Using Artificial Neural Network*

14.3.5.1 Architecture of ANN

The input layer takes in all the variables given to it. Here, the 37 risk factors are fed as the input variables for the analysis. The significant step involved in the analysis is importing the required data. The data obtained from the RII analysis is fed as input for this ANN analysis. The output obtained provides the severity of each risk factor obtained, which is the input layer. Collected data are split into three subsets: training, testing, and validating.

- Training set: For the 37 risks identified, a set of data is used for learning to fit the parameters of the network.
- Validation set: A set of examples used to tune the network's parameters (i.e., architecture). For example, it can choose the number of hidden units in a neural network.
- Test set: A set of examples used only to assess a fully specified network's performance (generalization) or to apply successfully in predicting output whose input is known.

As per the research process, the data obtained from the RII analysis was used as input. The inputs were randomly provided with weights, and they were trained accordingly. The neural network has been trained up to its fifth epoch. The best performance is obtained at the third epoch with a validation performance 5.6624e-06. The network has been trained until when the MSE becomes negligible. The value of Mu obtained at the fifth epoch is 1e-08. The model has an MSE of 0.0003. The R-value obtained for the model is 0.99509.

Figures 14.2, 14.3 and 14.4 depicting the validation performance up to the fifth epoch is shown below:

The factor that has the highest value gets the highest rank, which signifies that risk is the maximum influencing factor, whereas the factor which has the lowest rank (the one with the lowest ANN value) signifies that it has the most negligible impact on the project. Based on the training, the ANN value for the 37 factors obtained and its ranking are tabulated below (Table 14.2).

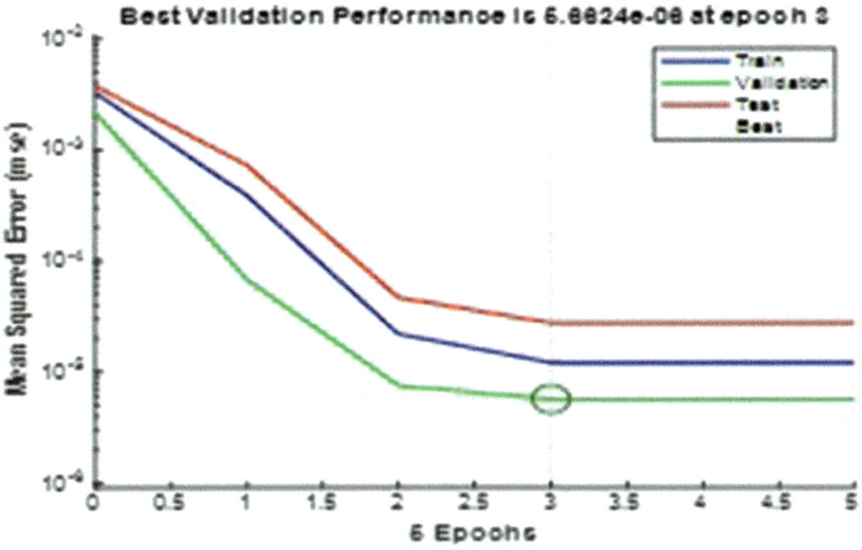

Fig. 14.2 Plot for best validation performance

Fig. 14.3 Regression plot for the test set

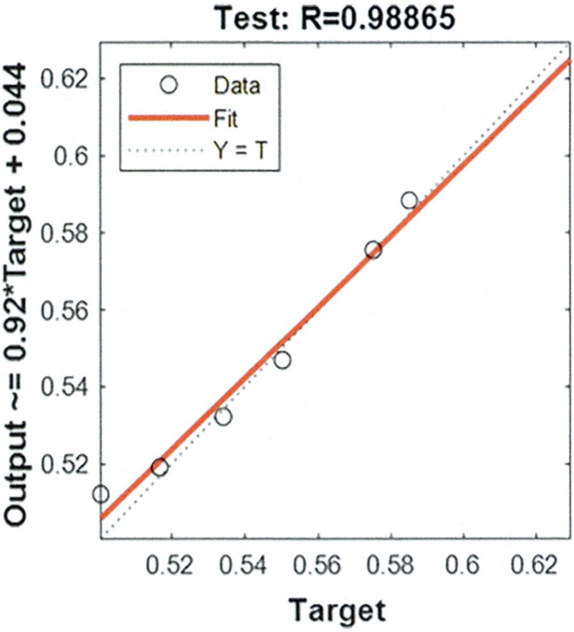

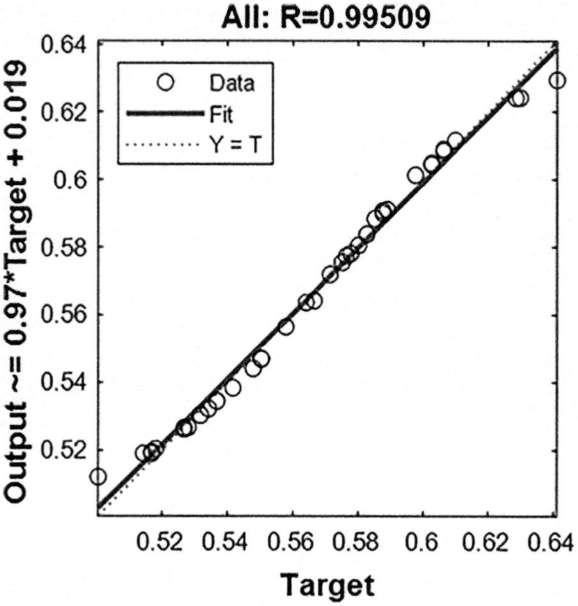

Fig. 14.4 Regression plot for the data set

14.4 Risk Response Strategy

Risk response is one deliberate option meant to reduce the project risk and enhance project profitability. Researchers have defined it in diverse ways, with the vital objective of ensuring the project's profitability. Risk response is identifying, selecting, evaluating, and implementing effective measures to diminish the probability of incidence of risks and minimize the adverse impact of these risks. The strategies play a dynamic part in mitigating risk on the project. The strategies based on risk response include four kinds of actions, namely risk avoidance, reduction, retention, and transfer, that decrease the chances of risk occurrence and its negative impact. The study looks at the management actions that should be taken through risk response strategies. Detailed strategies are proposed to alleviate the project risks. Table 14.3 shows the selection, controllability, and type of risk for risk response.

The following risks and their response strategies will assist in developing policies and guidelines to minimize the risk impacts. As it delves into construction, traffic, time, design, market, management, legal, financial, and safety and environmental risks, it provides an insight to the issues faced in developing cities. With further adaption with digital techniques data can be collected using sensors and IoT of the existing structures to identify the current issues that can be diagnosed by data mining and solutions to mitigate them can be worked on. This will benefit in mitigated issues on existing road infrastructures and provide insight to develop sustainable infrastructure as a planned city should be developed.

Table 14.2 ANN predicted output and rank

Risk ID	Risk factor	ANN value	Rank
R1	Lack of availability of continuous ROW/desired width	0.603	6
R2	Property demolition/land acquisition, including private, railway, and defence property	0.604	4
R3	Stoppage of work/rework due to lack of clarity on alignment	0.542	25
R4	Rock and soil drill ability/suitability	0.595	9
R5	Improper soil nailing/filling retaining structures	0.574	17
R6	Risks associated with sharp curves/gradient provision	0.635	2
R7	Unforeseen underground utilities	0.580	14
R8	Proactive governmental approach	0.599	8
R9	Rework due to poor quality of materials	0.554	22
R10	Accidents during construction	0.510	37
R11	Inadequate data collection/poor definition of project scope	0.525	30
R12	Labour disputes/availability/shortage	0.519	33
R13	Lack of efficiency/shortage of equipment	0.521	32
R14	Less productivity of workers	0.543	24
R15	Traffic intensity/network optimization	0.601	7
R16	Permissions including approvals and local traffic police	0.583	13
R17	Inadequate underpasses, service roads, and at-grade junctions	0.604	5
R18	Improper time allocation—Inability to execute the work within the specified period	0.638	1
R19	Delay in decision-making and mobilization	0.579	15
R20	Lack of proper site investigation	0.559	21
R21	Risks associated with cross-drainage works	0.548	23
R22	Incomplete/lack of coordination in design (structural, mechanical, electrical, etc.)	0.522	31
R23	Alteration in design	0.539	26
R24	Variations in material types, prices, and specifications during construction	0.563	20
R25	Change of ruling parties/top management	0.572	18
R26	Poor site management/inefficiency of subcontractor	0.585	11
R27	Unfairness/inexperience in tendering	0.525	29
R28	Lack of consistency between drawings, bills of quantities, and specifications	0.532	27
R29	Improper verification of contract documents/undocumented change orders	0.517	34
R30	Approval procedures linked with administrative government departments	0.585	12
R31	Payment-related problem from the contractor/owner side	0.634	3
R32	Change in bank formalities and lenders	0.531	28
R33	Fluctuation of material price/tax rate	0.578	16
R34	Theft of site materials (impact)	0.517	35

(continued)

Table 14.2 (continued)

Risk ID	Risk factor	ANN value	Rank
R35	Adverse weather conditions/terrain conditions of the site	0.586	10
R36	Problems with local people	0.569	19
R37	Natural disaster (flood, earthquake, etc.)	0.515	36

Table 14.3 Risk response strategy for the top risks

Type of risk	Controllability	Specific risk-handling actions	Type of risk-handling strategy
Improper time allocation—Inability to execute the work within the specified period	Medium	Select a strict subcontractor Increase productivity	Risk prevention Risk adaption
Risks associated with sharp curves/ gradient provision	High	Proper supervision Site investigation and field survey	Risk prevention
Payment-related problem from the contractor/owner side	Low	Lump-sum contracts Timely completion of bill	Risk prevention
Property demolition/land acquisition, including private, railway, and defence property	High	Site investigation of the field commence work after legal proceedings	Risk prevention
Inadequate underpasses, service roads, and at-grade junctions	Medium	Study ongoing and previous similar projects for more accuracy	Risk adaption Risk prevention
Lack of availability of continuous ROW/desired width	High	Proper cutting Land acquisition	Risk adaption
Traffic intensity/network optimization	Medium	Provide special provision Improvise intersections pathways	Risk prevention
Proactive governmental approach	Low	Clarify all ambiguity and approvals	Risk adaption
Rock and soil drill ability/ suitability	High	Increase workforce and equipment Increase working hours	Risk adaption
Adverse weather conditions/ terrain conditions of the site	Low	Purchase insurance coverage Secure machine materials	Risk adaption

14.5 Conclusion

Management of risks in highway projects is often recognized as an essential procedure to achieve the purposes of the project in terms of quality, safety, cost, etc. The risk events impact the choice of the risk response strategy. Risk response strategies and their mitigation measures are vital in mitigating the negative influence of risk over the project objectives. The manager's attitude towards risk, level of controllability, and project characteristics of risk can influence and vary the choice of the response strategy. In this research, 37 risk factors have been identified through literature surveys and expert interviews throughout the project. RII values obtained for each risk factor are fed as the input for ANN analysis. Several scenarios are trained and tested to figure out the best ANN model, producing a result with a mean-squared error of 0.003. Regression analysis is done on these factors, and the regression R-value obtained is 0.99509. The analysis identified the severity of various risk factors through the RII and ANN methods. Out of 37 risks incorporated into the questionnaire, the top 10 risks have been highlighted. Improper time allocation—inability to execute the work within the specified period, risks associated with sharp curves/gradient provision, problems from contractor/owner side, property demolition/land acquisition including private, railway, and defence property, inadequate underpasses, service roads, and at-grade junctions, lack of availability of continuous ROW/desired width, traffic intensity/network optimization, proactive governmental approach, rock and soil drill ability/suitability and adverse weather conditions/terrain condition of site are the top significant risk factors which effects highway constructions. The risk response strategy, along with the mitigation measures that are to be executed, have been provided. Proper initiation of the strategy and measures at every stage of the project could lead to the evacuation of various risks associated with the project. These could be used as a future reference for upcoming highway constructions, and appropriate measures must be taken to mitigate these risks.

References

Ahmadi, M., Behzadian, K., Ardeshir, A., & Kapelan, Z. (2017). Comprehensive risk management using fuzzy FMEA and MCDA techniques in highway construction projects. *Journal of Civil Engineering and Management, 23*(2), 300–310.

Aziz, R. F., & Abdel-Hakam, A. A. (2016). Exploring delay causes of road construction projects in Egypt. *Alexandria Engineering Journal, 55*(2), 1515–1539.

Cheraghi, E., Khalilzadeh, M., Shojaei, S., & Zohrehvandi, S. (2017). A mathematical model to select the risk response strategies of the construction projects: A case study of Saba tower. *Procedia Computer Science, 121*, 609–616.

De Marco, A., & Thaheem, J. (2014). Risk analysis in construction projects: A practical selection methodology. *American Journal of Applied Sciences, 11*, 74.

Dziadosz, A., & Rejment, M. (2015). Risk analysis in construction project-chosen methods. *Procedia Engineering, 122*, 258–265.

Gupta, V. K., & Thakkar, J. J. (2018). A quantitative risk assessment methodology for construction project. *Sādhanā, 43*(7), 116.

Institue, P.M. and PMBOK, A. (2013). *Guide to the project management body of knowledge* (5th ed.). Project Management Institute.

Rezakhani, P. (2012). Classifying key risk factors in construction projects. *Buletinul Institutului Politehnic din lasi Sectia Constructii, Arhitectura, 58*(2), 27.

Vishwakarma, A., Thakur, A., Singh, S., & Salunkhe, A. (2016). Risk assessment in the construction of highway projects. *International Journal of Engineering Research & Technology, 5*(2), 637–640.

Waziri, B. S., Bala, K., & Bustani, S. A. (2017). Artificial neural networks in construction engineering and management. *International Journal of Architecture, Engineering and Construction, 6*(1), 50–60.

Yu, J. H., Jeon, M. Y., & Kim, T. W. (2015). Fuzzy-based composite indicator development methodology for evaluating overall project performance. *Journal of Civil Engineering and Management, 21*(3), 343–355.

Zhao, J., Fu, X., & Zhang, Y. (2016). Research on risk assessment and safety management of highway maintenance projects. *Procedia Engineering, 137*, 434–441.

Chapter 15
Eco-Friendly Transportation Solutions for Urban Areas

M. Punitham and S. K. Udhaya Jothi

Abstract The chapter examines the need for environmentally friendly transportation options that limit negative effects on the environment, emphasizing developments in public transportation, electric vehicles (EVs), alternative fuels, and creative mobility ideas. It offers a thorough examination of the benefits and drawbacks of environmentally friendly transportation options, emphasizing the productivity increases of well-thought-out public transportation networks and the revolutionary advancements made possible by electric cars that run on rechargeable batteries. But there are obstacles in the way of sustainable transportation, such the need for behavioral changes, budgetary constraints, and limited infrastructure. The chapter breaks down these difficulties, illuminating the nuances that require consideration from stakeholders, politicians, and scholars alike. Building and sustaining broad public transit networks, EV charging infrastructure, and alternative fuel production facilities require deliberate investment and cooperation, which makes infrastructure development a crucial cornerstone. Cost is a significant obstacle, especially as the initial expenses of alternative fuel and electric cars are higher than those of their conventional equivalents. The chapter promotes financial gap-bridging government incentives and research funding to make sustainable transportation solutions available to a wider range of people. Advances in technology, from battery science to self-driving car capabilities, are what propel the development of environmentally friendly transportation. The chapter emphasizes the necessity of ongoing research and development projects to improve these technologies' cost, range, and efficiency. Changes in behavior provide a special issue that calls for successful public awareness campaigns and governmental initiatives. The chapter looks at the ways that activism, education, and innovative legislation might help to bring about this cultural change.

M. Punitham
Department of Commerce (A&F), PSG College of Arts & Science, Coimbatore, India

S. K. Udhaya Jothi (✉)
Full Time Research Scholar, Department of Commerce, PSG College of Arts & Science, Coimbatore, India

V. Baskaran et al. (eds.), *EcoTech Urbanism*, The Urban Book Series,
https://doi.org/10.1007/978-3-031-83833-0_15

Keywords Sustainable transportation · Public transit · Electric vehicles · Alternative fuels · Shared mobility · Smart cities · Green infrastructure · Policy interventions · Social equity

15.1 Introduction

The global transportation industry is at a turning point that calls for quick response and revolutionary change (Araújo et al., 2017; Chakwizira et al., 2014). It is impossible to overestimate the influence of conventional, fossil fuel-powered transportation on climate change and air quality, as it accounts for a startling 29% of global greenhouse gas emissions (Olivier et al., 2017; Kennedy et al., 2009). The need for a paradigm change toward sustainable transportation solutions is more important than ever as the globe struggles with the effects of human activity on the environment (Petrov et al., 2023; Bisht et al., 2020). Long the mainstay of international transportation networks, fossil fuel-powered cars are now recognized as one of the main causes of climate change (Romero-Ocaño et al., 2022; Arboleda et al., 2022). The objective of this chapter is to present a thorough analysis of sustainable transportation options by examining several aspects of innovation, difficulties, and potential paths forward. The trip starts with a review of the transportation industry's current situation, an analysis of its effects on the environment, and a determination of the urgent need for reform. We will analyze the obstacles to widespread implementation of these solutions, including infrastructural constraints, financial issues, and changes in behavior, in order to identify the underlying complications. There are chances to advance beyond the obstacles. With the goal of laying the groundwork for a more sustainable transportation future, this paper will highlight important avenues for future study and development. Through tackling the various facets of this transformation, ranging from legislative changes and technology breakthroughs to cultural transformations, we want to support the joint efforts in creating a society that is greener, cleaner, and more just.

15.1.1 Environmentally Sustainable Transport (EST) Forum

High-level governmental decision-makers from 22 Asian nations have gathered annually at the Regional Environmentally Sustainable Transport (EST) Forums in Asia since 2005 to explore ways to handle transportation-related issues in an environmentally sustainable manner. Representatives from the United Nations and other international organizations, NGOs, scientific and research institutions, the commercial sector, and national and municipal governments attend the EST Forum. It convenes line ministries and agencies from many sectors, including energy, urban development, health, environment, and transportation, to talk about multisectoral policy concerns pertaining to the transportation industry. The Bangkok 2020

Declaration 348 (2010–2020) was approved by the Forum in 2010 and includes 20 goals as well as a set of benchmarks that Asian nations should use when making general choices on transportation planning, policy, and development. The countries concerned have demonstrated a level of voluntary engagement never seen before. In addition to tackling unsustainable transportation habits, avoid, shift, and improve will help the region's cities and communities become safer, more livable, resilient, and sustainable. In order to guide the development of the transport sector in Asia and the Pacific, the policy discussions at the Forum are in line with the 2030 Agenda, the Paris Agreement, the New Urban Agenda, and other pertinent international agreements that have established goals and/or targets for enhancing national connectivity, access to rural and urban areas, and economic, social, and environmental sustainability.

15.2 Green Bonds

Global climate funds, such as the Global Environment Facility, the Green Climate Fund, and the Clean Technology Fund, are crucial in promoting climate action, which includes the transition to low- or zero-emission vehicles. Green bonds are a growingly popular choice for institutional investors seeking longer-term sustainable investments. Investing in transportation is one of these bonds' most popular uses. A worldwide green bond fund with an emphasis on emerging economies is called the Real Economic Green Investment Opportunity (REGIO) fund. 357 Bond interest rates in certain markets are determined by the environmental, social, and governance (ESG) ratings of the issuer. Governments may also offer tax breaks to promote the issue of green bonds.

15.3 Sustainable Transportation

Transportation that is inexpensive, low- or zero-emission, and energy-efficient is referred to as sustainable transportation (Shah et al., 2021; Pamucar et al., 2021). Reducing adverse environmental effects while facilitating societal mobility demands in a way that preserves such requirements for future generations is the goal of sustainable transportation. Low-emission public transit, local fuels, electric and alternative-fuel cars, and other strategies may all contribute to sustainable mobility. Reducing greenhouse gas emissions, improving air quality, and saving money on gasoline and automobiles are some advantages of sustainable mobility (Patil, 2021; Liu et al., 2019). Around the world, groups and governments are moving to support environmentally friendly transportation. For instance, the 2021 United Nations Sustainable Transport Conference in Beijing, China, sought to address the global climate issue and promote sustainable transport initiatives (Gao & Zhu, 2022). In order to protect the environment and secure the welfare of future generations, it is critical to implement sustainable transportation methods. The dynamic road toward sustainable transportation is characterized by technologies that transform the way we move and navigate our environment. This section explores a wide range of environmentally friendly transportation options, each offering a special strategy to solve the problems with conventional means of transportation. The aspects and categories of sustainable transport are displayed in Fig. 15.1.

15.4 Public Transit

As a strong substitute for private automobiles, public transportation is an essential part of sustainable urban development (Cong et al., 2022). Public transportation, including buses, trains, and other vehicles, is essential for increasing productivity, easing traffic, and lowering pollution (Yannis & Chaziris, 2022). Globally, effective public transportation systems are now essential to sustainable urban growth (Liu et al., 2023). A key component of the effort to develop sustainable urban settings is public transportation, which tackles the wider social and environmental issues related to the use of private vehicles. Reduction of traffic, environmental effect, and efficiency are some of the main benefits of strong public transportation systems (Magalhães & Santos, 2022).

A great example of the transformational potential of a well-thought-out public transportation system is Singapore's well-functioning bus system. Strategic route design that takes into account the demands of various groups is the first step toward Singapore's public transport success (Tedjopurnomo et al., 2022). The TransMilenio fast transit system in Colombia, South America, is a prime example of how creative public transportation solutions may be used to meet the problems caused by urbanization (Casa Nova et al., 2023; Cabrera-Moya & Prieto-Rodríguez, 2022). According to Diaz et al. (2023), the BRT system offers a quick and dependable substitute for private automobiles, therefore bringing about a paradigm change in public transportation. The system offers travelers a smooth transit experience by utilizing a wide network, dedicated bus lanes, and effective boarding procedures. Innovation and the use of cutting-edge technologies are key components of public transportation's future (Ji et al., 2022). Public transportation operations may be further optimized by projects like the creation of electric buses, intelligent transportation systems, and real-time data analytics. Particularly, electric buses help cut emissions and support the overall objective of switching to greener energy sources (Rodrigues & Seixas, 2022). Public-private partnerships are another example of collaboration since it may bring in more resources and knowledge to improve public transportation services and infrastructure. Inclusionary public transportation system design promotes fair access to mobility choices and helps create a more socially and economically viable metropolitan environment.

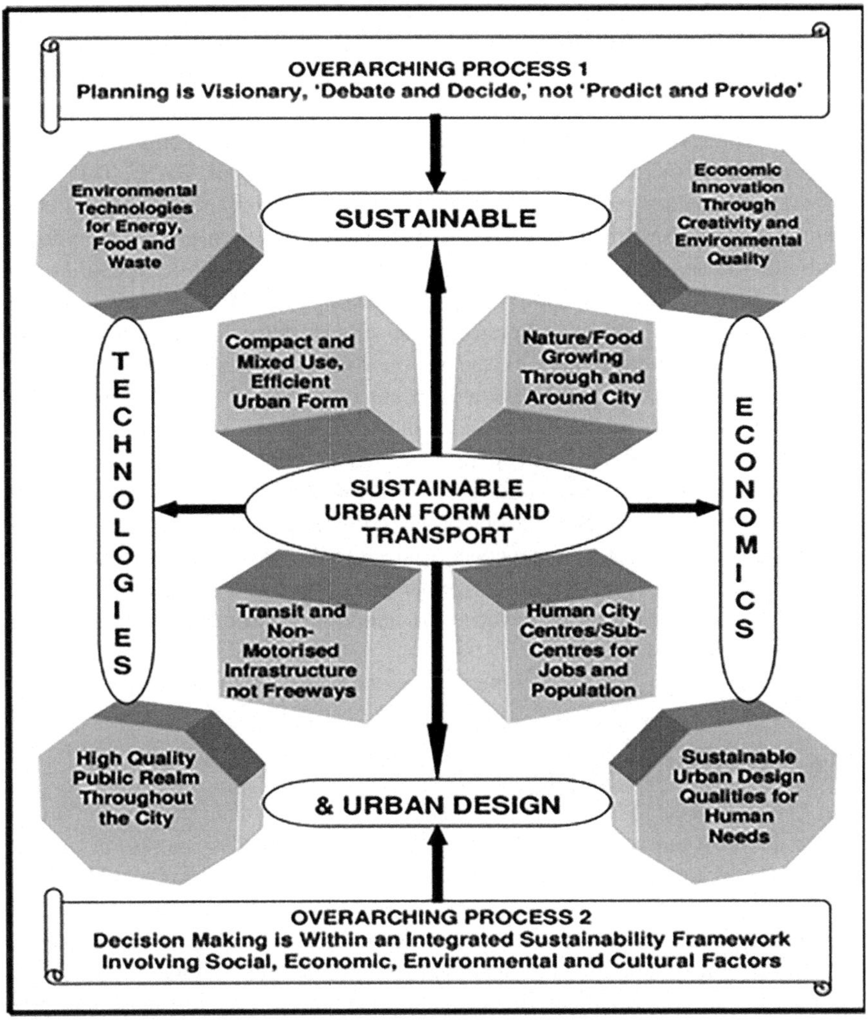

Fig. 15.1 Schematic of sustainable transport dimensions and categories

15.5 Electric Vehicles

The global adoption of electric vehicles (EVs) is being propelled by developments in infrastructure and battery technology, which are completely changing the transportation scene. As governments, corporations, and consumers become more aware of the environmental advantages of electric mobility, cities around the world are seeing a movement toward fleets that are predominately electric vehicles (EVs) (Roberts, 2022). This change represents a paradigm shift toward a more robust and sustainable transportation system rather than merely a technology fad (Hou et al.,

2023). With its successful transformation into the "Electric Vehicle Capital" because too smart incentives, extensive infrastructure development, and progressive legislative support, Oslo, Norway, is a leader in the global shift toward electric mobility (Mega, 2022). Using electric boats driven by renewable energy sources, Amsterdam, the Netherlands, offers a cutting-edge method of environmentally friendly transportation on its famous canals (Minak, 2023; Chidolue & Iqbal, 2023). This action encourages a cleaner and healthier urban environment by lowering water pollution, a major problem for urban waterways. Amsterdam's adoption of electric mobility highlights the adaptability of electric mobility solutions, which let localities customize their sustainable transportation plans outside of the city limits. It emphasizes how crucial it is to go beyond conventional transportation options and embrace technology that supports more expansive environmental goals. Cities are able to customize their sustainable transportation plans because to the adaptability of electric mobility options, such as buses and boats. Campaigns for education and public awareness are essential to the effective adoption of electric cars. Oslo's success may have been aided by clear explanations of the advantages of EVs, prompt resolution of issues, and encouragement of a favorable attitude toward electric vehicles. The global character of environmental concerns necessitates international collaboration. Sharing best practices and working together on research and development projects may help cities and countries advance the adoption of sustainable transportation options on a larger scale.

15.6 Alternative Fuels

One of the most important ways to diversify the sustainable transportation environment is to investigate cleaner fuel sources (Khan et al., 2023; Ukoba et al., 2019). Biodiesel, hydrogen fuel cells, and synthetic fuels are examples of alternative fuels that provide technical innovation, environmental advantages, and energy source diversity (Halder et al., 2023; Uddin et al., 2022). One state that has led the way in the construction of a hydrogen highway is California, which has established a vast network of hydrogen filling stations to service fuel cell vehicles. This program delivers major infrastructural investment while addressing the range restrictions commonly associated with battery electric cars. Without needing a whole redesign of the transportation system, biodiesel offers a greener option that integrates in smoothly with the current diesel infrastructure. Germany's biodiesel programs are noteworthy for their ability to seamlessly integrate biodiesel into the current infrastructure, rendering it a feasible and easily available substitute for a diverse array of cars. The link between the three main sustainability challenges in transportation—social, economic, and environmental—is depicted in Fig. 15.2. The success Germany has had in promoting biodiesel shows that alternative fuels can coexist peacefully with conventional modes of transportation without sacrificing effectiveness or convenience. The construction of infrastructure necessitates significant planning and investment, and adopting cars that run on alternative fuels has greater

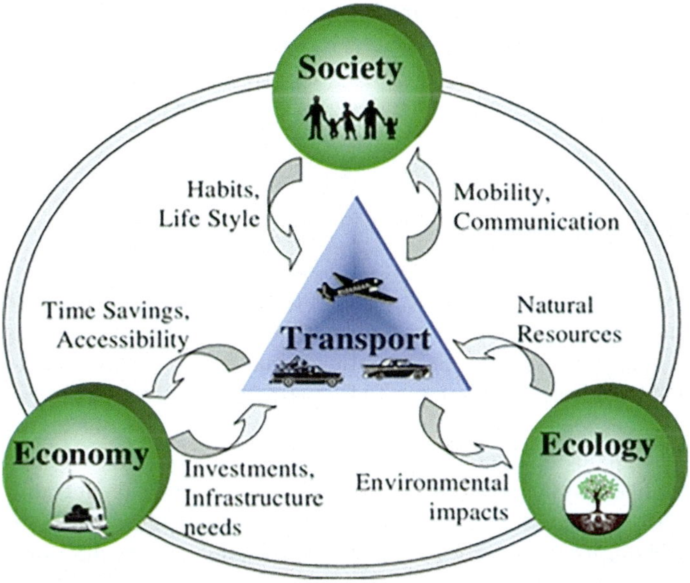

Fig. 15.2 Schematic of relationship of key sustainability issues in transportation

upfront costs. The success of alternative fuels depends on ongoing technological developments, such as enhanced fuel cell technology, more efficient biofuel production methods, and the investigation of novel synthetic fuel possibilities. The secret to a successful adoption of alternative fuels is raising customer acceptance and knowledge.

15.7 Shared Mobility

By maximizing resource usage and lowering the ownership of private vehicles, shared mobility and micromobility are cutting-edge ideas that seek to build more resilient and environmentally friendly urban settings. These solutions help to create more resilient and environmentally friendly urban settings by meeting the rising need for convenient, flexible, and sustainable transportation choices. Micromobility and shared mobility have several benefits, such as better quality of life and sustainability, less environmental effect, increased urban mobility, and efficient use of resources. In order to encourage its citizens to share cars and bikes, Barcelona, Spain, has accepted shared mobility services as essential elements of its urban mobility policy. Eco-friendly transportation alternatives are also important for Berlin, as these micromobility solutions improve air quality and contribute to the city's overall sustainability objectives by producing less emissions than traditional cars (Göddeke et al., 2022; Ikwuagwu et al., 2020). Strong and well-maintained infrastructure is necessary for infrastructure growth, and regulatory frameworks must strike a balance between innovation and public safety. With last-mile connection, shared and micromobility solutions may significantly improve connectivity and become an integrated part of the larger transportation network.

15.8 Green Infrastructure

Building sustainable transportation infrastructure is a big task that needs careful planning and large financial outlays. Governments are powerful change agents who can effect change by strategically investing, something that may be accomplished through public-private partnerships. These partnerships can close the gap between public resources and private sector innovation by accelerating the construction of facilities for producing alternative fuels and EV charging infrastructure. By combining intelligent systems like real-time data analytics, intelligent traffic management, and integrated ticketing systems, smart city technologies provide a chance to maximize the use of currently installed infrastructure. This method simplifies the entire traveling experience and increases the effectiveness of public transit networks. For public transportation networks to be efficient and accessible, strategic investments are essential. Infrastructure investments for EV charging are essential to the general adoption of electric vehicles. Establishing charging stations may be made easier by public-private partnerships, which guarantee that the infrastructure keeps up with the rising demand for electric vehicles. Investing in facilities that produce alternative fuels, such hydrogen or biodiesel, helps to diversify the energy mix used in transportation. To enhance the production and distribution of alternative fuels, governments might sponsor research projects and offer financial incentives. The incorporation of intelligent city technology presents a revolutionary method for infrastructure advancement, augmenting the overall efficacy of transportation networks. Governments and technology companies may work together to put these ideas into practice, resulting in a more flexible and responsive urban transportation system.

15.9 Policy Interventions

The adoption of sustainable transportation is beset by substantial obstacles since alternative fuel and electric vehicle (EV) beginning prices are greater than those of their conventional equivalents (Aijaz & Ahmad, 2022). Through focused incentives and research funding, governments play a crucial role in mitigating these costs (Muzir et al., 2022; Adebukola et al., 2022). Governments may create a more competitive and cost-effective sustainable transportation market by enacting comprehensive policies. Targeted government incentives, such tax breaks, refunds, and subsidies, can increase customers' ability to afford EVs and other alternative fuel cars. Customers are incentivized by these financial gains to select eco-friendly choices, which accelerates the uptake of sustainable mobility. Tax credits are a useful tool for encouraging customers to invest in environmentally and energy-conscious automobiles by lowering their cost and supporting more general environmental and energy policy objectives. Governments have the ability to set aside money for at-the-point refunds or subsidies, which increase the appeal of EVs and alternative fuel cars to people on a budget. Funding for cost-effective research is essential for fostering breakthroughs that lower the production and operating costs of environmentally friendly transportation solutions. Early adoption can be accelerated and the cost-benefit realization sped up with the aid of government support and incentives.

15.10 Social Equity

Ensuring social fairness and accessibility in sustainable transportation is a critical challenge to prevent the escalation of preexisting inequities. In order to do this, it is crucial to develop inclusive policies and include communities, taking precautions to avoid the possibility of sustaining current inequalities. Numerous avenues exist for promoting social justice and accessibility in the context of sustainable mobility (Bocarejo & Urrego, 2022; Brown, 2022). Accessibility for all groups may be given top priority in inclusive policy design, guaranteeing that the advantages of sustainable mobility are shared fairly. Thorough evaluations and focused approaches can assist in identifying marginalized communities and places. They can also address particular needs by modifying public transportation routes, offering financial aid to low-income people, or putting in place shared mobility options that are customized to meet their needs. Including the community is essential to creating transportation policy that put social fairness first. Another crucial component of social justice is accessibility evaluations, which pinpoint obstacles to accessibility and put in place solutions that are tailored to the particular needs of certain demographic groups. In order to comprehend and overcome obstacles to sustainable mobility, tailored solutions and universal design are critical first steps. To prevent making already existing disparities worse, a commitment to fair resource allocation is also required. Targeted investments in underprivileged neighborhoods can take the form of improving the infrastructure for bicyclists and pedestrians, creating new public transportation lines, or establishing accessible and reasonably priced shared mobility services. Fair resource distribution makes ensuring that infrastructure and financial investments take into account the particular difficulties that various communities confront, which helps to create a transportation system that is more socially and inclusive.

15.11 Conclusion

Moving toward more environmentally friendly transportation is a path of profound change that goes beyond simple technical improvements. It necessitates a multipronged strategy that includes regulatory changes, technical advancements, infrastructural development, and behavioral adjustments. The foundation of change is infrastructure development, which includes building and maintaining large networks of public transit, setting up infrastructure for electric vehicle (EV) charging, and setting up facilities for the generation of alternative fuels. But these difficulties also present never-before-seen chances for societal change. Governments may distribute funding for public transportation network expansion and improvement via strategic investment and collaboration, and public-private partnerships can expedite the implementation of electric vehicle charging infrastructure. By incorporating smart city technology, current infrastructure is further optimized to guarantee effectiveness and customer happiness. Real and noticeable obstacles include the cost of alternative fuel and electric vehicle (EV) options. By enacting targeted incentives like tax credits and subsidies, government financing for research may help remove these obstacles and increase the accessibility of EVs and alternative fuel cars to a wider audience. Funding for research may also spur the creation of affordable technologies, which helps to inevitably lower prices as technology advances. The engine of transition is technological improvement; further development in the areas of autonomous vehicle capabilities, alternative fuel generation, and battery technology is critical for the future. These technical developments are mostly the result of partnerships between government agencies, commercial businesses, and academic institutions supporting research and development projects. Researchers and inventors alike are drawn to the frontiers of enhanced battery technology, alternate fuel generation techniques, and the incorporation of autonomous vehicle capabilities into mainstream transportation. Perhaps the most significant change needed for sustainable transportation is to encourage a shift in behavior away from a culture that is centered around cars and toward public transportation, shared mobility, and active travel. Public awareness campaigns emphasizing the advantages of sustainable transportation for the environment and human health become powerful tools, and education, lobbying, and policy innovation play a key role. Our priorities should be on inclusive policies that put accessibility for marginalized people first. Governments, businesses, communities, and individuals must come together with a common goal of creating a future that is greener, cleaner, and more egalitarian. Accepting the complex strategy presented in this investigation paves the way for a day when environmental responsibility and mobility coexist.

References

Adebukola, A. A., Navya, A. N., Jordan, F. J., Jenifer, N. J., & Begley, R. D. (2022). Cyber security as a threat to health care. *Journal of Technology and Systems, 4*(1), 32–64.

Aijaz, I., & Ahmad, A. (2022). Electric vehicles for environmental sustainability. *Smart Technologies for Energy and Environmental Sustainability*, 131–145.

Araújo, K., Mahajan, D., Kerr, R., & da Silva, M. (2017). Global biofuels at the crossroads: An overview of technical, policy, and investment complexities in the sustainability of biofuel development. *Agriculture, 7*(4), 32.

Arboleda, F. J. M., Romero, J. E. P., & Szczęsna, A. (2022). A serious game for the responsible use of fossil fuel powered vehicles. *International Journal of Computer Applications in Technology, 69*(1), 62–72.

Bisht, A., Kamboj, N., Kamboj, V., & Bisht, A. (2020). A review on the role of emerging anthropogenic activities in environmental degradation and emphasis on their mitigation. *Archives of Agriculture and Environmental Science, 5*(3), 419–425.

Bocarejo, J. P., & Urrego, L. F. (2022). The impacts of formalization and integration of public transport in social equity: The case of Bogota. *Research in Transportation Business & Management, 42*, 100560.

Brown, A. (2022). From aspiration to operation: Ensuring equity in transportation. *Transport Reviews, 42*(4), 409–414.

Cabrera-Moya, D. R. R., & Prieto-Rodríguez, G. A. (2022). On the need for structuring of the integrated public transport system-IPTS of Bogotá, Colombia as a viable system. *Revista de Investigaciones Universidad del Quindío, 34*(2), 440–461.

Casa Nova, S. P. D. C., Patino-Jacinto, R. A., & Macias, H. A. (2023). Diversity and equity in accounting education during COVID-19: Insights from Colombia. *Issues in Accounting Education*, 1–17.

Chakwizira, J., Bikam, P., & Adeboyejo, T. A. (2014). Sustainable transport solutions at a crossroads in developing countries: Insights and perspectives. *WIT Transactions on the Built Environment, 138*.

Chidolue, O., & Iqbal, M. T. (2023). Design and performance analysis of an oil pump powered by solar for a remote site in Nigeria. *European Journal of Electrical Engineering and Computer Science, 7*(1), 62–69.

Cong, C., Kwak, Y., & Deal, B. (2022). Incorporating active transportation modes in large scale urban modeling to inform sustainable urban development. *Computers, Environment and Urban Systems, 91*, 101726.

Diaz, A. J. O., Cantillo, V., & Arellana, J. (2023). Understanding how individuals perceive changes in the built environment and the transport system after implementing a BRT system. The case of Barranquilla, Colombia. *Journal of Transport Geography, 110*, 103623.

Gao, Y., & Zhu, J. (2022). Characteristics, impacts and trends of urban transportation. *Encyclopedia, 2*(2), 1168–1182.

Göddeke, D., Krauss, K., & Gnann, T. (2022). What is the role of carsharing toward a more sustainable transport behavior? Analysis of data from 80 major German cities. *International Journal of Sustainable Transportation, 16*(9), 861–873.

Halder, P., Babaie, M., Salek, F., Haque, N., Savage, R., Stevanovic, S., Bodisco, T. A., & Zare, A. (2023). Advancements in hydrogen production, storage, distribution and refuelling for a sustainable transport sector: Hydrogen fuel cell vehicles. *International Journal of Hydrogen Energy*.

Hou, X., Lv, T., Xu, J., Deng, X., Liu, F., & Lam, J. S. L. (2023). Electrification transition and carbon emission reduction of urban passenger transportation systems—A case study of Shenzhen, China. *Sustainable Cities and Society, 93*, 104511.

Ikwuagwu, C. V., Ajahb, S. A., Uchennab, N., Uzomab, N., Anutaa, U. J., Sa, O. C., & Emmanuela, O. (2020). Development of an Arduino-controlled convective heat dryer. In *UNN International Conference: Technological Innovation for Holistic Sustainable Development (TECHISD2020)* (pp. 180–195).

Ji, J., Bie, Y., Zeng, Z., & Wang, L. (2022). Trip energy consumption estimation for electric buses. *Communications in Transportation Research, 2*, 100069.

Kennedy, C., Steinberger, J., Gasson, B., Hansen, Y., Hillman, T., Havranek, M., Pataki, D., Phdungsilp, A., Ramaswami, A., & Mendez, G. V. (2009). Greenhouse gas emissions from global cities.

Khan, M. Z. A., Khan, H. A., Ravi, S. S., Turner, J. W., & Aziz, M. (2023). Potential of clean liquid fuels in decarbonizing transportation–An overlooked net-zero pathway? *Renewable and Sustainable Energy Reviews, 183*, 113483.

Liu, Y. H., Liao, W. Y., Li, L., Huang, Y. T., Xu, W. J., & Zeng, X. L. (2019). Reduction measures for air pollutants and greenhouse gas in the transportation sector: A cost-benefit analysis. *Journal of Cleaner Production, 207*, 1023–1032.

Liu, C., Bardaka, E., & Paschalidis, E. (2023). Sustainable transport choices in public transit access: Travel behavior differences between university students and other young adults. *International Journal of Sustainable Transportation, 17*(6), 679–695.

Magalhães, I., & Santos, E. (2022). Evaluating the potential of mobility plans for achieving sustainable urban development. *Research in Transportation Business & Management, 43*, 100743.

Mega, V. (2022). The energy race to Decarbonisation. In *Human sustainable cities: Towards the SDGs and green, just, smart and inclusive transitions* (pp. 105–141). Springer International Publishing.

Minak, G. (2023). Solar energy-powered boats: State of the art and perspectives. *Journal of Marine Science and Engineering, 11*(8), 1519.

Muzir, N. A. Q., Mojumder, M. R. H., Hasanuzzaman, M., & Selvaraj, J. (2022). Challenges of electric vehicles and their prospects in Malaysia: A comprehensive review. *Sustainability, 14*(14), 8320.

Olivier, J. G., Schure, K. M., & Peters, J. A. H. W. (2017). Trends in global CO2 and total greenhouse gas emissions. *PBL Netherlands Environmental Assessment Agency, 5*, 1–11.

Pamucar, D., Deveci, M., Canıtez, F., Paksoy, T., & Lukovac, V. (2021). A novel methodology for prioritizing zero carbon measures for sustainable transport. *Sustainable Production and Consumption, 27*, 1093–1112.

Patil, P. (2021). Sustainable transportation planning: Strategies for reducing greenhouse gas emissions in urban areas. *Empirical Quests for Management Essences, 1*(1), 116–129.

Petrov, M., Nikolaeva, Z., & Dimitrov, A. (2023). The impact of anthropogenic activity on the global environment. *Science Business Society, 8*(2), 59–64.

Roberts, C. (2022). Easy street for low-carbon mobility? The political economy of mass electric car adoption. In *Electrifying mobility: Realising a sustainable future for the car* (pp. 13–31). Emerald Publishing Limited.

Rodrigues, A. L., & Seixas, S. R. (2022). Battery-electric buses and their implementation barriers: Analysis and prospects for sustainability. *Sustainable Energy Technologies and Assessments, 51*, 101896.

Romero-Ocaño, A. D., Cosío-León, M. A., Valenzuela-Alcaraz, V. M., & Brizuela, C. A. (2022). The impact of gradually replacing fossil fuel-powered vehicles with electric ones: A bi-objective optimisation approach. *Expert Systems with Applications, 194*, 116546.

Shah, K. J., Pan, S. Y., Lee, I., Kim, H., You, Z., Zheng, J. M., & Chiang, P. C. (2021). Green transportation for sustainability: Review of current barriers, strategies, and innovative technologies. *Journal of Cleaner Production, 326*, 129392.

Tedjopurnomo, D., Bao, Z., Choudhury, F., Luo, H., & Qin, A. K. (2022, June). Equitable public bus network optimization for social good: A case study of Singapore. In *Proceedings of the 2022 ACM Conference on Fairness, Accountability, and Transparency* (pp. 278–288).

Uddin, S. U., Chidolue, O., Azeez, A., & Iqbal, T. (2022, June). Design and analysis of a solar powered water filtration system for a Community in Black Tickle-Domino. In *2022 IEEE International IOT, Electronics and Mechatronics Conference (IEMTRONICS)* (pp. 1–6). IEEE.

Ukoba, K., Fadare, O., & Jen, T. C. (2019, December). Powering Africa using an off-grid, stand-alone, solar photovoltaic model. In *Journal of Physics: Conference Series* (Vol. 1378, No. 2, p. 022031). IOP Publishing.

Yannis, G., & Chaziris, A. (2022). Transport system and infrastructure. *Transportation Research Procedia, 60*, 6–11.

Chapter 16
Building Sustainable Cities: Leveraging IoT and Analytics for Urban Automation and Management

R. Parkavi, R M. Soundar Arunachalam, S. Praveen, and J. E. Arunkumar

Abstract The rapid urbanization witnessed in recent decades has led to a surge in population density within urban areas, necessitating advanced automation and management systems to operate and manage cities sustainably. This study examines the crucial role of Internet of Things (IoT) devices and analytics in establishing automated solutions across diverse urban sectors. IoT technologies offer a versatile framework capable of automating and managing crucial aspects of urban life such as electricity consumption, pollution control, vehicle parking, health monitoring, and traffic control. This research investigates the use of IoT in critical areas such as smart water management, smart waste management, smart traffic management, and smart buildings.

Through a comprehensive examination of these domains, it elucidates how IoT-enabled solutions facilitate efficient resource utilization, enhance environmental sustainability, and get better on the whole quality of life in urban environments. Furthermore, the combination of analytics with IoT data facilitates predictive modeling, real-time monitoring, and data-driven decision-making, thereby optimizing city operations and fostering a more sustainable future. By analyzing case studies and existing implementations, this work highlights the tangible benefits and challenges associated with deploying IOT and analytics-driven solutions in urban settings. Additionally, it offers insights into emerging trends, potential barriers, and opportunities for further advancements in leveraging IoT and analytics for sustainable urban development. Ultimately, this work aims to present a broad understanding of the transformative potential of IoT and analytics in shaping the cities of tomorrow towards greater resilience, efficiency, and sustainability.

Keywords Intelligent traffic control · Renewable energy integration · Real-time monitoring · IoT-enabled smart sensors · Automated fault detection and isolation · Bidirectional power flow · Predictive analytics

R. Parkavi (✉) · R. M. Soundar Arunachalam · S. Praveen · J. E. Arunkumar
Department of Information Technology, Thiagarajar College of Engineering, Madurai, India
e-mail: rpit@tce.edu; soundar@student.tce.edu; spraveen1@student.tce.edu; arunkumarje@student.tce.edu

V. Baskaran et al. (eds.), *EcoTech Urbanism*, The Urban Book Series,
https://doi.org/10.1007/978-3-031-83833-0_16

16.1 Introduction

In the twenty-first century, urbanization has surged dramatically, with more than half of the global population now residing in urban areas. Rapid city growth presents both substantial opportunities and challenges, making cities centers of economic activity, innovation, and cultural diversity. However, the escalating demands placed on urban infrastructure and resources raise concerns about sustainability, resilience, and livability.

In response to these challenges, there is a growing imperative to embrace technological innovations that can revolutionize how cities are planned, managed, and operated. Among these innovations, the Internet of Things (IoT) has emerged as a transformative force, providing numerous opportunities to develop smarter, more efficient, and sustainable urban environments. Fundamentally, IoT involves a network of interconnected devices, sensors, and systems that can collect, exchange, and analyze vast amounts of data in real time. This interconnected ecosystem lays the groundwork for implementing automated solutions across various urban domains, ranging from energy management and environmental monitoring to transportation and public services.

The integration of IoT with advanced analytics further enhances its potential, enabling cities to harness data-driven insights for predictive modeling, proactive decision-making, and optimization of resource allocation. By leveraging IoT and analytics, cities cannot only address immediate challenges such as traffic congestion, waste management, and air quality but also anticipate future needs and adapt to evolving urban dynamics.

In this context, this work aims to provide a complete overview of the role of IoT and analytics in driving sustainable urban development. Specifically, it explores how IoT technologies can revolutionize key aspects of urban life, including water management, waste management, traffic management, and building operations. By integrating theoretical analysis, case studies, and practical insights, this research aims to clarify the transformative potential of IoT and analytics in reshaping the cities of the future.

By investigating current trends, challenges, and opportunities in leveraging IoT and analytics for sustainable urban development, this work aims to educate policymakers, urban planners, and stakeholders about the advantages and impacts of adopting these technologies. Ultimately, it advocates for a holistic approach to urban planning and management that prioritizes innovation, efficiency, and sustainability, ensuring that cities remain vibrant, resilient, and livable for future generations.

16.2 IoT Components and Architecture

Internet of Things (IoT) architecture is the blueprint that outlines how devices, sensors, and systems communicate and interact within the IoT ecosystem. At its core, IoT architecture comprises three main components: devices/sensors, connectivity, and data processing/storage (Environmental Protection Agency, 2023a, b)

Devices/sensors: Devices and sensors form the foundational components of IoT, comprising physical objects equipped with sensors, actuators, or other mechanisms to gather data from the surrounding environment. Examples include sensors for temperature measurement, motion detection devices, and smart meters. These devices gather information from the physical environment and transform it into digital data for further processing.

Connectivity serves as the bridge enabling communication between IoT devices and the central system or network infrastructure. IoT utilizes various connectivity technologies such as connectivity options encompass Wi-Fi, Bluetooth, Zigbee, cellular networks (such as 3G, 4G, and 5G), and Low-Power Wide-Area Networks (LPWAN). Selection of the appropriate connectivity relies on factors including range, power consumption, data transfer speed, and deployment requirements.

Data processing and storage are essential steps once data is collected from IoT devices. Data processing involves tasks like filtering, aggregating, and analyzing raw data to derive meaningful insights. This process can occur via edge computing, where data is processed locally on devices or nearby gateways to reduce latency and bandwidth usage. Alternatively, cloud computing entails handling data on remote servers, offering scalability and accessibility for extensive data analysis and storage needs.

IoT abstract architecture is shown in Fig. 16.1 and the IoT burst from the year 2015 to 2025 is shown in Fig. 16.2.

16.2.1 IoT Architecture Layers

IoT architecture layers are classified into four categories. They are:

1. Perception layer: This tier includes IoT devices and sensors responsible for gathering and transmitting information as of the physical environment.
2. Network layer: The network layer facilitates message between IoT devices and the central system or network infrastructure.
3. Middleware layer: This layer offers critical services such as data processing, security protocols, and device management functionalities.
4. Application layer: At the top level, the application layer encompasses IoT applications and services that utilize processed data to deliver specific functionalities or solutions.

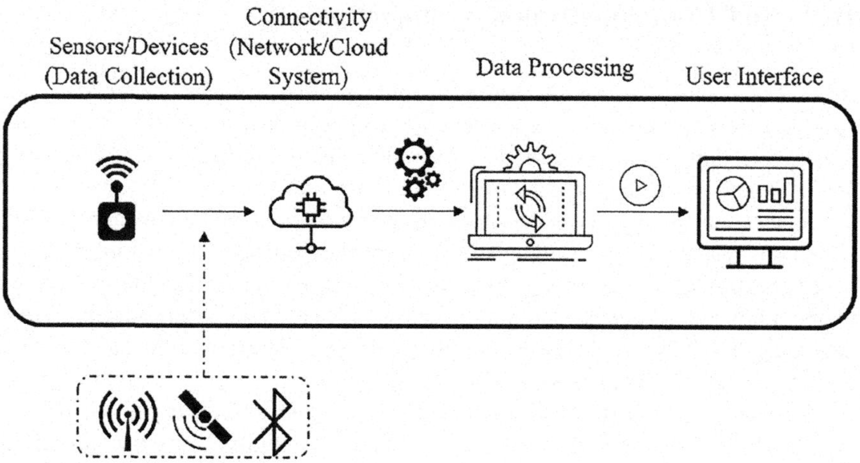

Fig. 16.1 IoT abstract architecture

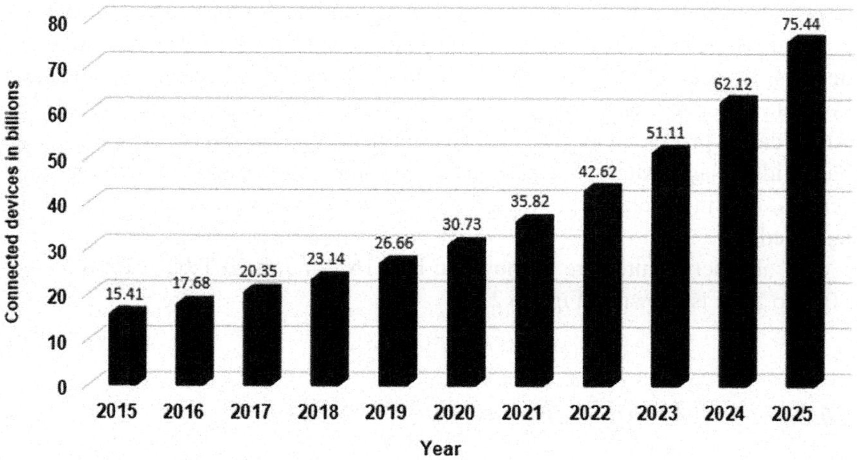

Fig. 16.2 IoT burst

16.3 Smart Water Management

The innovative smart water management techniques used in urban areas like Singapore and Barcelona. These cities have implemented advanced technologies to optimize water usage and ensure sustainable management of resources.

Advanced smart water management plays a pivotal task in the growth of sustainable smart cities. These cities face significant challenges such as water scarcity and aging infrastructure. Traditional approaches to water management often lack real-time data, hindering effective decision-making. However, with the emergence of advanced technologies such as Internet of Things (IoT) sensors and analytics, cities can revolutionize their water management strategies (Smith, 2020)

One key component of advanced smart water management is the deployment of IOT sensors and monitoring systems. These sensors facilitate real-time monitoring of water quality, flow rates, and levels within water infrastructure. By integrating with Geographic Information Systems (GIS), cities can perform spatial analysis and mapping of water resources, facilitating more informed decision-making processes.

Data analytics and predictive modeling are another crucial aspect of advanced smart water management. Machine learning algorithms can examine historical data to predict future water demand and usage patterns. Furthermore, predictive maintenance models can detect and resolve potential issues in water infrastructure before they become severe, ensuring the reliability and longevity of water systems (Wang, 2021).

Smart water distribution networks further enhance efficiency and resilience in water management. Implementation of smart meters and leak detection systems enables cities to detect and mitigate water leaks in distribution networks, reducing water loss and conserving valuable resources. Demand-responsive controls and pressure management technologies optimize water distribution systems, improving operational efficiency and reducing costs for water utilities.

In addition to infrastructure improvements, advanced smart water management also focuses on water conservation and demand management. Smart irrigation systems and demand management strategies, including tiered pricing and water conservation campaigns, encourage responsible water use among residents and businesses, contributing to overall sustainability efforts.

Several successful case studies highlight the effectiveness of advanced smart water management in smart cities. For instance, Singapore's Smart Water Grid utilizes real-time monitoring and control systems to achieve one of the lowest water leakage rates globally. Similarly, Barcelona's Smart Water Management Platform employs IOT sensors and data analytics to detect pollution events and facilitate rapid response measures, safeguarding public health and the environment.

The benefits of advanced smart water management extend beyond environmental sustainability to include economic efficiency and social equity. By reducing water wastage and pollution, cities can achieve cost savings and operational efficiency for water utilities. Moreover, improved access to clean and reliable water services enhances public health and quality of life, particularly in underserved communities.

Looking ahead, the integration of smart water management systems with other smart city initiatives is crucial for achieving holistic urban development. Cooperation among government agencies, utilities, and technology providers is crucial for tackling challenges connected to privacy and security of data, and integration, ensuring the effective deployment of advanced smart water management solutions.

Advanced smart water management holds immense potential for sustainable urban development. By utilizing IoT sensors, data analytics, and automation technologies, cities can improve the efficiency of water use, minimize wastage, and promote fair access to clean water resources, thereby establishing a foundation for a resilient and sustainable future.

16.4 Smart Traffic Control System

Navigating through bustling streets has become increasingly challenging. Traffic congestion, accidents, and safety concerns pose daily obstacles. The rising number of vehicles not only exacerbates traffic jams but also contributes to environmental pollution. Traditional traffic management systems often struggle to cope with urbanization and evolving mobility patterns. The solution must address these challenges more effectively than conventional systems. An optimal solution lies in an IoT-based smart traffic management system. IoT sensors, adaptable to various real-life situations, offer potential for safer, more efficient, and smoother traffic flow. This work delves into IoT-based smart traffic control systems. A smart traffic control system utilizes real-time data from traffic cameras, sensors, and GPS devices to overcome limitations of conventional traffic management systems. It also examines essential technologies utilized in IoT-driven smart traffic control systems.

16.4.1 Technologies

There are several technologies integral to smart traffic control systems. They are:

Traffic lights and IoT control Smart traffic signals, resembling conventional stoplights, utilize an array of sensors to monitor real-time traffic. The goal is to minimize vehicle stoppage time, thereby reducing carbon emissions. These lights can communicate with each other, adjusting to changing traffic conditions. Necessary devices include wireless sensors, RFID tags and readers, and geographic information systems (GIS) (Johnson, 2019)

Parking-enabled through IoT Smart meters and mobile apps facilitate convenient street parking by providing real-time notifications on parking availability. Drivers receive alerts and can reserve parking spots instantly, ensuring safer parking. Implementation requires hardware such as microcontrollers and distance-measuring sensors, cloud infrastructure like AWS IoT, and user interfaces like web or mobile applications (Kim, 2018).

Emergency assistance through IoT IoT-based traffic management systems expedite emergency response by detecting accidents via road sensors and alerting relevant authorities promptly.

Autonomous vehicles as traffic nodes In this system, vehicles act as intelligent nodes, sharing vital information on speed, location, and hazards. Vehicular Ad-hoc Networks (VANETs) are essential for gathering and distributing this information.

Fleet management Fleet management systems, vital for efficient goods transportation, focus on ensuring safe, cost-effective vehicle operations. They address concerns such as hazardous goods transportation, environmental impact, position tracking and navigation, traffic management, and optimizing fleet operations.

16.5 Challenges and Future Concerns

The challenges and future considerations for smart traffic control systems are:

Improvement in data collection: Enhanced sensing methods and automatic data-capturing techniques are essential for improving data accuracy and completeness.

Integration of nonhomogeneous data sources: Developing common standards is crucial for efficiently integrating diverse devices with different standards, thereby reducing costs and improving data transfer efficiency.

Managing huge amounts of data: Cutting-edge technologies like cloud and edge computing, along with big data analytics, are necessary for efficiently managing the exponential growth of IoT-generated data.

Privacy concerns: Ensuring data privacy and protection against hackers require implementing ethical AI and ML practices, advanced encryption, and decentralized data processing (Kim, 2018).

Policy formulations: Integrating traffic-related data into policymaking processes can lead to more effective traffic rules and regulations.

Smart traffic system schematic diagram is shown in Fig. 16.3.

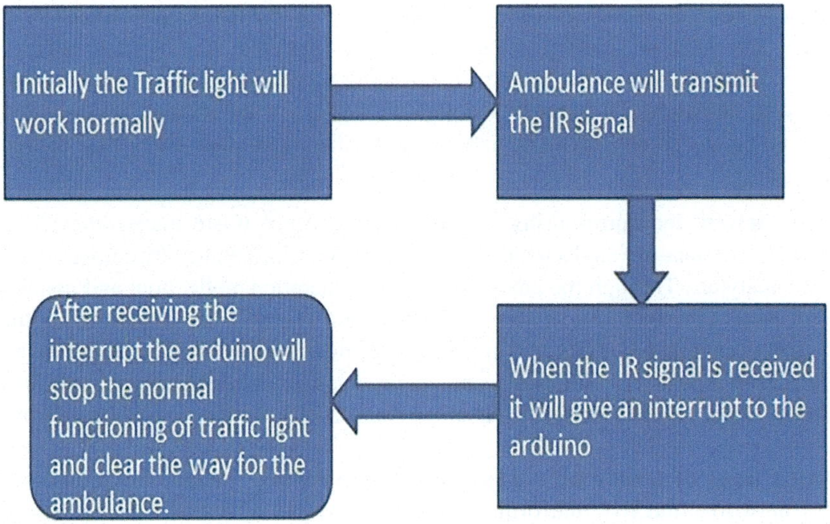

Fig. 16.3 Smart traffic system schematic diagram

16.6 Smart Waste Management for Urban Sustainability

In the pursuit of sustainable urban development, smart waste management systems have emerged as indispensable tools, harnessing cutting-edge technologies to revolutionize waste collection processes. By integrating using IoT sensors, data analytics, and automation, these systems enhance resource utilization, reduce operational costs, and mitigate environmental impact. This paper delves into the technical intricacies and innovative solutions driving smart waste management in modern cities, drawing insights from both theoretical research and practical implementations (Martinez, 2020).

Optimized waste collection routes Central to smart waste management is the utilization of IoT sensors installed in waste bins to track fill levels in real-time. By leveraging data analytics and dynamic routing algorithms, these systems optimize waste collection routes dynamically. This approach minimizes travel distances and fuel consumption while maximizing operational efficiency, ensuring timely waste removal without unnecessary resource expenditure.

Efficient resource allocation Moreover, smart waste management systems provide invaluable insights into waste generation patterns and collection requirements, enabling precise resource allocation in real time. Advanced data analytics facilitate dynamic resource deployment, optimizing workforce scheduling, equipment utilization, and route planning. By aligning resources with actual demand, waste management authorities can enhance service levels while reducing costs, thereby fostering sustainable practices (Lee, 2018).

Automation plays an essential role in enhancing waste sorting and recycling processes within smart waste management frameworks. By integrating advanced sensors and machine learning algorithms, these systems identify and segregate recyclable materials at the source, streamlining recycling workflows and minimizing contamination. This approach fosters transparency and traceability throughout the recycling supply chain, bolstering recycling rates and promoting resource recovery.

Data-driven decision-making Fundamental to smart waste management is the adoption of data-driven decision-making strategies, underpinned by comprehensive data analytics. Through the analysis of waste generation, collection, and composition data, authorities gain actionable insights into system performance and optimization opportunities. Informed by data analytics, decision-makers can fine-tune collection schedules, route plans, and operational workflows to enhance efficiency, reduce costs, and elevate service quality (Patel, 2019).

Integration of sensor networks and IOT platforms The combination of sensor networks and IoT platforms forms the foundation of smart waste management systems, allowing for real-time data collection, transmission, and analysis. IOT-enabled waste bins and collection vehicles continuously monitor critical parameters, transmitting information to centralized platforms for analysis and decision-making. Integration with GIS and mapping technologies further enhances operational visibility and control, empowering waste management authorities to proactively manage collection activities and respond swiftly to emerging challenges.

The flowchart for the smart traffic control system is shown in Fig. 16.4.

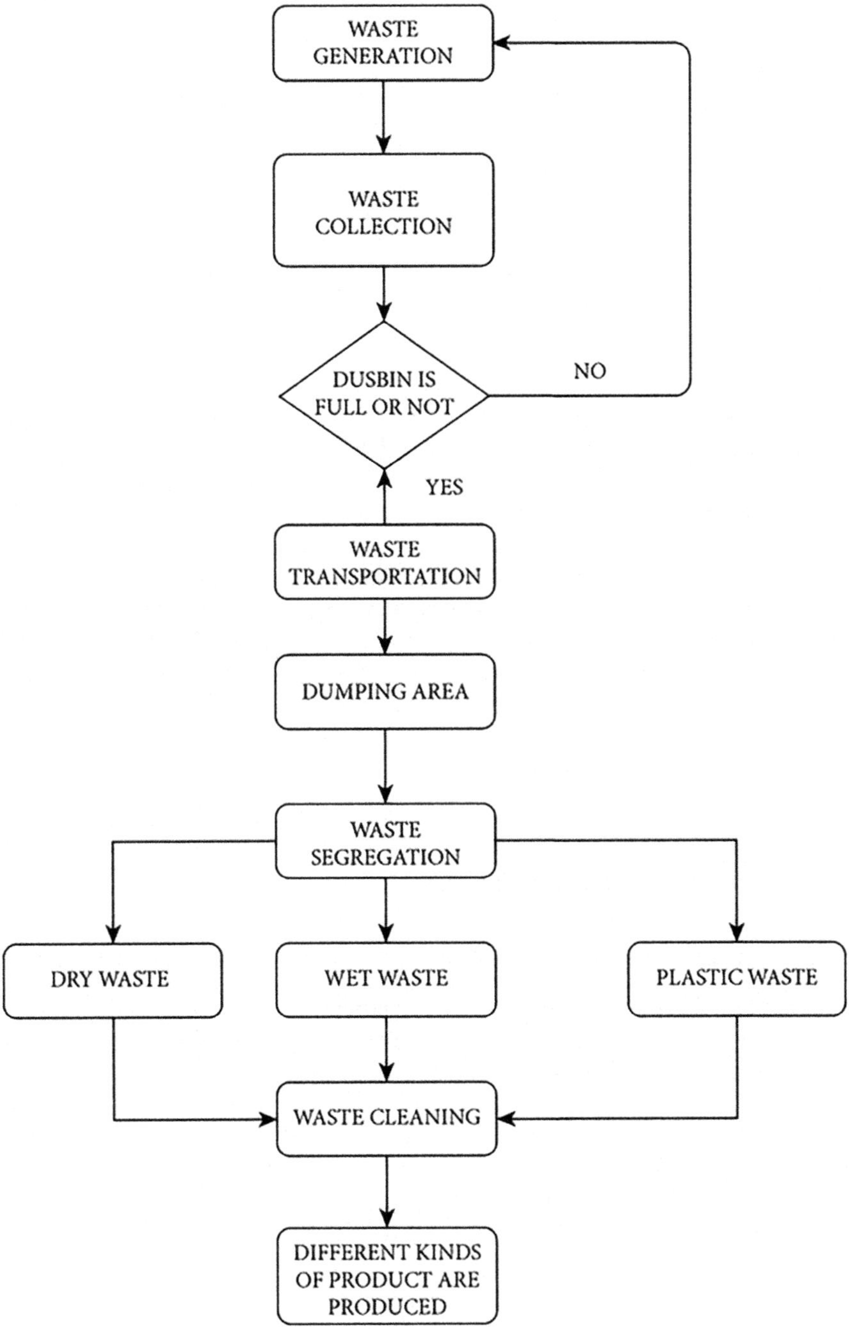

Fig. 16.4 Smart waste management

16.7 Smart Agriculture

Smart agriculture is all about using some tech stuff to make farming better. The technologies are nothing but sensors, drones, and artificial intelligence to assist farmers in growing crops more effectively and sustainably. Now the key principles behind smart agriculture are precision, sustainability, automation, and being data-driven (Qiu & Li, 2021).

16.7.1 Established Smart Agriculture Systems

Cities around the world have already implemented various smart agriculture systems for growing crops more efficiently and sustainably. The recent technologies used in smart agriculture systems are:

- Precision hawk: This company offers drone-powered aerial data collection and analytics for agriculture. They provide farmers with detailed insights into crop health, weed pressure, and stand counts, helping them to optimize the management strategies.
- Vertical farming and hydroponic systems employ advanced technologies like LED lighting, climate control, and precise nutrient delivery to grow crops indoors in controlled environments. These techniques allow for year-round cultivation and decrease water usage, and achieve higher crop yields compared to conventional farming practices.
- Precision farming tools, including GPS-guided tractors and automated irrigation systems, are vital to smart agriculture. These technologies enable farmers to apply resources like water, fertilizers, and pesticides with accuracy, reducing waste and enhancing crop yields.

16.7.2 Future Smart Agriculture Innovations

Future agricultural solutions that innovate and streamline operations include:

Autonomous farming equipment IoT technology facilitates autonomous tractors, drones, and robots designed for tasks such as planting, weeding, and harvesting. This reduces reliance on manual labor, enhancing operational efficiency.

Blockchain technology Block chain enhances traceability, transparency, and trust within the food supply chain. It enables farmers to authenticate the origin and quality of agricultural products, thereby minimizing fraud and enhancing market integrity access.

Augmented reality Augmented reality provides farmers with real-time data and insights, enabling them to make well-informed decisions about crop management, irrigation, and harvesting based on data-driven analysis (Tzounis & Kagianas, 2020).

16.7.3 Benefits of Smart Agriculture

By harnessing IoT, big data, artificial intelligence, and other innovations, farmers can enhance their operations, minimize waste, and boost profitability. Here are primary advantages of smart agriculture:

Enhanced decision-making Smart agriculture technologies deliver real-time data and insights, allowing farmers to make informed decisions regarding crop management, resource allocation, and market trends. Utilizing data analytics and machine learning, farmers can forecast crop yields, detect patterns in crop health and growth, and optimize operations for maximum profitability.

Increased productivity Smart agriculture practices contribute to higher crop yields and overall productivity. Through robotics and automation, farmers achieve more efficient and precise task execution, reducing reliance on manual labor and elevating output.

Improved traceability Smart agriculture technologies improve the traceability and transparency of the food supply chain. Block chain technology allows farmers to record and authenticate the cause and eminence of agricultural products, mitigating fraud and ensuring food safety.

16.7.4 Challenges and Constraints

While using these technologies we need to face many challenges and constraints. Some basic challenges and constraints in smart agriculture are:

- Data privacy and security: Protecting data privacy and security is essential when collecting and analyzing extensive agricultural data. Implementing strong data protection measures and complying with rigorous privacy regulations are essential to safeguarding sensitive information.
- Limited Funding and Resources: Limited funding and resources for research and development can hold back the adoption and implementation of new agricultural technologies. By Investing in public-private partnerships, grant programs, and other funding sources can help support research and development efforts and drive innovation.

- Limited skilled labor and training: Limited availability of skilled labor and training opportunities in precision agriculture technologies may hinder adoption and effective use. By Providing training and education programs for farmers and agricultural workers can help build the necessary skills and expertise for successful adoption and implementation of smart agriculture technologies.

16.8 Smart Environmental Monitoring and Management

16.8.1 Smart Environmental Monitoring

Smart environmental monitoring is crucial for assessing environmental health, identifying pollution sources, and informing policies aimed at reducing environmental risks and promoting public health.

16.8.1.1 Principles

Data-driven: Relies on real-time data collection and analysis for accurate environmental condition monitoring (Zhao et al., 2022a, b).

Proactive: Detects environmental changes and pollution events early to prevent adverse effects on health and ecosystems.

Collaborative: Involves cooperation among government agencies, research institutions, industries, and communities to tackle environmental challenges.

16.8.1.2 Technologies in Smart Environmental Monitoring

Key technologies include:

IoT sensors: Measure air quality (e.g., PM2.5, CO_2), water quality (e.g., pH, dissolved oxygen), noise levels, and temperature at various monitoring stations.

Remote sensing: Utilizes satellite images, aerial drones, and on-ground sensors to observe large-scale environmental changes such as deforestation and natural disasters.

Machine learning and AI: Analyzes extensive environmental data to identify trends, predict pollution events, and generate actionable insights for environmental management.

16.8.1.3 Benefits of Smart Environmental Monitoring

Early warning: Provides timely alerts about environmental hazards like air pollution, water contamination, and extreme weather, enabling prompt responses.

Ecosystem protection: Identifies threats to biodiversity, habitat loss, and ecosystem degradation, guiding conservation and sustainable land management efforts.

Citizen engagement: Encourages public participation in environmental monitoring through citizen science initiatives and access to environmental data.

Policy support: Offers scientific data and insights to support evidence-based policy decisions and assess the effectiveness of environmental regulations.

16.8.1.4 Challenges of Smart Environmental Monitoring

Data quality: Ensuring the accuracy, reliability, and consistency of data collected from diverse sources.

Sensor calibration: Regular calibration and maintenance of sensors to ensure precise measurements and minimize errors.

Privacy and security: Protecting sensitive environmental data with strong privacy measures and security protocols.

Accessibility and equity: Ensuring equitable access to environmental monitoring data and technologies, especially in underserved communities.

16.8.1.5 Future Directions of Smart Environmental Monitoring

Sensor miniaturization: Developing smaller, cheaper, and more portable sensors for widespread and real-time monitoring.

Citizen science: Expanding initiatives to involve citizens in data collection and monitoring efforts.

Data analytics: Developing machine learning and AI to derive valuable insights from large environmental datasets.

International collaboration: Strengthening global cooperation and knowledge sharing to address environmental challenges.

Policy integration: Incorporating environmental monitoring data into broader policy frameworks, such as urban planning and climate change adaptation strategies.

16.8.2 Smart Environment Management

The integration of IoT technology into environmental monitoring and management systems marks a significant advancement in addressing global environmental challenges. IoT facilitates comprehensive strategies to manage issues like air and decline in water quality, loss of biodiversity, and effects of climate change (Zheng & Luo, 2021).

16.8.2.1 Evolution of IoT in Environmental Monitoring

IoT technology has transformed environmental monitoring from traditional static methods to dynamic, real-time systems. This shift has enhanced data accuracy, accessibility, and the ability to generate actionable insights (Ahmed & Zhang, 2020).

16.8.2.2 Early Developments and Foundation

The origins of IoT in environmental monitoring originate from the development of sensor networks and telemetry systems in the late twentieth century. These early systems enabled remote data collection and transmission, initially focusing on basic parameters like temperature, humidity, and soil moisture.

16.8.2.3 Advancements in Sensor Technology

Improvements in sensor technology have led to smaller, more cost-effective, and accurate sensors designed to measure a broader range of environmental parameters. This progress has broadened the scope of monitoring to include complex indicators like greenhouse gas emissions and particulate matter concentrations.

16.8.2.4 Integration of Wireless Communication Protocols

The shift from wired to wireless communication protocols has improved the scalability and flexibility of IoT-enabled monitoring systems. Standards such as Wi-Fi, Bluetooth, Zigbee, and LoRaWAN facilitate seamless connectivity and real-time data transmission over large areas.

16.8.2.5 Applications of IoT in Smart Environmental Monitoring

Water quality monitoring: IoT sensors track parameters like pH, dissolved oxygen, turbidity, and temperature in freshwater and marine ecosystems to identify pollution sources and assess ecological conditions.

Precision agriculture: IoT technologies monitor soil conditions, crop health, and microclimatic variables to optimize resource use, reduce environmental impacts, and enhance crop yields.

Wildlife and habitat monitoring: IoT devices collect data on wildlife behavior, habitat conditions, and biodiversity indicators to support conservation and ecological research efforts.

16.8.2.6 Challenges and Considerations

Data security and privacy: Ensuring the protection of sensitive data from breaches, unauthorized access, and cyberattacks through strong encryption, secures storage, and access control.

Interoperability and standards: Overcoming challenges related to integrating diverse IoT technologies, devices, and platforms by establishing universal standards and protocols to ensure seamless connectivity and system scalability.

16.9 Smart Grids

In the pursuit of a more sustainable and efficient energy landscape, smart grid technologies have emerged as transformative solutions for modernizing power generation, distribution, and consumption. This paper delves into the technical intricacies and advancements driving the development of smart grid systems, exploring their role in enhancing reliability, resiliency, and sustainability in the energy sector (Zhang, 2022)

16.9.1 Components of Smart Grids

Smart grids consist of several essential components:

Advanced metering infrastructure (AMI) AMI forms the backbone of smart grid systems, enabling two-way communication between utility providers and consumers. Smart meters, equipped with real-time monitoring capabilities, allow for accurate measurement of energy consumption. This technology facilitates demand response programs and dynamic pricing, helping consumers manage their energy usage more effectively.

Distribution automation Distribution automation technologies improve the reliability and efficiency of electricity distribution networks within smart grids. By integrating sensors, actuators, and control systems, these technologies monitor and manage distribution infrastructure in real time. Automated fault detection, isolation, and restoration capabilities reduce downtime, enhancing system resilience and minimizing disruptions for consumers and businesses.

Renewable energy integration Smart grids support the seamless integration of renewable energy sources, such as solar and wind power, into the electricity grid. Advanced forecasting algorithms and grid management techniques optimize the integration of intermittent renewable generation, maintaining grid stability and reliability. Bidirectional power flow and grid-scale energy storage facilitate the incorporation of distributed energy resources (DERs) and microgrids, promoting sustainable energy practices.

Demand response and energy management Demand response programs utilize smart grid capabilities to manage electricity demand efficiently. Smart meters and home energy management systems enable consumers to participate in demand response initiatives by adjusting their energy usage in response to dynamic pricing or incentive schemes. Optimization algorithms ensure real-time balance between supply and demand, reducing peak loads and improving overall grid efficiency.

16.9.2 Advanced Grid Analytics

Predictive Maintenance Advanced analytics tools enable predictive maintenance of grid infrastructure, using machine learning algorithms to predict equipment failures and prioritize maintenance tasks. By analyzing historical data and sensor readings, these models identify early signs of equipment deterioration, allowing utilities to address issues proactively and minimize operational disruptions.

Grid Optimization Grid optimization algorithms enhance the efficiency and reliability of smart grid systems. These algorithms use real-time data on energy supply, demand, and grid conditions to optimize operations, minimize transmission losses, and reduce system costs. Techniques such as optimal power flow, voltage control, and load balancing strategies ensure optimal performance under varying operating conditions.

16.9.3 Cyber Security and Resilience

Cyber security Measures Smart grid systems incorporate robust cyber security measures to protect against cyber threats and vulnerabilities. These measures include encryption protocols, intrusion detection systems, and network segmentation strategies to safeguard critical infrastructure and data from unauthorized access or manipulation.

Resilience Enhancements Smart grid technologies enhance grid resilience through redundancy, flexibility, and adaptive control mechanisms. Grid resiliency features include self-healing capabilities, decentralized control architectures, and islanding functionalities to isolate and restore critical grid segments during emergencies or disturbances. Additionally, micro grid integration and distributed energy resources enhance local resilience and provide backup power during grid outages.

16.10 Conclusion

The integration of IOT, analytics, and automation technologies has significantly transformed various aspects of urban infrastructure, ranging from water management to environmental monitoring. Smart solutions have enabled more efficient resource usage, reduced waste, and enhanced sustainability across multiple sectors. Smart water management systems, driven by IOT sensors and data analytics, have optimized distribution networks and minimized water waste, contributing to more sustainable water usage practices. Similarly, intelligent traffic control systems have mitigated congestion and pollution, improving both mobility and air quality within urban areas. In waste management, smart solutions have streamlined collection, recycling, and disposal processes, fostering circular economies and environmental conservation. Additionally, smart grid technologies have revolutionized energy management and renewable integration, promoting grid reliability and sustainability. Smart agriculture holds tremendous potential to transform farming practices, boost productivity, and enhance sustainability. By utilizing IoT sensors, data analytics, and automation, these systems optimize resource use and monitor crop health. This enables precision farming techniques that reduce environmental impact.

Furthermore, smart environmental monitoring is essential for assessing and safeguarding environmental health, informing evidence-based policies, and addressing environmental challenges. Ongoing investment in research, innovation, and collaboration is essential to unlocking the full potential of smart solutions and meeting sustainable urban development goals. In summary, the integration of IoT, analytics, and automation provides cities with the opportunity to tackle challenges, enhance quality of life, and foster more sustainable urban environments. Ongoing innovation, collaboration, and investment are crucial for overcoming obstacles and fully harnessing the potential of smart technologies in shaping the cities of the future.

References

Ahmed, S., & Zhang, Y. (2020). Smart environmental monitoring systems: Technologies, applications, and challenges. *IEEE Access, 8*, 119652–119670.

Al-Ali, M., Al-Anazi, A., & Al-Said, A. (2021). A survey on IoT-based smart agriculture: Applications, technologies, and challenges. *IEEE Access, 9*, 119899–120012.

Chen, D. (2021). Smart waste management: A case study of city X. *International Conference on Environmental Engineering*, 112–125.

Environmental Protection Agency. (2023a). *Smart cities and sustainable development: A policy brief*. EPA. www.epa.gov/smart-cities-policy. Accessed 25 Mar 2024.

Environmental Protection Agency. (2023b). *Smart environmental monitoring for sustainable development: A policy brief*. EPA. www.epa.gov/smart-environmental-monitoring-policy. Accessed 25 Mar 2024

Garcia, J. (2020). *Smart traffic control: Case studies and best practices*. Urban Dynamics Press.

International Energy Agency. (2023). *Smart grids: A global perspective*. IEA. www.iea.org/smart-grids-report. Accessed 20 Mar 2024.

Johnson, E. (2019). *Advancements in traffic control systems*. UrbanTech Publishers.

Kim, D. (2018). *Advanced analytics for traffic management: Models and algorithms*. Springer.

Lee, M. (2018). Smart grid Technologies for Sustainable Energy Management. *IEEE Transactions on Power Systems, 30*(4), 789–802.

Li, K. (2023). Renewable energy integration in smart grids: Challenges and opportunities. *IEEE Power and Energy Magazine, 40*(2), 45–58.

Martinez, S. (2020). Community engagement in smart waste management: Lessons learned from city Y. *International Conference on Sustainable Cities*, 225–238.

Patel, P. (2019). Advancements in smart waste management technologies. *Waste Management & Recycling, 15*(1), 75–88.

Qiu, X., & Li, X. (2021). A review of IoT-based smart agriculture: From the perspective of IoT-based agriculture data management. *Sustainability, 13*(10), 5269.

Smith, J. (2020). Optimizing water distribution in smart cities. *Journal of Sustainable Urban Development, 5*(2), 45–58.

Tzounis, A., & Kagianas, N. (2020). Sustainable agriculture and smart farming: A review of the role of ICT in precision agriculture and animal farming. *Sustainability, 12*(12), 5019.

United Nations Environment Programme. (2022). *Smart cities: Challenges and opportunities*. UNEP. www.unep.org/smart-cities-report. Accessed 15 Mar 2024.

Wang, L. (2021). Integration of IoT and analytics in smart water management systems. *Water Resources Management, 25*(3), 150–165.

Zhang, W. (2022). Predictive maintenance in smart grids: A review of current trends and future directions. *IEEE Transactions on Smart Grid, 10*(3), 112–125.

Zhao, Y., Hu, H., & Zhang, Y. (2022a). A review of IoT-based smart agriculture: From the perspective of agriculture-big-data-driven decision-making. *Computers and Electronics in Agriculture, 194*, 106424.

Zhao, Z., Chen, Y., Wang, Y., & Wang, X. (2022b). A review of IoT-based smart environmental monitoring systems: From the perspective of agriculture environmental monitoring. *Sustainability, 14*(12), 7214.

Zheng, Y., & Luo, J. (2021). A review of IoT-based smart environmental monitoring systems: From the perspective of agriculture environmental monitoring. *IEEE Access, 9*, 119729–119750.

Chapter 17
Evaluation of Wastewater Quality Index (WWQI) for Domestic Sewage Disposal in Mandovi River, Goa

Shaivi Chanekar, Sameer Mahato, Devendra Patidar, Bright Singh Seeni, and Yogeshwar Boddu

Abstract The development of a wastewater quality index (WWQI) is essential to address environmental concerns, comply with regulations, protect public health, support resource management objectives, and optimize economic considerations. Such a study can provide valuable insights for decision-makers, policymakers, and stakeholders involved in wastewater management, and contribute to more effective and sustainable wastewater disposal methods. This project presents the development of an index that represents the degree of pollution or contamination in the wastewater. The wastewater quality index (WWQI) is developed using weighted arithmetic function by considering the relevant criteria pollutants based on standards and assigning weightage to these parameters. An artificial neural network (ANN) model is developed to predict WWQI values and a WWQI predictor web page is created.

Keyword Wastewater quality index (WWQI) · Water pollution · Wastewater disposal · RNN · LSTM

S. Chanekar · S. Mahato · D. Patidar · Y. Boddu
Department of Civil Engineering, National Institute of Technology Goa, Cuncolim, India
e-mail: chanekarshaivi@nitgoa.ac.in; sam0905@nitgoa.ac.in; devendra01@nitgoa.ac.in; bodduyogeshwar3@nitgoa.ac.in

B. S. Seeni (✉)
Department of Civil Engineering, National Institute of Technology Goa, Cuncolim, India

TIET-Virginia Tech Center of Excellence in Emerging Materials (CEEMS), Thapar Institute of Engineering and Technology, Patiala, India
e-mail: bright_ps@thapar.edu, bsingh@nitgoa.ac.in

V. Baskaran et al. (eds.), *EcoTech Urbanism*, The Urban Book Series,
https://doi.org/10.1007/978-3-031-83833-0_17

17.1 Introduction

Water quality parameter analysis plays a significant role in understanding the quality of water in surface and subsurface sources (Jose et al. 2021). Disposal of wastes such as industrial effluents, residential zone-based effluents, agricultural runoffs, dumping of solid waste, etc. through unsuitable practices affects the quality of water (Yongo et al. 2023). The wastewater quality index (WWQI) is a well-known and reliable scientific method for tracking the regional and temporal change of wastewater quality. It combines all the distinctive values of a complicated set of wastewater quality data into a single value. Numerous research use both water quality indices and wastewater quality indices to determine the overall water quality (Singh et al., 2008; Shah & Longsheng, 2020; Vijay et al., 2016; Raut et al., 2017; Jamshidzadeh & Barzi, 2020). The idea of a water quality index and a mathematical expression for the aggregation function were pioneered by Horton in 1965. The water quality and wastewater quality indices over the previous four decades are then determined using a number of mathematical expressions developed for aggregation functions (Swamee & Tyagi, 2007).

When determining the water quality index (WQI) or wastewater quality index (WWQI), the choice of an aggregation function is crucial. According to analyses of the literature, the majority of aggregation functions have issues with ambiguity (overestimation) and eclipsing (underestimation) (Swamee & Tyagi, 2007, Singh et al., 2008; Jamshidzadeh & Barzi, 2020). Additionally, these issues could result in inaccurate results. A suitable aggregate function should therefore be free of the dual issues of ambiguity or eclipsing, or at the very least, minimize their effects. Additionally, these issues could result in inaccurate results. A suitable aggregate function should therefore be free of the dual issues of ambiguity or eclipsing, or at the very least, minimize their effects. The low water quality subindex must be treated fairly by the ambiguity-free aggregation function, which must also affect the subindices of all water quality variables. The water quality subindex should not be biased by the eclipsing-free aggregating function. The subindex of each individual quality parameter should be sensitive to changes in an appropriate aggregation function.

The study aims at developing wastewater quality index (WWQI) to determine the disposal quality of effluent disposed into inland surface water and on land for irrigation in Goa. Further, it focuses on predicting WWQI values using ANN model and developing a web page of WWQI predictor. Prediction of the WWQI for the future years can contribute to the holistic sustainability of the city which will improve the ecological and technological improvement of the cities.

17.2 Methodology

17.2.1 Study Area

Panjim (Panaji), the capital of the Indian state of Goa, lies on the banks of the Mandovi River estuary. With a population of 114,759 in the metropolitan area, Panaji is Goa's largest urban agglomeration. There are three sewage treatment

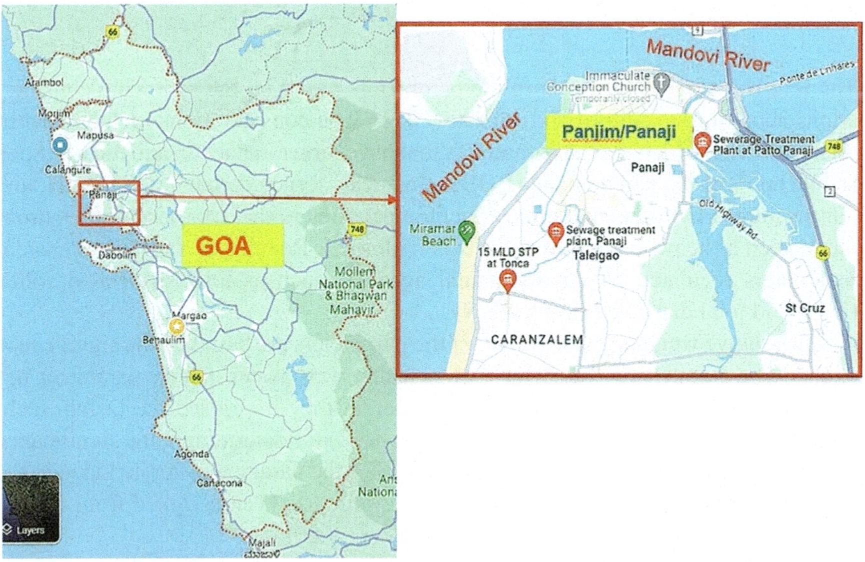

Fig. 17.1 Study area

plants (STPs) in Panjim, which are Patto plant having 2 MLD (million liters per day) capacity and Tonca plants having 15 MLD and 12.5 MLD capacities (Fig. 17.1). Each of these plants are based on the sequencing batch reactor (SBR) technology. The sequencing batch reactor (SBR) is a fill-and-draw-activated sludge system for wastewater treatment where wastewater is added to a single batch reactor, treated to remove undesirable components, and then discharged. The treated effluent from all the three STPs is disposed of in the Mandovi River.

The test results of 2 years obtained from the Tonca STP of 15 MLD capacity for influent and effluent wastewater are for the parameters COD (chemical oxygen demand), BOD (biochemical oxygen demand), pH, temperature, chloride, TDS (total dissolved solids), and TSS (total suspended solids).

17.2.2 Selection of Wastewater Quality Parameters

Based on the objectives, appropriate wastewater quality parameters were selected for analysis. These parameters include physical, chemical, and biological parameters. The selection of parameters is based on their relevance to the project objectives, availability of data and resources, and adherence to relevant standards or guidelines of Central Pollution Control Board (CPCB).

17.2.3 Assignment of Relative Weights

The relative weights are given to the parameters based on the following considerations according to Abbasi and Abbasi (2012). The considerations include health-related effects, which are the parameters that have direct impact on aquatic life, such as biochemical oxygen demand (BOD), chemical oxygen demand (COD), pH, and temperature. Environmental effects include parameters that can harm the environment, such as heavy metals and nutrients (chloride). Regulatory standards consider parameters regulated by environmental agencies such as total suspended solids (TSS) and total dissolved solids (TDS).

The relative weights given to each of the parameters are based on the significance values. The significance value for each selected wastewater quality parameter has been assigned from Singh et al. (2008), wherein they have used the Delphi technique (based upon opinions collected from experts) for determining the significance values for each wastewater quality parameter. For the parameters not included in the study of Singh et al. (2008), the significance value has been assigned from Abbasi and Abbasi (2012) and Jamshidzadeh and Barzi (2020).

The guidelines for effluent quality standards are taken from the Indian standards issued by the Environmental Protection Act rules (EPA, 1986). The disposal quality of effluent has been considered for two cases, which are disposal in inland surface water and disposal on land for irrigation.

The effluent quality standards and relative weights for surface disposal and disposal on irrigational land are given in Tables 17.1 and 17.2, respectively.

17.2.4 WWQI Development

There are several methods for estimating the wastewater quality index. The method chosen for determining WWQI is weighted arithmetic function as it exhibits less eclipsing problems, a uniform and true linear behavior with variations in subindex values for extreme pollutants, and ranks second in sensitivity among six aggregation functions (Singh et al., 2008). The formula used for developing WWQI is shown in eq. (17.1).

Table 17.1 Relative weights and effluent standards of selected parameters for surface disposal

Parameters	Permissible limits for disposal in inland surface water	Relative weight (w_i)
pH	5.5–9	0.1456
BOD (mg/L)	30	0.1552
COD (mg/L)	250	0.1897
Temperature (°C)	40	0.1533
SS (mg/L)	100	0.1149
DS (mg/L)	2100	0.1149
Chloride (mg/L)	1000	0.1264

Table 17.2 Relative weights and effluent standards of selected parameters for disposal on irrigational land

Parameters	Permissible limits for disposal in inland surface water	Relative weight (w_i)
pH	5.5–9	0.2216
BOD (mg/L)	100	0.2362
SS (mg/L)	200	0.1749
DS (mg/L)	2100	0.1749
Chloride (mg/L)	600	0.1924

Table 17.3 Rating and grade of the wastewater quality for corresponding levels of WWQI (WHO, 2018)

WWQI	Rating/class
0–25	Excellent
25–50	Good
50–75	Fair
75–100	Poor
>100	Very poor

$$\mathrm{WWQI} = \sum_{i=1}^{n} w_i q_i \tag{17.1}$$

Where.
w_i = the relative weight of the ith parameter and q_i = quality rating given by:

$$q_i = \left(\frac{v_n - v_{id}}{s_n - v_{id}} \right) \times 100 \tag{17.2}$$

Where.

v_n = estimated value of water quality parameter.
s_n = standard permissible value of water quality parameter
v_{id} = ideal value of water quality parameter (0 for all values except pH = 7 and temperature = 26 °C)

The final rating of WWQI is an ascending scale that ranges from 0 to 100 (higher scores indicate higher pollution). The quality of the water body is categorized also into five classes: "Excellent," "Good," "Fair," "Poor," and "Very poor". These classes are shown in Table 17.3.

17.2.5 WWQI Distribution

Upon calculation of the WWQI values for the entire dataset, the data distribution for raw sewage and treated sewage was obtained. From the pie chart shown for raw sewage (Fig. 17.2), major percentage of the samples fall under "Very poor"

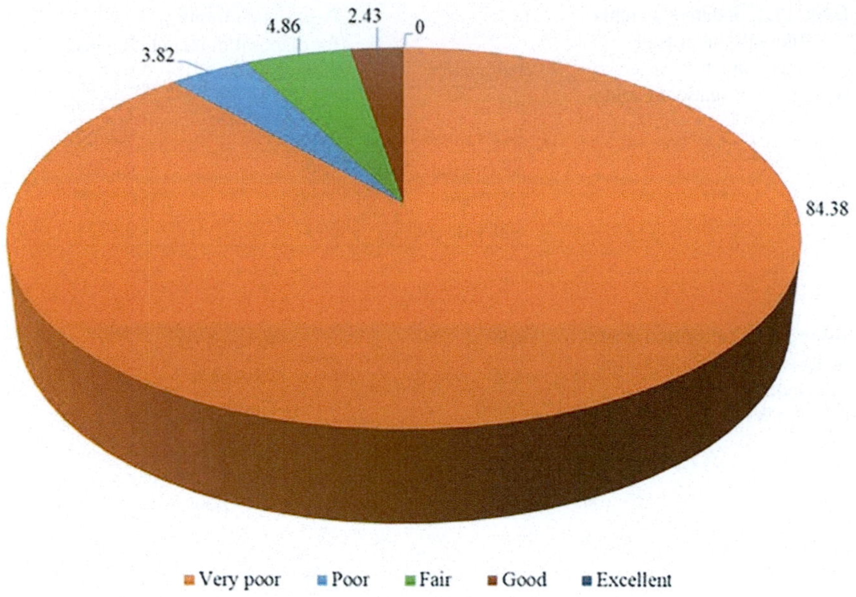

Fig. 17.2 Distribution of raw sewage (%)

category. The status of WWQI shows excellent results after treatment as very high percentage of the samples fall under "Excellent" category, as shown in Figs. 17.3 and 17.4 for disposal in surface water and on land for irrigation, respectively. This indicates that the sewage treatment plant has good efficiency in treating the raw sewage.

17.3 Ann Modelling

The focus of the project is on "Time Series Forecasting" using artificial neural networks (ANNs), specifically leveraging the power of long shortterm memory (LSTM) networks. The goal is to enhance prediction accuracy by exploiting the temporal dependencies within the data. The initial phase involves comprehensive data preprocessing, ensuring the time series is appropriately formatted for input into the LSTM network. Subsequently, the model is designed and trained, with careful consideration given to hyperparameter tuning to optimize performance. In addition to the LSTM-based ANN, the implementation of a recurrent neural network (RNN) model is done for Time Series Forecasting. This comparative analysis aims to evaluate the performance of the LSTM and RNN architectures, discerning their strengths and weaknesses in capturing temporal patterns within the data. The RNN undergoes a similar process of data preprocessing, model design, and hyperparameter tuning.

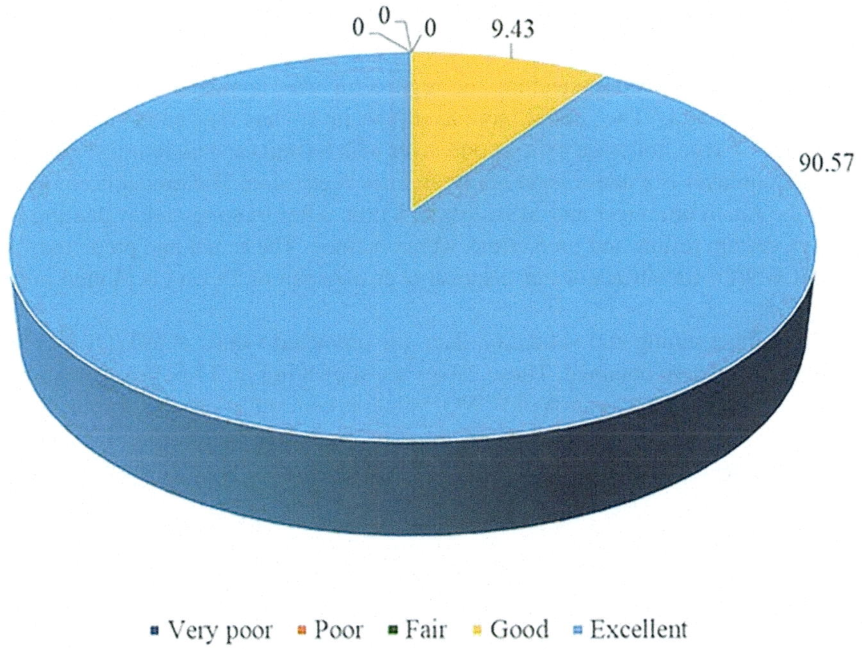

Fig. 17.3 Distribution of treated data to be disposed in surface water (%)

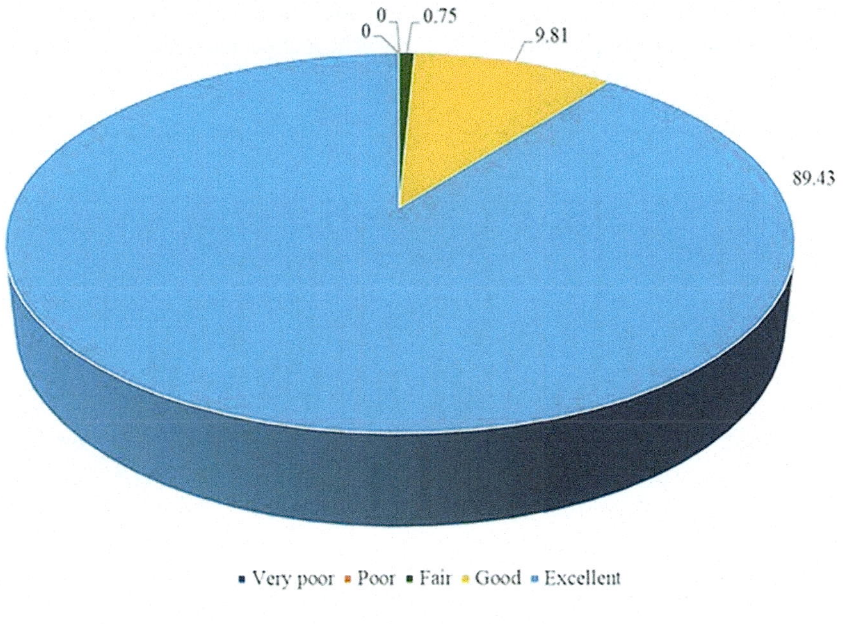

Fig. 17.4 Distribution of treated data to be disposed on irrigational land (%)

17.3.1 RNN (Recurrent Neural Network) Model

A two-layered RNN architecture for regression was implemented using Keras sequential model. The initial simple RNN layer has 16 units with return sequences = True, followed by a second layer with 64 units and no return sequences. The output layer is a dense layer configured for regression. The model is compiled with the Adam optimizer and MSE loss function. After training and evaluating the model, certain results and predictions were obtained. The actual and predicted values of WWQI for surface water disposal over the span of 2 years is plotted in the Fig. 17.5.

From this training and validation data, the predicted values of WWQI over the next 10 years were obtained. These values are shown in Fig. 17.6. From the graph, it is observed that the variation of WWQI values is constantly varying. Similarly, the actual vs predicted values were obtained for disposal on irrigational land, which are shown in Fig. 17.7. The predicted values of WWQI over the next 10 years are shown in Fig. 17.8.

17.3.2 LSTM (Long Short-Term Memory) Model

In the designed LSTM architecture tailored for regression tasks, temporal dependencies within sequential data are emphasized. The model, optimized using Adam and trained with MSE (mean squared error) loss, incorporates an early stopping mechanism to mitigate the risk of overfitting. The architecture employs a sequential

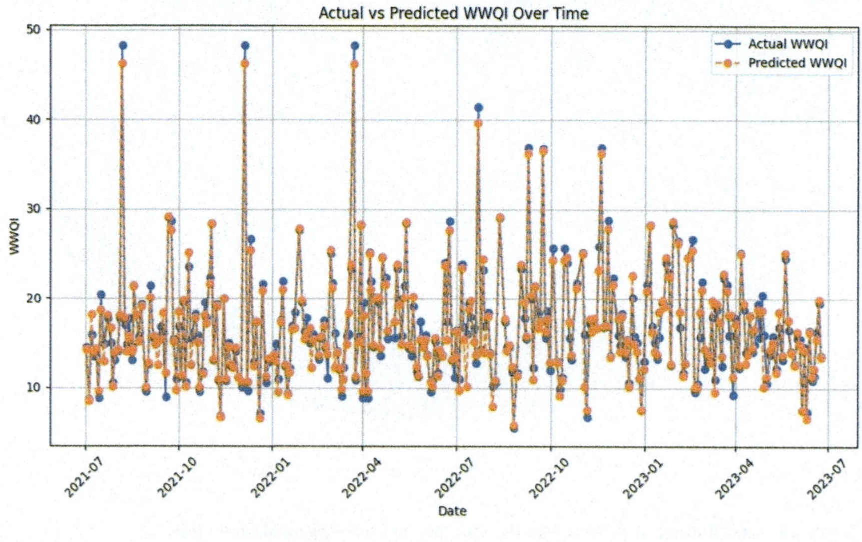

Fig. 17.5 Actual vs predicted WWQI over time for disposal in surface water using RNN model

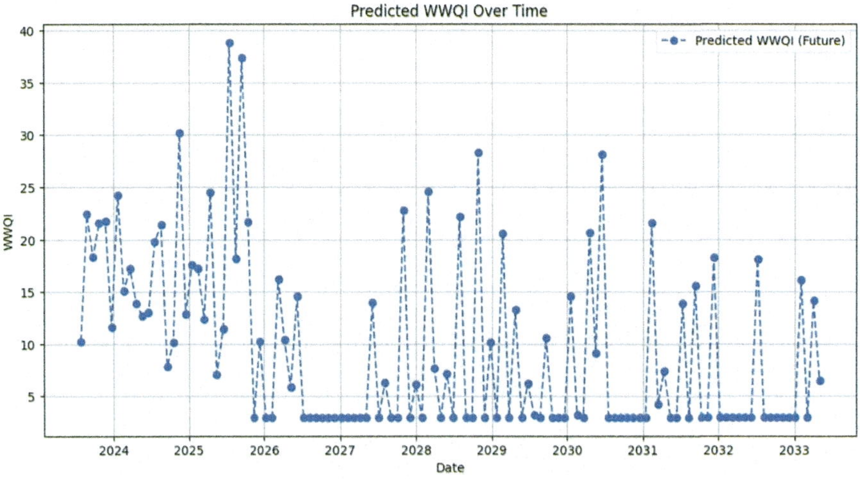

Fig. 17.6 Predicted values of WWQI over time for disposal in surface water using RNN model

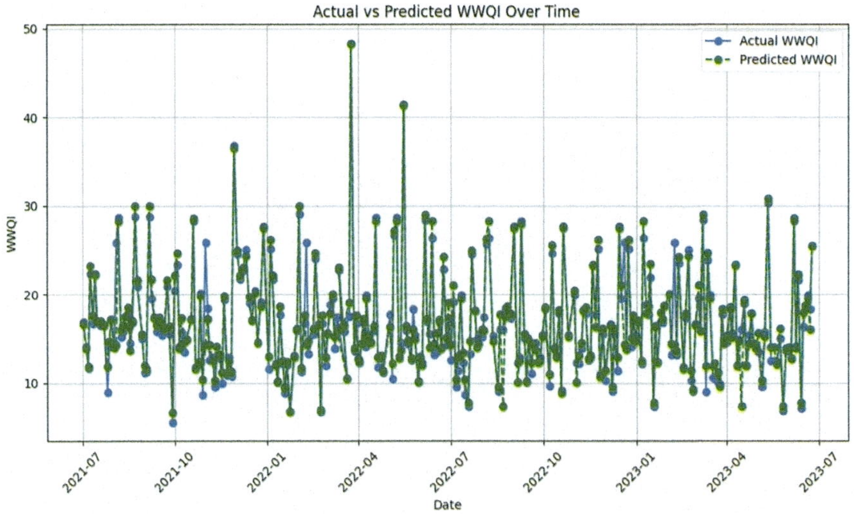

Fig. 17.7 Actual vs predicted WWQI over time for disposal on irrigational land using RNN model

model featuring two LSTM layers. The first layer, comprising 16 units, returns full sequences, enabling the capture of intricate temporal patterns. Subsequently, the second layer, consisting of 64 units, is designed to capture higher-level dependencies within the sequential data. The output layer is a dense layer configured for regression, with the number of units matching the output size.

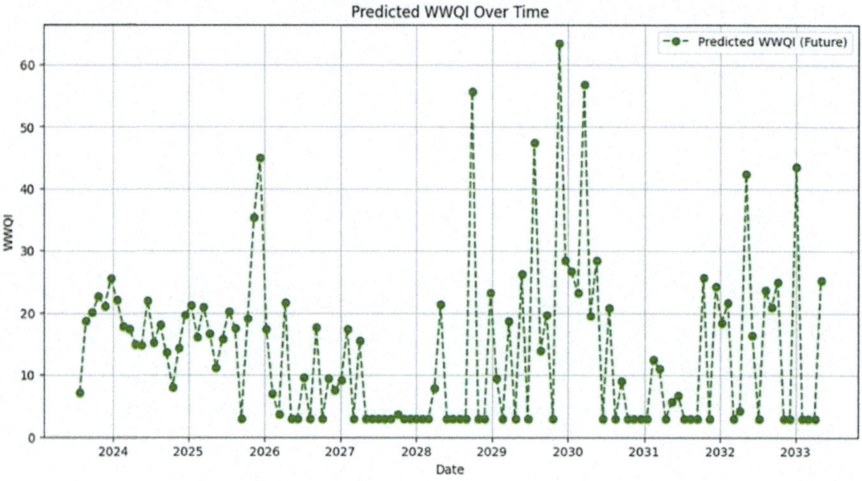

Fig. 17.8 Predicted values of WWQI over time for disposal on irrigational land using RNN model

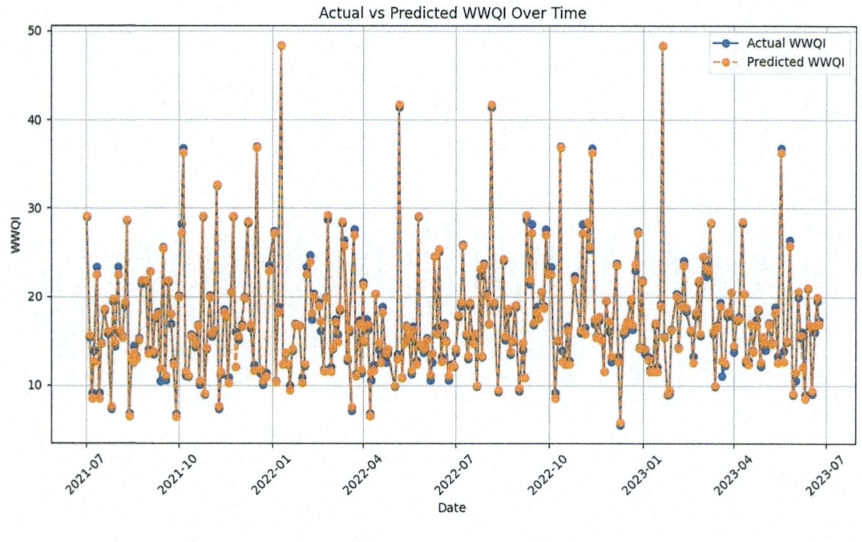

Fig. 17.9 Actual vs predicted WWQI over time for disposal in surface water using LSTM model

The graph showing the actual and predicted values of WWQI over time for disposal in surface water is shown in Fig. 17.9, while the predicted values over time for the next 10 years are shown in Fig. 17.10.

Similarly, the actual vs predicted values were obtained for disposal on irrigational land are shown in Fig. 17.11. The predicted values of WWQI over the next 10 years for disposal on irrigational land are shown in Fig. 17.12.

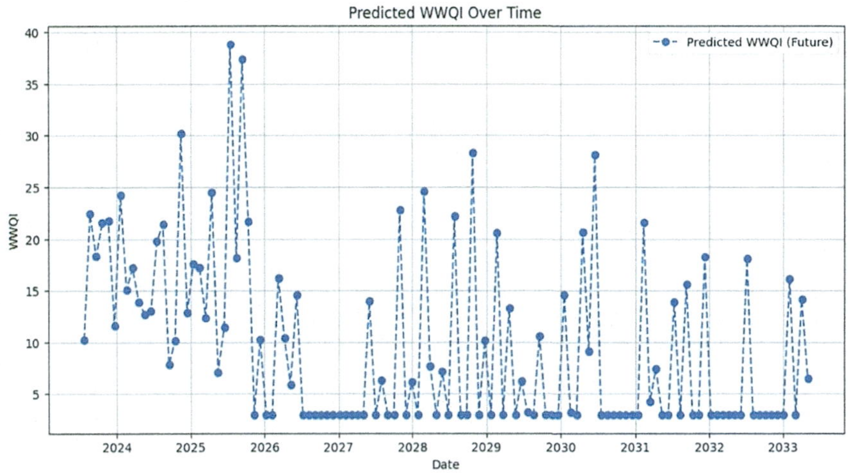

Fig. 17.10 Predicted values of WWQI over time for disposal in surface water using LSTM model

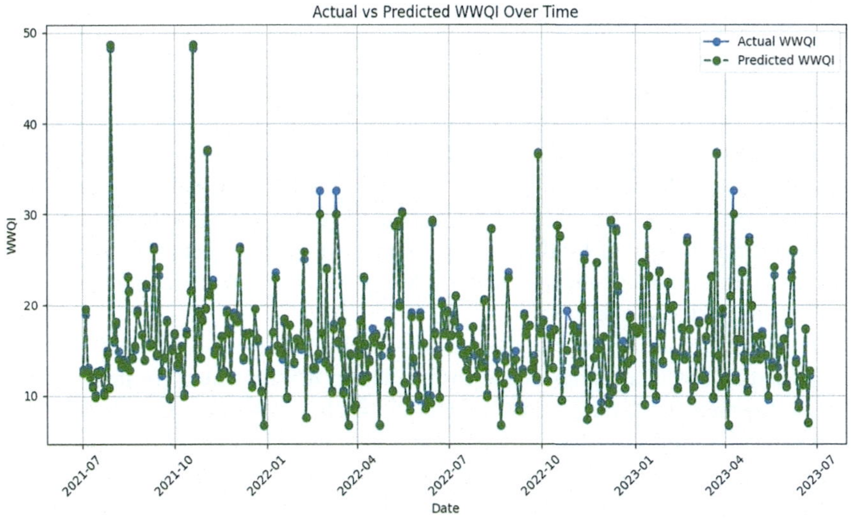

Fig. 17.11 Actual vs predicted WWQI over time for disposal on irrigational land using LSTM model

17.3.3 Performance Analysis

The performance analysis of the data is done using mean absolute error (MAE), root mean square error (RMSE), and R squared. RMSE provides an estimate of the standard deviation of the residuals, indicating how well the model's predictions match the observed data. Lower RMSE values indicate better model performance, with a

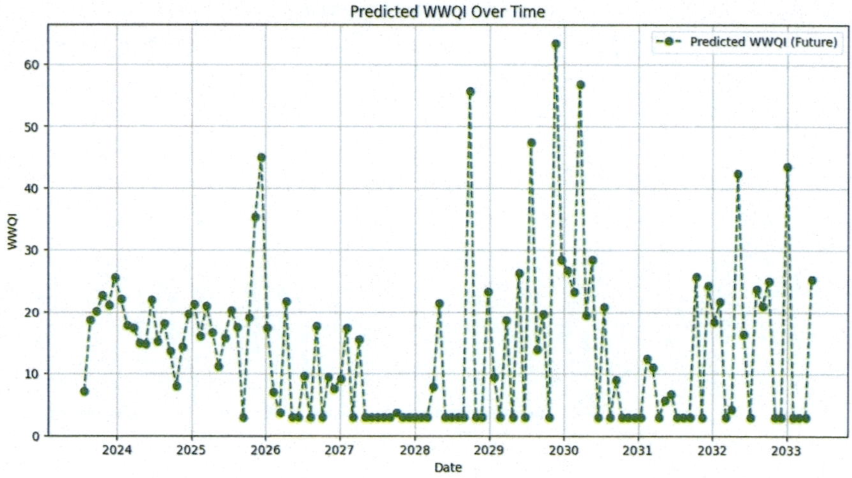

Fig. 17.12 Predicted values of WWQI over time for disposal on irrigational land using LSTM model

value of 0 indicating perfect predictions. Like RMSE, lower MAE values indicate better model performance, with a value of 0 indicating perfect predictions. R-squared measures the proportion of the variance in the dependent variable (target) that is explained by the independent variables (features) in the model. It ranges from 0 to 1, with higher values indicating a better fit of the model to the data. R-squared can be interpreted as the percentage of the response variable variation that is explained by the model. A value of 1 indicates that the model explains all the variability of the response data around its mean, while a value of 0 indicates that the model does not explain any of the variability.

Figure 17.13 shows the comparison of MAE, RMSE, and R-squared values of RNN and LSTM models developed for predicting WWQI values for treated water to be disposed in surface water. Figure 17.14 shows the comparison of MAE, RMSE, and R-squared values of RNN and LSTM models developed for predicting WWQI values for treated water to be disposed on irrigational land. The R-squared values, close to 1 and the lower MAE and RMSE values indicate high accuracy of the two models.

17.4 WWQI Predictor

A web page is developed to predict the values of WWQI when provided with a specific date in a span of 10 years. The interface of the website is displayed in Fig. 17.15. An example of the prediction done by the code is shown in Fig. 17.16. The WWQI

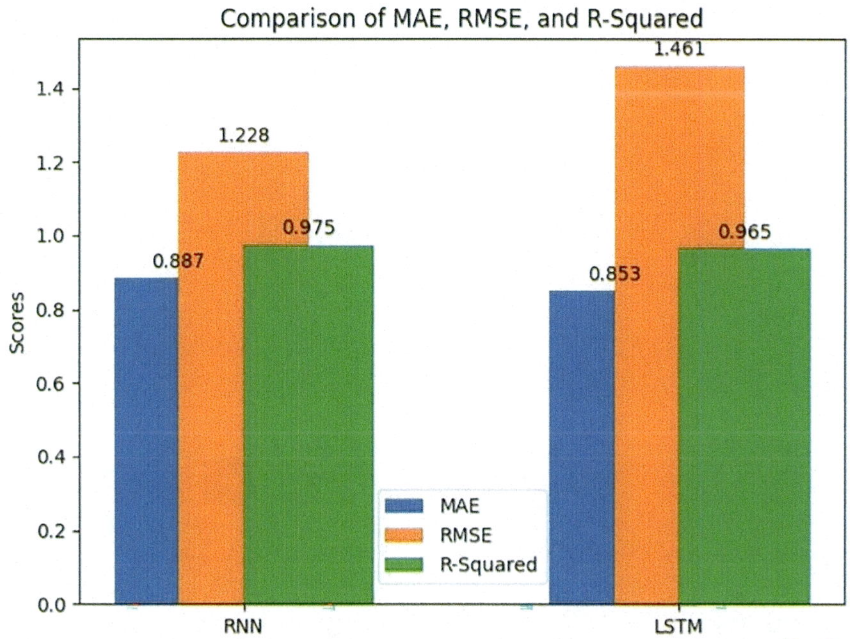

Fig. 17.13 Comparison of MAE, RMSE, and R-squared for disposal in surface water

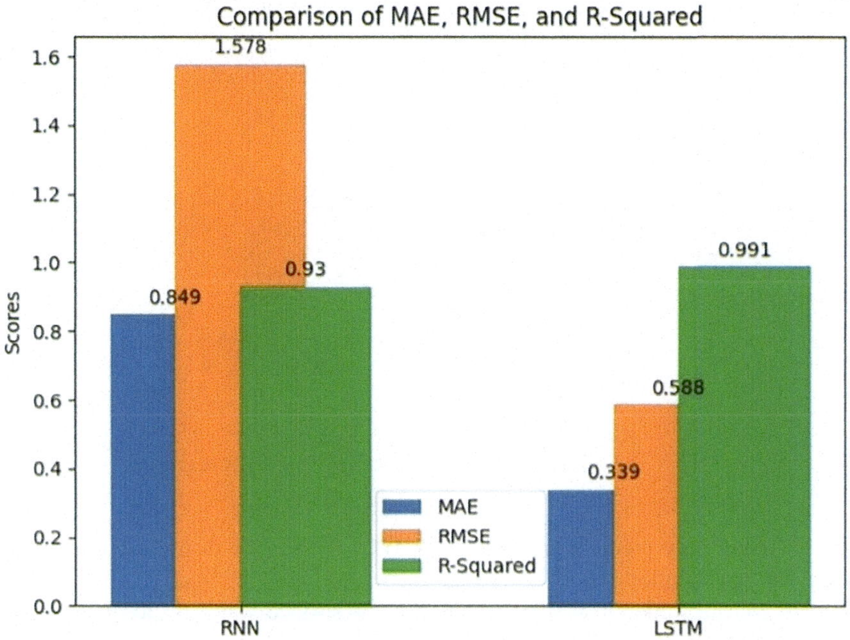

Fig. 17.14 Comparison of MAE, RMSE, and R-squared for disposal on irrigational land

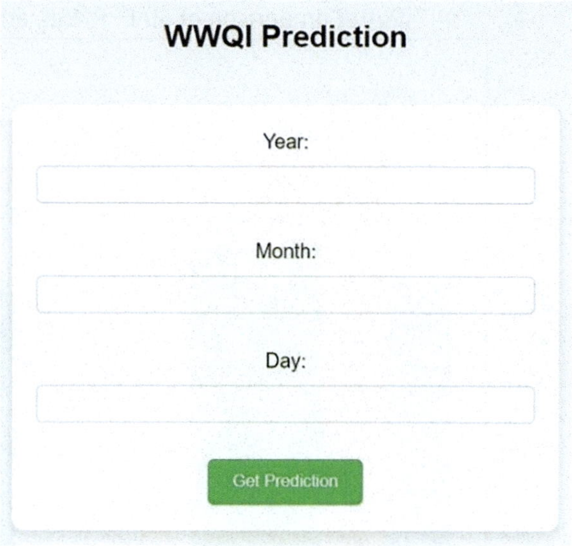

Fig. 17.15 Interface of the WWQI predictor web page

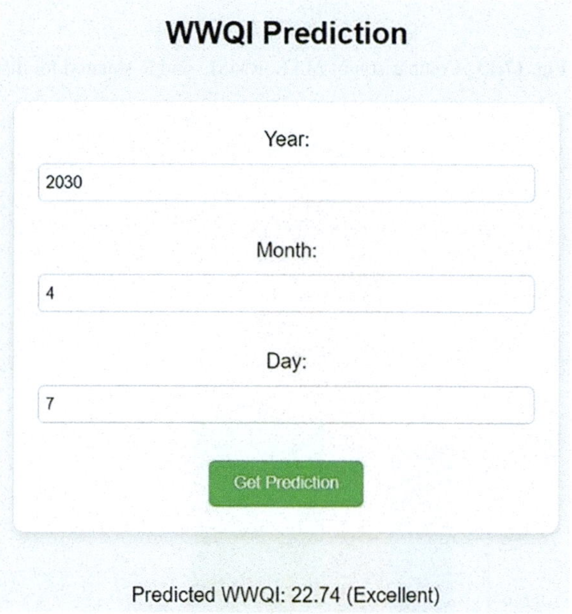

Fig. 17.16 Illustration showing the predicted WWQI value

predictor not only predicts the value of WWQI for surface disposal but also provides its classification status.

The website link for WWQI prediction for surface water disposal is https://sameer1232018.github.io/wwqi_web.

17.5 Conclusion

In conclusion, dealing with the increasing wastewater issues in Goa requires a thorough plan. There is an urgent need for a specific measure, the wastewater quality index (WWQI), tailored for local conditions like discharging treated water into rivers and fields. Using a weighted arithmetic function is a great way to create the WWQI, ensuring accurate results. Also, employing advanced tools like RNN and LSTM, helps predict WWQI values over time.

The models for treated sewage are to be made including measures like mean absolute error (MAE), root mean square error (RMSE), and R-squared in model assessment which provide clear insights into accuracy. By addressing these gaps and taking a comprehensive approach, Goa can better manage wastewater amidst growing urban challenges. The developed predictor can be modified for any city incorporating its own parameters, which can help in achieving the sustainable improvement of the cities.

References

Abbasi, T., & Abbasi, S. A. (2012). *Water quality indices*. Elsevier.

Jamshidzadeh, Z., & Tavangari Barzi, M. (2020). Wastewater quality index (WWQI) as an assessment tool of treated wastewater quality for agriculture: A case of North Wastewater Treatment Plant effluent of Isfahan. *Environmental Science and Pollution Research, 27*, 7366–7378.

Jose, J. M. A., Shinu, P. M. S., Gladis, E. H. E., Arthi, D., & Joseph, J. (2021). Preliminary assessment for physicochemical quality parameters of groundwater in Tirunelveli district. *Materials Today Proceedings, 45*, 1843–1848.

Raut, S. B., Anaokar, G. S., & Dharnaik, A. S. (2017). Determination of wastewater quality index of municipal wastewater treatment plant using fuzzy rule base. *European Journal of Advances in Engineering and Technology, 4*(10), 733–738.

Shah, S. A. A., & Longsheng, C. (2020). New environmental performance index for measuring sector-wise environmental performance: A case study of major economic sectors in Pakistan. *Environmental Science and Pollution Research, 27*, 41787–41802.

Singh, R. P., Nath, S., Prasad, S. C., & Nema, A. K. (2008). Selection of suitable aggregation function for estimation of aggregate pollution index for river Ganges in India. *Journal of Environmental Engineering, 134*(8), 689–701.

Swamee, P. K., & Tyagi, A. (2007). Improved method for aggregation of water quality subindices. *Journal of Environmental Engineering, 133*(2), 220–225.

Vijay, R., Mardikar, T., & Kumar, R. (2016). Impact of sewage discharges on coastal water quality of Mumbai, India: Present and future scenarios. *Environmental Monitoring and Assessment, 188*, 1–13.

Yongo, E., Mutethya, E., Zhang, P., Lek, S., Fu, Q., & Guo, Z. (2023). Comparing the performance of the water quality index and phytoplankton index of biotic integrity in assessing the ecological status of three urban rivers in Haikou City, China. *Ecological Indicators, 157*, 111286.

Chapter 18
Energy-Based Data-Driven Smart Sustainable Cities Using IoT, AI, and Big Data Analytics

M. Nirmala Devi, Venkatesh Baskaran, S. Arun Kumar, and A. Anish Pranav

Abstract The concerns of environmental degradation and climate change have recently been the focus of intense attempts to apply creative solutions from AI, IoT, data science, and big data. Effective rainfall prediction and classification are crucial for various sectors. However, traditional forecasting methods often struggle to accurately capture the complex dynamics of rainfall patterns, particularly in regions with diverse weather conditions. In response to these challenges, we propose a novel "Rainfall Classification for Agricultural Sustainability" (RCAS) approach, which utilizes advanced machine learning (ML) techniques, specifically XGBoost, to classify rainfall patterns based on historical meteorological data. By leveraging feature engineering and model optimization strategies, RCAS identifies and categorizes rainfall events into distinct classes, facilitating more informed decision-making for stakeholders. Predicting wind turbine energy is essential for optimizing renewable energy utilization and ensuring grid stability. Additionally, precise predictions support efficient grid management, allowing utilities to balance supply and demand in real time, ultimately enhancing energy reliability and sustainability. This study bridges the gap by exploring various machine learning (ML) and deep learning (DL) methodologies to enhance wind power forecasts. We emphasize the importance of accuracy in these predictions, aiming to overcome current standards. Our approach leverages these models to predict wind power generation for the next 15 days, utilizing the SCADA Turkey dataset and Tatapower Poolavadi Dataset. R^2 score is used alongside traditional metrics like mean absolute error (MAE) and root

M. Nirmala Devi (✉) · S. Arun Kumar
Department of Computer Science and Engineering, Thiagarajar College of Engineering, Madurai, Tamil Nadu, India

V. Baskaran
Department of Civil Engineering, Alagappa Chettiar Government College of Engineering and Technology, Karaikudi, Tamil Nadu, India

A. Anish Pranav
Department of Civil Engineering, Thiagarajar College of Engineering, Madurai, Tamil Nadu, India

V. Baskaran et al. (eds.), *EcoTech Urbanism*, The Urban Book Series,
https://doi.org/10.1007/978-3-031-83833-0_18

mean square error (RMSE) to evaluate model performance. Leveraging these methodologies aimed to enhance the accuracy of wind power forecasting, enabling more efficient utilization of renewable energy resources.

Keywords Rainfall prediction · XGBoost · Urban areas · Machine learning · Wind turbine energy prediction · Optimizing renewable energy utilization

18.1 Introduction

The evolving paradigm of urbanism, environmentally sustainable smart cities, represents a significant research gap in and of itself." It looks at how their underlying urbanization models have fragmented, merged, and transformed, as well as their convergent AI, IoT, and big data technologies and solutions. The evidence compiled suggests that the growing disapproval of the division between sustainable and smart cities, the broad adoption of the SDGs agenda, and the predominance of advanced ICT have all substantially impacted the development of environmentally sustainable smart cities, shaping their dynamics and landscape. These technologies do, however, come with environmental consequences and ethical and regulatory challenges. The findings can help policymakers create and carry out responsive environmental policies and educate academics and practitioners about the new data-driven technological solutions of smart cities.

In the realm of meteorology and environmental management, accurate prediction and classification of rainfall patterns play a pivotal role in various sectors. In light of the growing importance of leveraging technology to enhance weather forecasting and mitigate potential risks, the concept of "Rainfall Classification for Agricultural Sustainability" (RCAS) emerges as a vital solution to address these challenges. RCAS aims to revolutionize rainfall prediction and classification, particularly by using XGBoost to classify rainfall events based on historical meteorological data. By incorporating feature engineering and model optimization strategies, RCAS endeavors to categorize rainfall patterns into distinct classes, enabling stakeholders to make informed agricultural planning and management decisions.

Traditionally, rainfall forecasting methods have primarily focused on predicting precipitation amounts or timings without considering the nuanced classifications of rainfall events. While these methods have provided valuable insights, they often lack the granularity required to address the diverse needs of stakeholders. Existing solutions fail to account for the specific requirements of agricultural sustainability, water management, and disaster preparedness, leaving stakeholders vulnerable to the impacts of inaccurate or unreliable forecasts. In conclusion, RCAS represents a groundbreaking initiative in the field of meteorology and environmental management, offering a transformative solution to the challenges associated with traditional rainfall forecasting methods. By prioritizing accuracy, reliability, and usability, RCAS aims to empower stakeholders with the insights and tools needed to navigate the complexities of rainfall patterns and make informed decisions that promote sustainability and resilience in the face of changing weather conditions.

With the global population expanding rapidly and fossil fuel reserves dwindling, the imperative for clean, sustainable energy sources has never been more pressing. In this context, wind energy emerges as a viable solution, offering a renewable alternative to traditional energy sources. Harnessing the power of wind has been integral to human civilization for centuries, driving windmills to pump water and grind grain. However, despite its potential, wind energy has historically faced economic challenges, often being deemed costly and unpredictable compared to conventional fuels like petroleum. The former utilizes real-time meteorological data to construct simulation models, while the latter leverages historical data and AI algorithms for prediction.

In this study, we delve into machine learning (ML) and deep learning (DL) methodologies for wind power prediction, aiming to contribute to advancing renewable energy integration and grid management. By exploring various modeling techniques and evaluation metrics, we endeavor to enhance the reliability and efficiency of wind power forecasting, thereby facilitating the transition toward a more sustainable energy future.

18.2 Literature Survey

Ali et al. (2024) and Aboelmagd et al. (2023), illustrated that artificial intelligence (AI) technology, with artificial neural networks (ANNs) being the most popular technique, has shown great promise in environmental and renewable energy applications. A number of solar radiation forecasting techniques have been developed to address the lack of solar measurement in different parts of the world (GSR). In light of this, the goal of this work is to create temperature-based GSR models by utilizing artificial neural networks (ANNs), a method often used in machine learning approaches, to predict GSR from temperature data alone. Additionally, it contrasts these models' performance with the widely accepted empirical method. It also creates accurate GSR models for five other sites as well as the overall region, which does not yet have AI-based models even though there are planned solar energy facilities nearby.

Wang et al. (2024) in their study they look at something that hasn't been done much: how different validation dataset lengths affect the accuracy and prediction of solar radiation models. Additionally, it looks into various ANN architectures for GSR estimation and presents a thorough comparison analysis. The results show that the most sophisticated models using both approaches predict GSR with a high degree of accuracy, with coefficient of determination (R^2) values between 96% and 98%. Furthermore, for non-coastal locales, the empirical model's general and local formulas perform similarly. On the other hand, the general and local ANN-based models function nearly in the same way and have a great capacity to predict GSR in any place, even in the winter. Furthermore, single hidden layer ANN topologies with fewer neurones typically perform better than those with more.

This study also shows that meteorological factors like rain and clouds, particularly at coastal sites, had a major impact on how well the empirical models performed. The ANN-based models, on the other hand, performed about 7% better at coastal sites than the empirical ones and were less affected by such weather changes. Thus, the most evolved models—especially the ANN-based models—are strongly advised. The temperature data may be simply and continually collected for a variety of uses, and they facilitate the accurate and quick forecasting of GSR, which is helpful in the design and performance evaluation of different solar applications.

Alvarez et al. (2024) and Huangfu et al. (2022) discussed that clean energy solutions are necessary to combat climate change, and renewable energy sources like solar and wind are excellent options. However, integrating them into large-scale power systems is difficult because to their variability. In order to overcome this problem, a novel hybrid mathematical model is proposed in this study. The plan takes into account practical limitations including system demand, reserves, and transmission dynamics while integrating both fossil and renewable energy sources. The model integrates multiple methodologies. The model can be applied to large-scale systems due to its reduced computational complexity, which is achieved through the use of an innovative block composition technique. In order to better forecast the output of renewable energy, which is essential for controlling its intermittency, a neural network is also built.

Tharun et al. (2022) compared the accuracy of statistical modeling and regression techniques for rainfall prediction. The study found that regression techniques, such as random forest (RF), outperformed statistical modeling techniques, such as support vector machine (SVM) and decision tree (DT). The RF model performed best, with an accuracy of 90.5%.

Sarker (2023) used machine learning algorithms in his study. The study found that the SVM algorithm performed best, with an accuracy of 85.6%. However, the study did not address the impact of different environmental features on rainfall prediction. They compared the performance of deep learning models with other machine learning algorithms for rainfall prediction.

Another study compared the accuracy of statistical modeling and regression techniques, finding that regression techniques, such as random forest (RF), outperformed statistical modeling techniques, with the RF model achieving an accuracy of 90.5%. Researchers have also explored the use of support vector machine (SVM), support vector regression (SVR), and K-nearest neighbor (KNN) algorithms for daily rainfall prediction, with the SVM algorithm performing best at an accuracy of 85.6%. However, these studies did not address the impact of environmental features on rainfall prediction.

Huangfu et al. (2022) explained that about wind speed forecasting, which may offer thorough future knowledge on wind speed uncertainty, can help ease these problems. Interval forecasting is a more effective way to measure uncertainty in the production of renewable energy than typical point forecasting techniques. The concept of prediction intervals (PIs) is used in this work to represent wind speed uncertainties. In particular, a new interval forecast model utilizing long short-term memory (LSTM) neural networks is suggested to build PIs using the potent non-parametric forecasting technique known as the lower and upper bound estimation (LUBE) method.

Aboelmagd et al. (2023) and Iqbal et al. (2023) demonstrated that one of the most important integrated and impactful events for managing the system of new and renewable energy sources in the power grid is accurate forecasting in photovoltaic energy. The study offers a useful method for forecasting photovoltaic energy generation using neural networks. It is difficult to use neural networks for forecasting PV power generation because of the unpredictability of variations in temperature, humidity, wind speed, and solar radiation. High prediction performance is achieved by neural network methods, which set them apart from traditional mathematical models that struggle to handle intricate nonlinear interactions and produce poor predictions. The suggested forecasting model based on neural networks provides a useful method for raising the accuracy of PV generation estimation. Because it can recognize intricate relationships in the data and adjust to changing conditions, it holds enormous potential to improve the dependability and integration of solar energy in the existing electrical environment.

In addition to machine learning techniques, researchers have applied data mining, big data analysis, and hydrological models for rainfall prediction, with promising results. However, there is a need for further research to improve the accuracy of rainfall prediction across these various approaches. LSTM can capture complex temporal dependencies inherent in wind speed data. LSTM can remember past information for an extended period, which allows it to capture these complex temporal dependencies and make accurate predictions for the short term. LSTM can also handle nonlinearity and complex patterns in data very well. Recent studies delve into utilizing LSTMs for wind prediction, frequently implementing hybrid models with other techniques to enhance accuracy. Concurrently, researchers strive to refine the optimization process and enhance data preprocessing, both contributing to improved wind prediction accuracy.

Iqbal et al. (2023) discussed that wind energy has become more and more popular as a practical renewable energy source because of its many advantages, such as its low cost of production and real estate friendliness. However, because wind speeds are unpredictable, it experiences volatility in the production of electricity. Therefore, in order to meet the energy demands, it is essential to have a good estimate of wind energy forecasting.

Chen et al. (2024) demonstrated that wind energy provides the benefit of clean and pollution-free power generation as a renewable energy source. Because of its plentiful resources, wind power is currently the most popular and rapidly expanding form of electricity.

Song et al. (2023) explained that this sequential hybrid architecture utilizes LSTM for initial predictions, followed by SFS-PSO optimization to refine the results and elevate evaluation criteria. The algorithm specifically incorporates SFS to enhance the exploitation phase, leading to improved performance.

Joseph et al. (2023) demonstrated LSTM and TCN model accuracy by incorporating the CEEMDAN algorithm. CEEMDAN (complete ensemble empirical mode decomposition with adaptive noise) This technique decomposes the original wind energy data into multiple intrinsic mode functions (IMFs) representing different timescales and frequency components. The combination of TCN and LSTM allows the model to capture both short-term and long-term trends in the wind energy data, potentially leading to more accurate predictions. In this model, LSTM predicts the short-term components (IMF1 and IMF2) extracted by CEEMDAN and TCN predicts the long-term components (IMF3 to IMFn) extracted by CEEMDAN. CEEMDAN-TCN and CEEMDAN-LSTM have close accuracy of 91.2% and 92.5%, respectively, while the accuracy of CEEMDAN-LSTM-TCN is 99.8%.

Prathibha et al. (2023) proposed a dilated convolution, which serves as an effective tool for extracting features from lengthy sequential data. In the model's encoder component, it handles historical wind power and NWP data. Leveraging multihead attention and a gating mechanism, the encoder adeptly captures crucial features. As a result, the encoder produces an attention weight map, which provides insights into the distribution of information within the data. The decoder predicts future wind power values by taking the output of the encoder and short historical wind power data as inputs and uses cross-attention to leverage information from the encoder.

Rahman et al. (2022) set a new benchmark for wind speed forecasting by achieving significantly lower prediction errors compared to previous methods. At 3, 6, 12, and 24 h, the model's average error (MAE) is 0.243, 0.290, 0.362, and 0.453, respectively, representing a substantial improvement over the state of the art. This EEMD-transformer model outperforms the EEMD-gated recurrent unit (GRU) model in mean absolute error (MAE), root mean square error (RMSE), and R; MAE decreases by 3.5%, RMSE decreases by 4.7%, and R improves by 0.0018, while mean absolute percentage error (MAPE) increases by 0.91.

18.3 Proposed Methodology

Rainfall prediction is a critical task with significant implications for agriculture, water, and disaster management. Machine learning techniques have emerged as a promising approach for rainfall prediction, and several studies have explored the application of these techniques for daily, monthly, and annual rainfall prediction. The proposed methodology of "Rainfall Classification for Agricultural Sustainability" (RCAS) represents a groundbreaking approach to rainfall prediction and classification, leveraging advanced machine learning techniques to address the challenges associated with traditional forecasting methods. RCAS undergoes rigorous preprocessing, including data cleaning, feature engineering, and labeling, to extract relevant features and classify rainfall events into distinct categories. Utilizing the powerful XGBoost algorithm, RCAS develops a robust classification model trained on the preprocessed data, optimizing model parameters and evaluating performance through cross-validation techniques. The advantages of RCAS lie in its accuracy, reliability, customization, and adaptability, offering socioeconomic benefits such as enhanced agricultural productivity, optimized water resource allocation, and better preparedness for extreme weather events. In conclusion, RCAS represents a data-driven and innovative solution to the challenges of rainfall classification, empowering stakeholders with the insights and tools needed to make informed decisions and promote sustainability in agricultural and environmental management practices.

18.3.1 Rainfall Prediction Dataset

The dataset comprises 2906 entries and consists of 33 columns, each representing specific attributes related to meteorological conditions. Key columns include "name" and "datetime" to identify the observation and time stamp. Meteorological variables such as "tempmax," "tempmin," and "temp" denote maximum, minimum, and average temperatures, while "feelslikemax," "feelslikemin," and "feelslike" represent perceived temperatures. Other columns include "dew" for dew point, "humidity" for humidity levels, "precip" for precipitation amount, "windspeed" and "windgust" for wind speeds, "cloudcover" for cloud cover percentage, and "visibility" for visibility range. Additionally, the dataset contains columns for snow-related metrics like "snow" and "snowdepth," as well as atmospheric pressure ("sealevelpressure"), solar radiation ("solarradiation"), and UV index ("uvindex"). Furthermore, there are columns capturing sunrise and sunset times ("sunrise," "sunset"), moon phase ("moonphase"), and descriptions of weather conditions ("conditions," "description," "icon"). The dataset also includes several categorical columns such as "preciptype" indicating the type of precipitation, "severerisk" for severe weather risk, and "stations" for meteorological station identifiers. These diverse attributes offer comprehensive insight into meteorological phenomena, enabling sophisticated analysis and modeling for various environmental monitoring, agriculture, and disaster management applications.

18.3.2 Feature Extraction

In Table 18.1, the description of the rainfall prediction features in the dataset are given, After feature extraction only seven features which are influential features for decision making are selected from the original dataset

Table 18.1 Description of the features in the dataset

Sl.No	Feature name	Description	Data type
1.	datetime	Datetime	Nominal
2.	tempmax	Maximum temperature	Numeric
3.	tempmin	Minimum temperature	Numeric
4.	temp	Temperature	Numeric
5.	feelslikemax	Maximum feels like	Numeric
6.	feelslikemin	Minimum feels like	Numeric
7.	feelslike	Feels like	Numeric
8.	dew	Dew	Numeric
9.	humidity	Humidity	Numeric
10.	precip	Precipitation	Numeric
11.	precipprob	Precipitation prob.	Numeric
12.	precipcover	Precipitation cover	Numeric
13.	preciptype	Precipitation type	Nominal
14.	snow	Snow	Numeric
15.	snowdepth	Snow depth	Numeric
16.	windgust	Wind gust	Numeric
17.	windspeed	Wind speed	Numeric
18.	winddir	Wind direction	Numeric
19.	sealevelpressure	Sea level pressure	Numeric
20.	cloudcover	Cloud cover	Numeric
21.	visibility	Visibility	Numeric
22.	solarradiation	Solar radiation	Numeric
23.	solarenergy	Solar energy	Numeric
24.	uvindex	UV index	Numeric
25.	severerisk	Severe risk	Numeric
26.	sunrise	Sunrise	Nominal
27.	sunset	Sunset	Nominal
28.	moonphase	Moon phase	Numeric
29.	conditions	Conditions	Nominal
30.	sescription	Description	Nominal
31.	icon	Icon	Nominal
32..	stations	Stations	Nominal

18.3.3 Visualization of Data (Fig. 18.1)

18.3.4 Data Preprocessing

Data preprocessing includes dropping unnecessary columns, rearranging and renaming columns for clarity, handling missing values through mean and mode imputation, detecting and treating outliers using the interquartile range (IQR) method, and visualizing data distribution for insights. Categorical variables are label-encoded, and the dataset is split into features and target variables. Class imbalance is addressed using NearMiss undersampling and ADASYN oversampling

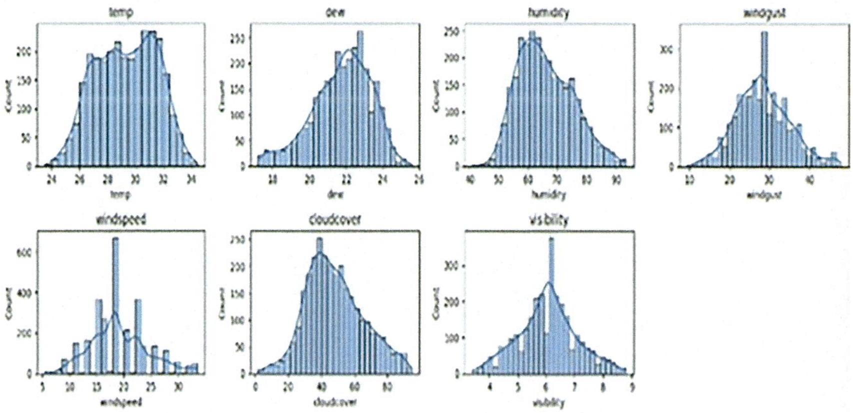

Fig. 18.1 Distribution of numerical data

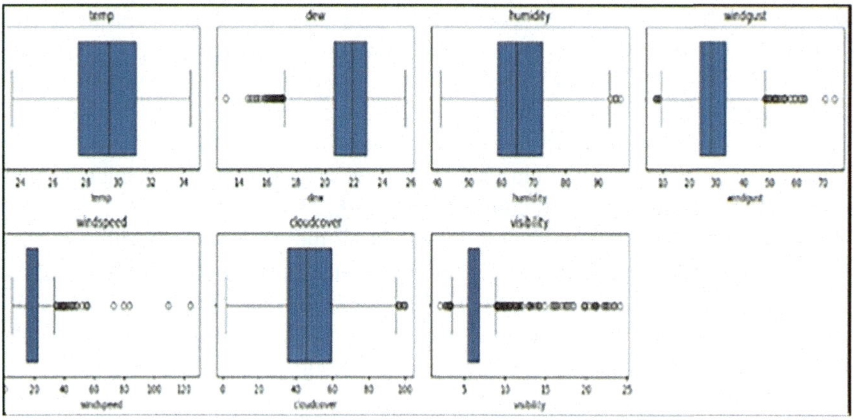

Fig. 18.2 Data before preprocessing

techniques. Subsequently, various machine learning algorithms are trained and evaluated on the preprocessed data, and the best-performing XGBoost model is saved for future use. These preprocessing steps ensure the dataset is appropriately prepared, enhancing the accuracy and reliability of the rainfall prediction model for Madurai (Figs. 18.2 and 18.3).

18.3.5 Machine Learning Techniques XG Boost

The XGBoost code uses the provided algorithm to develop a rainfall prediction model tailored for Madurai. By employing a gradient boosting framework, XGBoost sequentially constructs an ensemble of decision trees to minimize a predefined loss

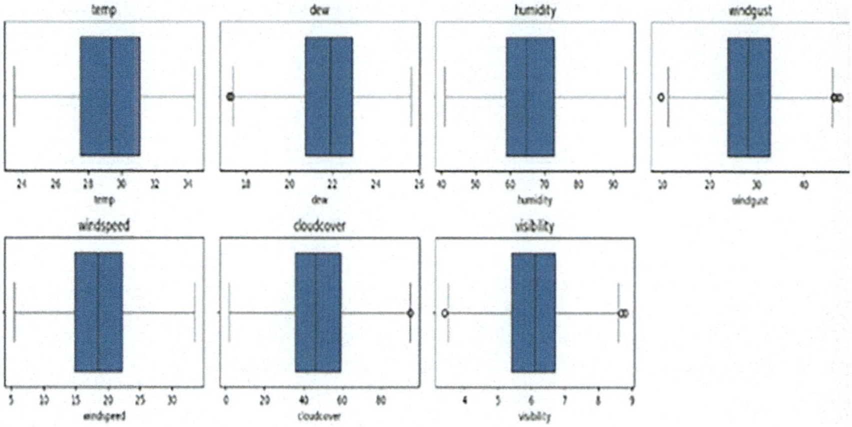

Fig. 18.3 Data after preprocessing

function during training. The dataset undergoes preprocessing and splitting into training and testing subsets, followed by model training using hyperparameter tuning via GridSearchCV and cross-validation for robust evaluation. The trained XGBoost model is saved for future use, and during inference, input data is collected, preprocessed, and utilized for prediction. The algorithm's efficiency, scalability, and ability to capture intricate patterns ensure accurate and reliable rainfall predictions, making it an ideal choice for this task.

In the rainfall prediction project for Madurai, the XGBoost algorithm plays a pivotal role in accurately forecasting rainfall events. Initially, the dataset, comprising various meteorological features, is preprocessed to handle missing values and outliers. Subsequently, the dataset is split into training and testing sets for model development and evaluation. Leveraging the powerful gradient boosting technique, XGBoost constructs an ensemble of decision trees iteratively, optimizing a predefined loss function at each step. The model undergoes hyperparameter tuning using GridSearchCV and cross-validation to enhance performance and generalization. Once trained, the XGBoost model is capable of effectively capturing complex relationships between meteorological variables and rainfall occurrence. During inference, the model efficiently processes input data to predict rainfall events, providing valuable insights for planning and decision-making. Overall, the XGBoost algorithm serves as a robust and reliable tool for accurate rainfall prediction in the Madurai region, contributing to improved disaster preparedness and resource management (Fig. 18.4).

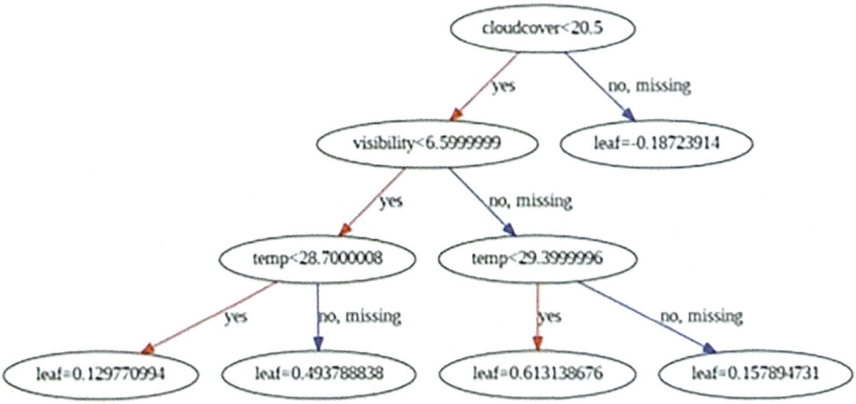

Fig. 18.4 XGBoost tree

18.3.6 Wind Power Prediction

One important factor to consider when researching wind energy resources is wind speed. The main focus of this research is on wind speed prediction models for wind energy production. Wind power generation is hampered by the extreme unpredictability, uncontrollability, and intermittency of wind speeds, which raises development costs and puts power systems' stability at risk. Thus, it is essential to predict wind speed variations scientifically in order to guarantee grid integration of wind power, equipment safety, and steady and safe power system operation.

Owing to the complexity of wind speed, sustaining wind power integration systems, guaranteeing the safe and stable functioning of power systems, and assuring the safety of wind power equipment all depend on the scientific prediction of wind speed fluctuations. This method improves wind speed data processing efficiency and more effectively handles strong stochastic and nonlinear features. In addition, weights for each component are calculated using mathematical analytical techniques, and a dynamic neural network model is built to optimise wind speed time series modelling in order to achieve a more precise prediction of wind speed fluctuations. Lastly, wind speed forecasting in the Xinjiang region is conducted using the optimised HHT-NAR model, which shows notable gains in coefficient of determination and reduction of root mean square errors.

This model is a useful prediction tool for the wind energy generation industry, demonstrating both theoretical novelty and greater performance in real-world applications.It is a critical area of research to enhance the efficiency and reliability of renewable energy sources. This study proposes a novel methodology integrating Long Short-Term Memory (LSTM) networks and Transformer architectures for wind power prediction, leveraging the TataPower dataset. Our approach amalgamates the strengths of LSTM networks in capturing temporal dependencies with the transformative capabilities of Transformers, surpassing conventional methods such as Convolutional Neural Networks (CNN) and standalone LSTM models.

This research methodology is implemented by collecting the TataPower dataset, encompassing multivariate time-series data related to wind speed, wind direction, and power generation, among other parameters. Preliminary preprocessing steps involved handling missing values and ensuring uniformity in data formats. Date and time information was meticulously extracted and formatted to facilitate temporal analysis. Furthermore, numerical features were scaled to normalize the data distribution, easing subsequent model training.

To gain insights into the dataset's characteristics, we conducted an extensive exploratory data analysis (EDA) phase. Descriptive statistics and visualization techniques, including pair plots and correlation heatmaps, were employed to uncover underlying patterns and relationships between variables. Additionally, a wind rose plot was constructed to visualize the relationship between wind direction and wind speed, providing valuable insights into directional wind patterns.

Feature engineering played a pivotal role in enhancing the predictive capabilities of our model. Key steps involved in feature engineering included: Date and time information from the timestamp column were extracted and engineered to capture temporal patterns inherent in the data. Features such as year, month, day, hour, and minute were derived to provide the model with additional temporal context. Domain-specific knowledge was leveraged to engineer additional features that may impact wind power generation. Features related to weather conditions, geographical location, and turbine specifications were incorporated to enrich the dataset and improve predictive performance. Features that did not significantly influence the target variable (power generation) were identified and pruned from the dataset to reduce noise and improve model efficiency.

This process involved careful analysis and domain expertise to retain only the most relevant features for prediction. Numerical features were normalized and scaled to ensure uniformity in data distribution and prevent features with larger magnitudes from dominating the model training process. This step was essential for stabilizing the training process and improving convergence during model optimization. Techniques such as principal component analysis (PCA) or feature importance ranking were employed to select the most informative features and reduce the dimensionality of the dataset. We aimed to streamline model training and enhance predictive accuracy by focusing on the most relevant features.

A scatter matrix was plotted to visualize pairwise relationships between the numerical features like wind speed, theoretical power curve, wind direction, and LV ActivePower (kW). The scatter plot between wind speed and theoretical power curve shows that wind turbine systems cannot generate any power if the wind speed is less than 4 m/s. When the wind speed is larger than 4 m/s to 11 m/s, the relationship between them is linear, meaning that as the wind speed increases, turbines generate more power. However, after the wind speed passes 11 m/s, the power generated saturates at 3600 KWh.

The correlation coefficient between wind speed and the theoretical power curve is close to 1, indicating a very strong positive correlation. This means that the theoretical power curve prediction increases as wind speed increases. The correlation between LV ActivePower and Theoretical_Power_Curve is also positive and relatively strong, but not as strong as the correlation between wind speed and theoretical power curve. This suggests that the theoretical power curve provides a decent estimate of actual power output, but other factors besides wind speed also influence the LV ActivePower generation. New attributes like week, month, season, day, and hour are created from the existing "Date/Time" column to extract more meaningful features. Finally, categorical features such as "Seasons" are encoded using a dictionary so the machine learning models can better understand them. Using bar charts, we plot numerical features (power, wind speed, theoretical power curve, and LV ActivePower) against categorical features (week, month, season). This helps identify trends in power generation based on these factors.

18.3.7 Preprocessing Techniques for Wind Turbine Power Prediction

Machine learning algorithms generally perform best when the data they are trained on is clean and free of errors or inconsistencies. The preprocessing step plays a vital role in ensuring the quality of the data used for wind turbine power prediction. This step involves a series of techniques to prepare the raw data for model training.

18.3.8 Outlier Detection and Treatment

One important aspect of preprocessing is outlier detection and treatment. Outliers are data points that fall significantly outside the typical range of values for a particular feature. Their presence can negatively impact the performance of machine learning models. Outlier detection is implemented using the interquartile range (IQR) method. This method calculates the IQR, which represents the spread of data between the first quartile (Q1) and third quartile (Q3). Outlier thresholds are then set based on a factor (e.g., 1.5 times) of the IQR from the quartiles. Data points falling outside these thresholds are considered outliers.

The identified outliers are then replaced with the IQR thresholds, effectively capping their values within the expected range. This approach helps mitigate the influence of extreme values on the model's learning process.

18.3.9 Splitting Target and Feature Variables

Another crucial step in preprocessing is separating the target variable from the feature variables. The target variable is the "LV ActivePower (kW)," which represents the actual power output of the wind turbine. Feature variables encompass all other data points used to predict the target variable, such as wind speed, theoretical power curve, and wind direction. The code achieves this separation by creating distinct datasets for training and testing purposes. "X_train" and "X_test" represent the feature matrices, excluding the target variable, for training and testing, respectively. Similarly, "y_train" and "y_test" contain the target variable values for training and testing. Separating the target variable allows the model to learn the relationship between the features and the power output without being influenced by the power output values themselves during training.

By incorporating outlier treatment and target-feature variable separation, the preprocessing step ensures the data used for wind turbine power prediction is of high quality and suitable for machine learning model training. These techniques can significantly improve the accuracy and generalizability of the resulting models. Features are seperated into numerical (wind speed, etc.) and categorical (season) types. Numerical features are standardized (scaled) for better model training, while categorical features are converted to numerical labels using ordinal encoding. Numerical features are scaled using a StandardScaler. This technique standardizes the features by subtracting the mean and dividing by the standard deviation. Standardization helps ensure that all features are on a similar scale, preventing features with larger ranges from dominating the model's learning process. Categorical features are encoded using an OrdinalEncoder. This encoder assigns numerical values to the categories in a sequential order (e.g., Spring = 1 and Summer = 2). Encoding allows the model to handle categorical data effectively.

A ColumnTransformer efficiently applies these techniques to each feature group. The code transforms both training and testing data using the same methods to ensure consistency during model evaluation. This preprocessing step prepares the data for machine learning by addressing inconsistencies and ensuring compatibility with the model.

18.3.10 LSTM Model Development

Central to our methodology was the development of a robust LSTM-based predictive model tailored specifically for wind power prediction. Key steps involved in LSTM model development included: The dataset was transformed into a supervised learning problem, where historical observations served as input features, and the target variable (power generation) served as the output. This transformation enabled the LSTM model to learn from past observations and predict historical trends. The architecture of the LSTM model was meticulously designed, taking into account parameters such as the number of LSTM layers, the number of neurons in each layer, and the inclusion of additional components such as dropout layers to prevent overfitting. These architectural choices were optimized through iterative experimentation to strike a balance between model complexity and predictive performance. Hyperparameters such as batch size, learning rate, and number of epochs were tuned to optimize model performance. Through systematic experimentation and validation on a separate validation dataset, optimal hyperparameter values were determined to ensure robust model training and generalization.

$$ft = \text{Sigmoid}\left(\omega f\left[ht - 1mt\right] + bf\right) \tag{18.1}$$

$$it = \text{Sigmoid}\left(\omega i\left[ht - 1mt\right] + i\right) \tag{18.2}$$

$$C't = \text{Sigmoid}\left(\omega O\left[ht - 1mt\right] + bC\right) \tag{18.3}$$

$$Ct = ft \times Ct - 1 + it \times C't \tag{18.4}$$

$$Ot = \text{Sigmoid}\left(\omega O\left[ht - 1mt\right] + bO\right) \tag{18.5}$$

$$ht - 1 = Ot \times \tan h\left(Ct\right) \tag{18.6}$$

where at time t, mt represents the current input vector, while ht-1 represents the output vector of the hidden layer from the previous time step. The forgetting gate, input gate, and output gate are denoted as ft., it, and Ot, respectively. ωf, ωi, and ωO represent the weights associated with the forgetting gate, input gate, and output gate, respectively. bf, bi, and bO represent the biases of the forgetting gate, input gate, and output gate, respectively. C't represents the candidate state, and Ct represents the newly calculated state.

The LSTM model was trained on the preprocessed and scaled dataset using an appropriate optimization algorithm such as Adam or RMSprop. During training, the model iteratively adjusted its internal parameters to minimize a defined loss function, typically mean squared error (MSE), thereby improving its predictive accuracy. Once trained, the LSTM model's performance was evaluated using established metrics such as mean absolute percent error (MAPE), root mean square error (RMSE), and coefficient of determination (R-squared). These metrics provided insights into the model's accuracy, precision, and ability to capture variations in wind power generation.

Before applying deep learning techniques, the dataset underwent preprocessing to ensure its suitability for modeling. This involved handling missing values using forward and backward filling techniques to maintain temporal continuity in the dataset. Additionally, feature scaling was performed to normalize the data, ensuring that all features were on a similar scale, which aids in convergence during model training. Moreover, the date column underwent transformation into sine and cosine components to capture cyclic temporal patterns, such as daily and yearly seasonality, in the data.

$$\text{Scaled}_{\text{Feature}} = \frac{\left(\text{Original}_{\text{Feature}} - \min_{\text{Value}}\right)}{\left(\max\_\text{Value} - \min\_\text{Value}\right)} \tag{18.7}$$

$$\text{New}_{\text{Feature}_{\text{Sine}}} = \sin\left(2 \times \text{PI} \times \text{Frequency} \times \text{Date}\right) \text{New}_{\text{Feature}_{\text{Cosine}}}$$

$$= \text{Cos}\left(2 \times \text{PI} \times \text{Frequency} \times \text{Date}\right) \tag{18.8}$$

In preparation for model training, the dataset was windowed to create input-output pairs. Each window consisted of a sequence of consecutive data points, where the input window contained historical observations, and the output window contained the target values to be predicted. This windowing approach allowed the models to learn from past observations and make predictions for future time steps, facilitating temporal learning in the deep learning models.

Several deep learning architectures were explored for wind power forecasting, each designed to capture different temporal and spatial patterns in the data. Baseline models served as reference points for comparison, including simple repeat and single-step baseline models. The linear model consisted of a single dense layer with linear activation, suitable for making single-shot predictions. A more complex dense model comprised multiple dense layers with rectified linear unit (ReLU) activation functions, enabling the model to capture nonlinear relationships in the data. Additionally, a convolutional neural network (CNN) model was employed to extract spatial patterns from the time-series data. Furthermore, a recurrent neural network (RNN) model, specifically a long short-term memory (LSTM) network, was utilized to capture temporal dependencies and long-term patterns in the data, leveraging its ability to retain information over extended sequences.

$$\text{ReLU}(x) = \max(0, x) \tag{18.9}$$

$$\text{Output}[i, j] = \text{SUM}\left(\text{Filter}[m, n] \times \text{Input}[i + m, j + n]\right) + \text{Bias} \tag{18.10}$$

Model training was conducted using the Adam optimizer with a learning rate of 0.01, along with early stopping based on validation loss to prevent overfitting. The models were trained for a maximum of 20 epochs to ensure convergence and prevent excessive training time. During training, the performance of each model was evaluated using traditional evaluation metrics such as mean absolute error (MAE) and root mean squared error (RMSE) on both the validation and test datasets. Additionally, the coefficient of determination (R-squared) was computed to assess the goodness of fit of the models, providing insights into their predictive accuracy and explanatory power.

$$\text{MAE} = \frac{1}{N} \times \text{SUM}\left(\text{Predicted\_Value} - \text{Actual\_Value}\right)^2 \tag{18.11}$$

18.4 Results and Discussion

The existing hybrid CNN-LSTM model beats the accuracy of traditional single models like RNN, LSTM, and CNN models. MAE, RMSE, CC, and R^2 of CNN-LSTM are 0.4783, 0.6480, 0.9528, and 0.9070, respectively. While LSTMs excel at capturing temporal dependencies, they may struggle with long-term predictions (days or weeks) due to the vanishing gradient problem.

Various machine learning algorithms, including logistic regression, random forest, and XGBoost, were trained and evaluated using meteorological data. The XGBoost algorithm outperformed others, achieving the highest accuracy and F1-score. The precision-recall and ROC curves further confirmed the model's robust performance, showcasing its ability to balance precision and recall effectively. Additionally, feature importance analysis revealed that factors such as temperature, humidity, and wind speed significantly influence rainfall prediction. These findings underscore the importance of leveraging advanced machine learning techniques for accurate and reliable rainfall forecasting, which can have profound implications for disaster preparedness, agriculture, and water resource management in the Madurai region (Table 18.2).

Firstly, we acquire the wind power dataset, which includes various features such as wind speed, temperature, humidity, and time of day. Prior to model training, preprocessing is conducted on the dataset. This involves handling missing values by either imputation or removal of incomplete samples. Additionally, numerical features are normalized to ensure uniform scaling across the dataset. Feature selection techniques are applied to identify the most relevant features for wind power prediction, optimizing the model's performance.

Ensemble regression models are chosen for their ability to capture complex relationships within the dataset. The ensemble models selected for this study include Gradient Boosting Regressor, Support Vector Regressor (SVR), Random Forest Regressor, Linear Regression, Extra Trees Regressor, AdaBoost Regressor, Decision Tree Regressor, XGBoost Regressor, and XGBoost with Random Forest (XGBRF) Regressor. These models offer diverse approaches to capturing the underlying patterns in the data and mitigating overfitting.

In selecting ensemble regression models for wind power prediction, each model was chosen for its unique strengths in capturing different aspects of the dataset. Gradient Boosting Regressor offers sequential model fitting, reducing bias and variance in predictions. Support Vector Regressor (SVR) effectively handles high-dimensional data and captures complex nonlinear relationships. Random Forest Regressor is robust to overfitting and computationally efficient, suitable for large datasets. Linear Regression provides simplicity and interpretability, serving as a baseline for comparison.

Extra Trees Regressor utilizes ensemble learning for improved robustness and generalization. AdaBoost Regressor focuses on improving weak learners' performance through sequential training. Decision Tree Regressor captures nonlinear relationships in data through hierarchical partitioning. XGBoost Regressor offers

Table 18.2 Performance comparison between ML models

Machine learning model	Accuracy
Logistic regression	0.77
Random forest	0.7524071526822559
XG boost	**0.94**

scalability, speed, and regularization techniques for handling diverse datasets. XGBoost with Random Forest (XGBRF) Regressor combines strengths for enhanced performance and robustness. Each model contributes to a comprehensive approach to wind power prediction, leveraging their respective strengths to capture the dataset's complexity effectively.

Model training and evaluation are conducted iteratively for each ensemble regression model. The training data (`X_1`, `y_train`) are used to fit the model, while the testing data (`X_1_test`, `y_test`) are employed to assess model performance. Evaluation metrics such as the coefficient of determination (R^2 score) and Root Mean Squared Error (RMSE) are calculated to gauge the model's goodness of fit and prediction accuracy.

Hyperparameter tuning is essential for optimizing the performance of each regression model. We define a hyperparameter search space for each model, including parameters such as `n_estimators`, `max_depth`, `learning_rate`, `min_child_weight`, and `base_score`. Randomized search cross-validation is employed to efficiently search over the parameter grid and find the optimal hyperparameters that maximize the R^2 score.

After hyperparameter tuning, the best-performing model is selected based on the highest R^2 score obtained. The selected model's performance is further evaluated using k-fold cross-validation to assess its generalization capability. Visualization techniques are utilized to compare the cross-validation scores of different models, facilitating the identification of the most suitable model for wind power prediction.

In the results analysis phase, the performance of each ensemble regression model is comprehensively assessed. R^2 scores and RMSE values are presented and compared to determine the model's predictive accuracy. The implications of the findings are discussed in the context of renewable energy management, highlighting the potential applications of accurate wind power forecasting in optimizing energy generation and distribution systems.

To assess the performance of each regression model, randomized search cross-validation is used for hyperparameter tuning. This technique involves randomly sampling a subset of hyperparameters from predefined ranges and performing cross-validation to identify the optimal combination. Each regression model is trained on the training dataset using the optimized hyperparameters obtained from the randomized search cross-validation. The trained models are then evaluated on the testing dataset using two evaluation metrics. The evaluation metrics used are R^2 score and RMSE, which provide insights into each model's explanatory power and prediction accuracy.

Fine-tuning: A set of hyperparameters is defined to search over, encompassing various configurations that can influence the performance of the XGBoost regressor. The parameters include n_estimators, max_depth, learning_rate, min_child_weight, and base_score, which are crucial for controlling the complexity and generalization of the model. After the completion of the randomized search, the best parameters and corresponding R^2 score are inspected and printed.

The best parameters obtained from the search are used to instantiate a new XGBoost regressor model. This obtained new model is trained on the entire training

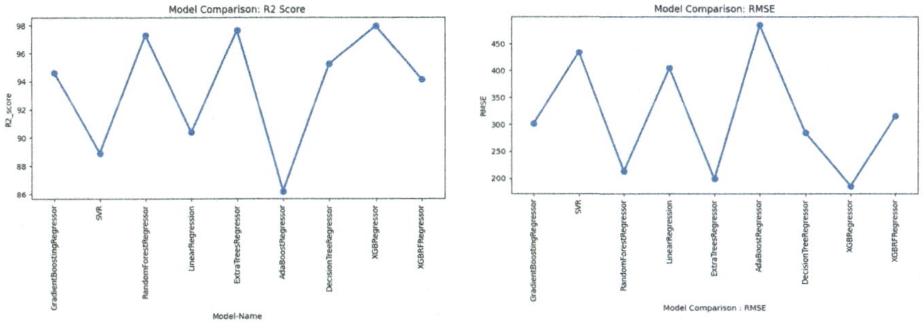

Fig. 18.5 Performance comparision of various models

dataset to ensure it captures the optimal configuration. The trained model is then utilized to make predictions on the test dataset, and the predicted values are obtained. The R^2 score is computed using the predicted and actual test labels, providing a measure of the model's predictive performance. It is shown in Fig. 18.5.

Cross-validation scores are computed for the best model using the cross_val_score function with 20 folds. k-fold cross-validation with a specified number of folds is employed during hyperparameter tuning to ensure robustness and generalization of results. Additional considerations such as random seed initialization and parallel processing are utilized to enhance reproducibility and computational efficiency.

18.5 Conclusion

In conclusion, the implementation of the XGBoost algorithm in the rainfall classification project for Madurai demonstrates promising results in accurately predicting rainfall conditions based on various meteorological factors. By leveraging historical weather data and advanced machine learning techniques, the model achieves a commendable level of accuracy in classifying rainfall conditions. The project highlights the potential of data-driven approaches in improving weather forecasting and providing valuable insights for disaster preparedness and mitigation efforts. However, there are areas for further refinement and enhancement, such as fine-tuning model parameters and incorporating additional features for better predictive performance. Overall, the project underscores the significance of data-driven methodologies in addressing complex environmental challenges and emphasizes the importance of ongoing research and innovation in meteorological forecasting and risk management.

In conclusion, this detailed methodology ensures a systematic approach to developing accurate wind power prediction models using ensemble regression techniques. By following this methodology, it aims to contribute valuable insights to the field of renewable energy forecasting and support decision-making processes in the renewable energy sector. A diverse set of regression models are chosen to compare between, these include traditional linear regression and more complex ensemble methods. The selected regression models include Gradient Boosting, Support Vector Regression (SVR), Random Forest, Linear Regression, Extra Trees, AdaBoost, Decision Tree, XGBoost, and XGBoost with Random Forest (XGBRF).

References

Aboelmagd, M. R., Selim, A., & Abdel-Akher, M. (2023). Predicting of photovoltaic power generation using similarity-based forecasting models. In *2023 24th International Middle East Power System Conference (MEPCON)* (pp. 1–8).

Ali, M. A., Elsayed, A., Elkabani, I., Akrami, M., Youssef, M., & Hassan, G. E. (2024). Artificial intelligence-based improvement of empirical methods for accurate global solar radiation forecast: Development and comparative analysis. *Energies, 17*.

Alvarez, G. E. (2024). Hybrid optimization model with neural network approach for renewable energy prediction and scheduling in large scale systems. *Management Science Letters, 14*, 247.

Chen, J., Guo, Z., Zhang, L., & Zhang, S. (2024). Short-term wind speed prediction based on improved Hilbert–Huang transform method coupled with NAR dynamic neural network model. *Scientific Reports, 14*.

Huangfu, J., Shi, Z., Liu, Y., & Li, K. (2022). Short-term interval prediction of wind speed using LSTM with improved grid search. In *2022 7th International Conference on Power and Renewable Energy (ICPRE)* (pp. 832–837).

Iqbal, U., Ramesh, J., & Al-Ali, A. R. (2023). Deep learning approach for wind energy time-series forecasting with environmental factors. In *2023 International Conference on Power and Renewable Energy Engineering (PREE)* (pp. 111–115).

Joseph, L. P., Deo, R. C., Prasad, R., Salcedo-Sanz, S., Raj, N., & Soar, J. (2023). Near real-time wind speed forecast model with bidirectional LSTM networks. *Renewable Energy, 204*(March), 39–58.

Prathibha, K., Rithvik Reddy, G., Kosre, H., Lohith Kumar, K., Rajak, A., & Tripathi, R. (2023). Rainfall prediction using machine learning. In D. S. Sisodia, L. Garg, R. B. Pachori, & M. Tanveer (Eds.), *Machine intelligence techniques for data analysis and signal processing* (Lecture Notes in Electrical Engineering) (Vol. 997). Springer. https://doi.org/10.1007/978-981-99-0085-5_37

Rahman, A. U., Abbas, S., Gollapalli, M., Ahmed, R., Aftab, S., Ahmad, M., Khan, M. A., & Mosavi, A. (2022). Rainfall prediction system using machine learning fusion for smart cities. *Sensors (Basel), 22*(9), 3504. https://doi.org/10.3390/s22093504. PMID: 35591194; PMCID: PMC9099780.

Sarker, S. (2023). A comparison of machine learning models for predicting rainfall in India. *Sustainability, 15*(18), 13724. https://www.mdpi.com/2071-1050/15/18/13724

Song, Y., Tang, D., Yu, J., Yu, Z., & Li, X. (2023). Short-term forecasting based on graph convolution networks and multiresolution convolution neural networks for wind power. *IEEE Transactions on Industrial Informatics, 19*(2), 1691–1702.

Tharun, E., et al. (2022). Rainfall prediction system using machine learning fusion for smart cities. *IEEE Access, 10*, 34176–34187.

Wang, S., Li, B., Li, G., Li, B., Li, H., Jiao, K., & Wang, C. (2024). A comprehensive review on the development of data-driven methods for wind power prediction and AGC performance evaluation in wind–thermal bundled power systems. *Energy and AI, 16*, 100336.

Chapter 19
Exploring the Application of Lean Tools for Sustainable Construction: A Strategy Based on Analytic Hierarchy Process Model

M. Durai Aravindh, G. Nakkeeran, G. PremKumar, L. Krishnaraj, R. Suresh Kumar, and D. Ashok Kumar

Abstract Recent contributions to the management and organizational literature necessitate research into how businesses can simultaneously implement lean tool improvement and waste reduction initiatives. Various lean tools are available to enhance the construction project delivery. However, it is difficult to implement in the modern environment. Therefore, understanding these implementations is critical to the lean tool's success and bridging the gap in its application. This study focuses on identifying the 10 most significant lean tools in the construction industry and selecting the optimal lean tool based on parameters such as time, cost, quality, and ease of application. One of the common multi-criteria decision-making (MCDM) approaches, analytic hierarchy process (AHP), is utilized in this study to select the optimal lean tool. The AHP model indicates that the attribute ease of application is given the highest priority and root cause analysis is preferred over other lean tools in construction projects. This study concludes by reflecting on the application of lean tool for optimizing construction management technology based on time, cost, quality, and ease of application. As a result, the simulator is an excellent resource for construction professionals to optimize project characteristics. This study provides original contributions to the field of construction operations management.

Keywords Lean tool · Analytic hierarchy process · Multi-criteria decision-making approach · Construction industry · Ease of application

M. Durai Aravindh · G. Nakkeeran · G. PremKumar · L. Krishnaraj (✉) · R. Suresh Kumar · D. Ashok Kumar
Department of Civil Engineering, College of Engineering and Technology, SRM Institute of Science and Technology, Kattankulathur, Chengalpattu, Tamil Nadu, India
e-mail: krishnal@srmist.edu.in

V. Baskaran et al. (eds.), *EcoTech Urbanism*, The Urban Book Series,
https://doi.org/10.1007/978-3-031-83833-0_19

19.1 Introduction

19.1.1 Lean and Analytic Hierarchy Process

Toyota's groundbreaking success with lean manufacturing principles gave birth to "Lean Construction" (Yilmaz et al., 2022), as shown in Fig. 19.1. In lean principles, reducing waste (Muda) and maximizing value are the goals. This study looks at the relationship between cost, time, quality, and ease of application function. The client's perceptions must be understood at the outset, supported by 14 principles and lean tools (Singh & Kumar, 2021).

The waste in lean product development affects the execution and delivery of complex projects (Belvedere et al., 2019). Lean project delivery differs from integrated project delivery based on its operating system (Mesa et al., 2019). The current condition and the future of lean construction were studied, and the means of implementation is rather a problem (Tezel et al., 2018). The application of lean principles in construction projects significantly improves project delivery characteristics such as enhanced job satisfaction and quicker project delivery (Aravindh et al., 2023). Multiple criteria decision-making methods allow the user to select the optimum or the most desired outcome from a decision question. To assess the sustainability of construction equipment, ways to enhance a novel hybrid MCDM methodology employing stepwise weight assessment ratio analysis, criteria importance through intercriteria correlation, and evaluation based on distance from the average solution are studied (Keshavarz Ghorabaee et al., 2018). The decision

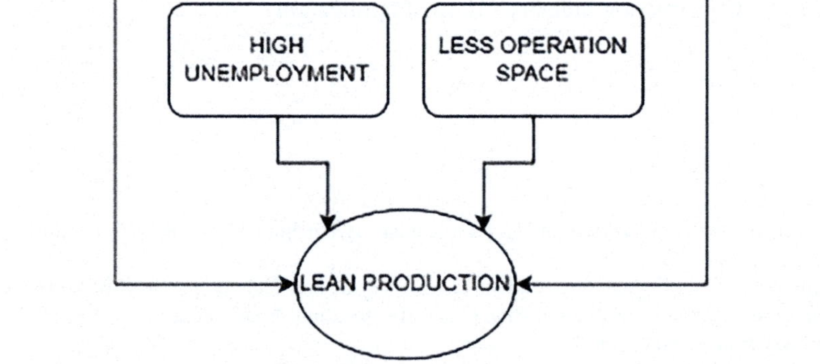

Fig. 19.1 Beginning of lean production (Aziz & Hafez, 2013)

criteria are evaluated using SWARA and COPRAS to rank the alternate projects based on environmental sustainability (Hashemkhani Zolfani et al., 2018). The lean tool should be selected based on its objectives and functionalities (Aslam et al., 2020). The AHP model is used in several industries to achieve the goal and determine the optimal solution using a hierarchy model (Awad & Jung, 2022; Benmoussa et al., 2019; Gupta et al., 2015; Improta et al., 2019; Mandavgade et al., 2021; Ranji et al., 2022; Zayed et al., 2008). AHP selects the contract with the highest discount in public works (Bertolini et al., 2006). AHP can reduce the number of comparisons using the hierarchy structure (Song & Kang, 2016). Furthermore, AHP is utilized to select the optimal project delivery method for the decision-maker (al Khalil, 2002). AHP is also used to develop a methodology to assess the important plant areas in Malaysia (Hamidah et al., 2022). AHP and preemptive goal programming (PGP) result in finding the ideal VSM tool and its implementation sequence. Productivity is important, and process activity mapping is identified as the optimal lean tool (Ramesh & Kodali, 2012). Comparative evaluation of the computerized maintenance management system (CMMS) is performed using the AHP model of 5 criteria and 16 sub-criteria in a synthetic foam production company (Meira et al., 2020). The best lean manufacturing concept for implementation in a manufacturing organization is selected using AHP (Vinodh et al., 2011). The combination of GIS and AHP identifies compact urban areas in Poland (Ogrodnik & Kolendo, 2021). Table 19.1 represents the lean tool and the literature studied for each tool.

Table 19.1 Literature for lean tool

Key	Lean Tool	Literature
A1	5S	Ogrodnik and Kolendo (2021), Jaca et al. (2014), Makwana and Patange (2019), Ablanedo-Rosas et al. (2010), Kanamori et al. (2015), Kanamori et al. (2016)
A2	PDCA (plan do check act)	Delisle and Freiberg (2014), Nsafon et al. (2020), Song and Fischer (2020), Santos et al. (2018), Prashar (2017), Garza-Reyes et al. (2018)
A3	FMEA (failure mode and effects analysis)	Liu et al. (2014), Alvand et al. (2021), Almannai et al. (2008), Chi et al. (2020), Reda and Dvivedi (2022), Kumru and Kumru (2013), Ramere and Laseinde (2021), Sutrisno et al. (2015), Meekhof and Bailey (2017)
A4	JIT (just-in-time)	Lux et al. (2016), Low and Wu (2014), Pun et al. (2014), Oral et al. (2003)
A5	LPS (last planner system)	
A6	Total productive maintenance	
A7	Kanban	
A8	5 Whys	
A9	VSM (value stream mapping)	
A10	Root cause analysis	

19.1.2 Literature Survey

A 5S model is an excellent tool for initiating and completing the process of continuous improvement. The effectiveness of the 5S implementation in Japan was determined by analyzing multiple case studies (Jaca et al., 2014). 5S assists organizations in building a framework and road map for effectively pursuing lean manufacturing for continuous improvement by exposing visible and crucial process waste. Additionally, 5S deployment has significantly improved employee work culture and morale (Makwana & Patange, 2019). Top management commitment, 5S practice inclusion in strategic planning, organization focus on how to keep the 5S practice going, measurement of 5S implementation's positive impact on organizational culture, and use of 5S as a foundation for advanced quality and continuous improvement philosophies are 5S implementation success factors (Ablanedo-Rosas et al., 2010). The 5S program improved the work environment by removing unwanted items, promoting orderliness, and labeling service units and directional markers. The 5S program improved the work environment by removing unwanted items, promoting orderliness, and labeling service units and directional markers (Kanamori et al., 2015). In low- and middle-income nations, the 5S might be proposed as a strategic alternative to improve healthcare service quality (Kanamori et al., 2016). Furthermore, the 5S framework has been demonstrated to be an effective and easy method for designing and improving processes (Delisle & Freiberg, 2014).

AHP-VIKOR and the PDCA cycle are integrated for continuous process improvement and the chosen hybrid energy system ensures a dependable system configuration, which is also the least expensive solution (Nsafon et al., 2020). A prototype with average usability of 95 was developed, which could be used by the foremen to generate a daily PDCA cycle (Song & Fischer, 2020). PDCA cycle process analysis was used to enhance a bus manufacturer's quality cost indication (Santos et al., 2018). EnMS combines technical and administrative energy-saving actions (ESAs) for constant energy efficiency and greener production (Prashar, 2017). Based on Deming's plan-do-check-act (PDCA) improvement cycle, a strategy for systematically performing Environmental-VSM (E-VSM) investigations is suggested (Garza-Reyes et al., 2018). After 3 months of employing the PDCA cycle, the department's running medical records have improved with 80% quality (Liu et al., 2014).

An integrated model of FMEA, SWARA, and WASPAS in a fuzzy environment is presented to detect and analyze construction project risks and to overcome specific FMEA flaws (Alvand et al., 2021). Quality function deployment (QFD) and FMEA are used to assess the best industrial automation solution and risk during system design and implementation (Almannai et al., 2008). The root cause analysis and FMEA may be utilized to reduce automobile recalls using orthogonal categorization, and this analysis may be used for various products or processes (Chi et al., 2020). Fuzzy QFD and FMEA quantify risk for lean application subelements. The future state plant structure and value stream map cut cycle time by 56.3%, lead time by 69.7%, materials transportation distance and activities by 75%, and employees by 202 to 200 (Reda & Dvivedi, 2022). A fuzzy-based FMEA is used to enhance public hospital buying, and the results show that applying fuzzy FMEA may efficiently find possible failure modes and consequences. It can also enable process assurance stability (Kumru & Kumru, 2013). FMEA is utilized to anticipate the condition of each component and assembly line. Predicting condition-based performance levels helps create maintenance plans and tactics that reduce unnecessary downtime (Ramere & Laseinde, 2021). Modified FMEA assesses maintenance waste criticality. An improved method for assessing the risk of maintenance waste mode utilizing waste priority number (WPN) is provided (Sutrisno et al., 2015). An overview of FMEA for process assessment describes a cataloging application, the tool's usage for error-proofing MARC record generation, and how FMEA may be utilized more effectively in an unstable setting (Meekhof & Bailey, 2017). The designer survey and research showed the need of extending the Process FMEA (P-FMEA) approach beyond quality and performance. Work situation FMEA (WS-FMEA), based on designer formalism, is recommended. Because the major stakeholders are already aligned with P-FMEA, this technique fits effortlessly into the design process, needing no additional resources or time (Lux et al., 2016).

The demand-pull technique was utilized for cement procurement and RMC delivery in Chongqing. For aggregates, sand, and concrete admixtures, economic order quantity was employed (Low & Wu, 2014). The application of JIT in building demolition project management was explored, and a Web-based waste exchange system is suggested. The system lets project participants connect before a demolition project starts, enabling waste products to be exchanged before they are created (Pun et al., 2014). A study with Turkish prefabrication enterprises was conducted, and the results revealed that inflation had little impact on Turkish prefabrication enterprises' supply-chain practices, preventing them from using JIT. Moreover, unlike in most undeveloped countries, material availability was adequate. The primary impediments to JIT implementation in the Turkish prefabrication business were financial and demand difficulties (Oral et al., 2003).

19.1.3 Research Gaps and Highlights

Research is about identifying 10 significant lean tools used in a construction project, to aid the successful completion of the project concerning cost, time, quality, and easy application and the top lean tool used in construction projects. The importance of the easy application of the lean tool is studied as an attribute as the optimal lean tool must be suitable for practical application in the site without difficulties that would deviate from the planned activities of the project.

The Indian construction industry needs lean tools for cost, time, and quality issues. First, identify lean tools to be used in construction projects and analyze the tools based on MCDM methods. Finally, suggest an important tool for optimizing time, cost, quality, and easy application in the construction industry.

19.2 Research Methodology

19.2.1 Questionnaire Design

The study is performed in two parts. In the first part, A questionnaire based on a 5-point Likert scale is used to find the 10 significant lean tools out of 30 lean tools based on their level of importance. The 5-point Likert scale utilized is (1) Insignificant, (2) Minor, (3) Moderate, (4) Major, and (5) Severe. In the next part, an AHP model is created for two levels with attributes in the upper level and alternatives in the lower level. To make the comparison matrix, the Saaty scale is utilized as follows: (1) Equally important, (2) Equally to moderately important; (3) Moderately important, (4) Moderately to strongly important, (5) Strongly important, (6) Strongly to very strongly important, (7) Very strongly important, (8) Very strongly to extremely important, (9) Extremely important. The element in each level of the hierarchy is subjected to a comparison matrix depending on the opinion of the evaluator. Table 19.2 represents the 30 lean tools selected for the study.

19.2.2 Data Collection

Private and public personnel in the Indian construction industry were surveyed for information. In the first part of the study, data was collected by issuing around 50 questionnaires, and a total of responses received were utilized to identify the significant 10 lean tools. For the second part, data was collected to create a comparison matrix between attributes and alternatives. A total of 6 comparisons were made for

Table 19.2 Identified lean tool

S.No	Lean tool	S.No	Lean tool
1	Andon	16	Visual management
2	Takt time	17	Poka-yoke
3	JIT (just-in-time)	18	Ishikawa diagram
4	VSM (value stream mapping)	19	Pareto analysis
5	Root cause analysis	20	5S
6	Kanban	21	Heijunka
7	LPS (last planner system)	22	A3 problem solving
8	Total productive maintenance	23	PDCA (plan-do-check-act)
9	FMEA (failure mode and effects analysis)	24	Muda walk
10	5 whys	25	Six big losses
11	Kaizen	26	SMART goals
12	Jidoka	27	Standardized work
13	Concurrent engineering	28	Line balancing
14	Bottleneck analysis	29	Overall equipment effectiveness
15	Gemba	30	Statistical process control

attributes, and 45 comparisons were made for alternatives by collecting data from 30 industry experts. The Eigen value and Eigen vector are calculated for the comparison matrix. The consistency of each matrix is evaluated to check the reliability of the received data. Table 19.3 represents an example questionnaire used in forming the comparison matrix for attributes in which the time attribute is compared with the cost, quality, and ease of application. Similarly, data was collected using three more questionnaires comparing other attributes with each other to form the comparison matrix for attributes.

Table 19.4 represents the example questionnaire utilized to form a comparison matrix in which A1 is compared with all the other alternatives A2 to A10. Similarly, data was collected using 9 more questionnaires comparing other alternatives with each other to form a comparison matrix for alternatives based on the time attribute.

19.2.3 Data Analysis

The AHP method involves several steps to create the hierarchy model. The proposed hierarchy model consists of three levels, where level 1 is the goal to select the optimum lean Tool in this study. The top level, i.e., level 2, represents attributes such as time, cost, quality, and Ease of application. The bottom level, that is, level 3, represents alternatives which are the 10 significant lean tools identified from the part 1 questionnaire study and are listed from A1 to A10. The AHP model is represented in Fig. 19.2.

In the AHP model, several elements in the hierarchy are compared, making the comparison difficult to remain consistent. For this, a consistency check is performed to evaluate the consistency of the comparison. Consistency is evaluated using the consistency index (C.I) and consistency ratio (C.R). The C.R should be less than 0.1 to deem the evaluation consistent. The C.I is calculated using the formula in Eq. (19.1).

$$C.I = \frac{\lambda_{\text{max}} - n}{n-1} \tag{19.1}$$

Where

λ_{max} = Maximum Eigen value
n = Number of evaluation criteria

The procedure for evaluating the optimal lean tool is as follows:

Table 19.3 Example questionnaire for attributes

Attribute	1	2	3	4	5	6	7	8	9
Cost									
Quality									
Ease of application									

Table 19.4 Example questionnaire for alternatives

	1	2	3	4	5	6	7	8	9
PDCA (plan-do-check-act)									
FMEA (failure mode and effects analysis)									
JIT (just-in-time)									
LPS (Last planner system)									
Total productive maintenance									
Kanban									
5 whys									
VSM (value stream mapping)									
Root cause analysis									

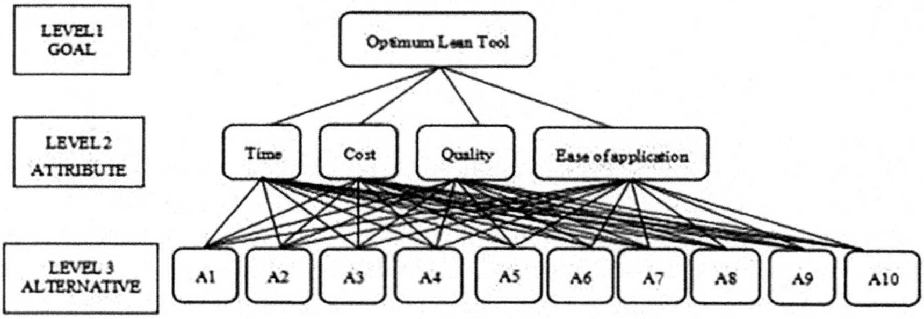

Fig. 19.2 AHP model

Step 1: Comparison Matrix

The first step is to create a pair-wise comparison matrix and the normalized matrix A*, which can be done by comparing the elements at each hierarchy level. The proposed AHP model consists of two hierarchy levels with 4 attributes and 10 alternatives. A total comparison is made. Table 19.5 represents the pair-wise comparison matrix for attributes.

The normalized matrix A*:

$$A* = \begin{bmatrix} a'_{11} & a'_{12} & a'_{13} & a'_{14} \\ a'_{21} & a'_{22} & a'_{23} & a'_{24} \\ a'_{31} & a'_{32} & a'_{33} & a'_{34} \\ a'_{n1} & a'_{n2} & a'_{n3} & a'_{nn} \end{bmatrix}$$

Where, $a'_{11} = \frac{a_{ij}}{\sum_{i-1}^{n} a_{ij}}$ for $i, j = 1,2,3,4$

Step 2: Eigen Vector

After forming the comparison matrix, the Eigenvalue and Eigenvector are computed as:

Table 19.5 Pair-wise comparison matrix for attributes

Attribute	Time	Cost	Quality	Ease of application
Time	1			
Cost		1		
Quality			1	
Ease of application				1

Table 19.6 Random index value

N	2	3	4	5	6	7	8	9	10
RI	0	0.52	0.89	1.11	1.25	1.35	1.4	1.45	1.49

$$W = \begin{bmatrix} W_1 \\ W_2 \\ W_3 \\ W_4 \end{bmatrix}$$

Where, $W_i = \frac{\sum_{i=1}^{n} a_{ij}'}{n}$

$$\text{for } i = 1, 2, 3,4;\ W' = AW = \begin{bmatrix} W_1' \\ W_2' \\ W_3' \\ W_4' \end{bmatrix}$$

$$\lambda_{\max} = \frac{1}{n}\left(\frac{W_1'}{W_1} + \frac{W_2'}{W_2} + \frac{W_3'}{W_3} + \frac{W_4'}{W_4}\right)$$

Where, W = Eigenvector

W_i = Eigenvalue
$\lambda_{\max}$ = Largest Eigenvalue of the comparison matrix

Step 3: Consistency Check
The next step is to check the consistency of the matrix formed using the responses collected from the evaluator. If the CR value exceeds 0.10, the matrix is inconsistent and unreliable. In this instance, the comparison matrix requires reevaluation. The R.I value observed from Saaty for matrix order is represented in Table 19.6.

The CR value is calculated using the formula in Eq. (19.2).

$$\mathrm{CR} = \frac{\mathrm{CI}}{\mathrm{RI}} \tag{19.2}$$

Where, $\mathrm{CI} = \frac{\lambda_{\max} - n}{n-1}$

Step 4: Global Priority

A Global priority table is formed. Then, the global priority value is calculated using the formula in Eq. (19.3):

$$GP_{ij} = \Sigma\left(LW_{ij} \times GW_{ij}\right) \tag{19.3}$$

Where,

GP_{ij} = Global priority value of the alternate tools
LW_{ij}= Local weight of tools concerning attribute
GW_{ij} = Global weight obtained by the product of local weights of attributes and alternatives

19.3 Result and Discussion

The relative importance placed on each criterion globally, and the relative priority of the various concept alternatives investigated. From the list of 30 lean tools, 10 lean tools are identified as significant by ranking using the relative importance index (RII). Table 19.7 illustrates the ten significant lean tools. In addition, the comparison matrix formed resulted in generating the global priority value of alternative lean tools identified in this study which is represented in Table 19.8.

Table 19.7 RII ranking of lean tools

Rank	Key	Lean tool	RII value	Rank	Key	Lean tool	RII value
1	*A1*	5S	0.838	*16*	*A16*	Statistical process control	0.538
2	*A2*	PDCA (plan-do-check-act)	0.819	*17*	*A17*	Bottleneck Analysis	0.519
3	*A3*	FMEA (failure mode and effects analysis)	0.813	*18*	*A18*	Visual management	0.513
4	*A4*	JIT (just-in-time)	0.806	*19*	*A19*	Gemba	0.5
5	*A5*	LPS (last planner system)	0.8	*20*	*A20*	Line balancing	0.494
6	*A6*	Total productive maintenance	0.794	*21*	*A21*	Takt time	0.488
7	*A7*	Kanban	0.794	*22*	*A22*	Andon	0.488
8	*A8*	5 whys	0.788	*23*	*A23*	Muda walk	0.488
9	*A9*	VSM (value stream mapping)	0.775	*24*	*A24*	Six big losses	0.488
10	*A10*	Root cause analysis	0.763	*25*	*A25*	Pareto analysis	0.481
11	*A11*	Kaizen	0.588	*26*	*A26*	SMART goals	0.481
12	*A12*	Jidoka	0.563	*27*	*A27*	Standardized work	0.481
13	*A13*	Ishikawa diagram	0.55	*28*	*A28*	Poka-yoke	0.45
14	*A14*	Concurrent engineering	0.538	*29*	*A29*	A3 problem solving	0.45
15	*A15*	Overall equipment effectiveness	0.538	*30*	*A30*	Heijunka	0.438

The procedure and model analysis for the attribute is given in Tables 19.8, 19.9, 19.10, and 19.11. The introduction of the 5S concept with the highest global priority score, as determined by the table will optimize workflow. As a result of conducting this study, the organization making efforts to implement root cause analysis controls in projects will save time and cost of a construction project. The decision-makers concluded that AHP permitted the selection of the optimal lean tool in this context, and additional actions were made to execute the implementation of the lean tool in their projects.

Table 19.8 Attribute-wise comparison matrix

Attribute	Time	Cost	Quality	Ease of application
Time	1.000	0.550	0.490	0.550
Cost	1.820	1.000	0.490	0.460
Quality	2.040	2.050	1.000	0.480
Ease of application	1.830	2.160	2.070	1.000

Table 19.9 Attribute-wise normalized matrix

Attribute	Time	Cost	Quality	Ease of application
Time	0.149	0.095	0.121	0.221
Cost	0.272	0.174	0.121	0.185
Quality	0.305	0.356	0.247	0.193
Ease of application	0.274	0.375	0.511	0.402

Table 19.10 Eigenvector and consistency ratio value

	W	W′	λ_{max}	CR
Time	0.147	0.600	4.134	CI = 0.044 RI = 0.89 CR = 0.05 CR < 0.1
Cost	0.188	0.769		
Quality	0.275	1.147		
Ease of application	0.390	1.634		

Table 19.11 Global priority value of alternatives

	Time (0.147)	Cost (0.188)	Quality (0.275)	Ease of application (0.390)	GP
A1	0.048	0.048	0.051	0.051	0.050
A2	0.054	0.054	0.058	0.062	0.058
A3	0.062	0.062	0.066	0.065	0.064
A4	0.075	0.076	0.077	0.076	0.076
A5	0.086	0.086	0.085	0.087	0.086
A6	0.098	0.095	0.093	0.096	0.095
A7	0.115	0.113	0.112	0.110	0.112
A8	0.130	0.130	0.131	0.130	0.130
A9	0.152	0.153	0.153	0.150	0.152
A10	0.178	0.183	0.175	0.173	0.176

Table 19.10 suggests that the attribute ease of application with a 0.390 weightage is given the highest priority over time, cost, and quality in selecting a lean tool that should be applied in a construction project. The time attribute is given the lowest weightage, followed by cost and quality.

Table 19.11 suggests that the lean tool A10 root cause analysis is preferred with a weightage of 0.176 over the other lean tools A1 to A9. In addition, the AHP evaluation quantified indicative lean tool selection values for the construction industry. Among them, it has been determined that ease of applying the lean tool on-site is a significant deciding factor in the selection of the lean tool. It also has the highest priority vector value, indicating that decision-makers deemed it to be significantly more important. Time (0.147), cost (0.188), quality (0.275), and ease of application (0.390) are the criteria priority values.

Implementation of lean tools preferring from the listed lean tool practices can help the construction industry determine which lean tool practices can aid in achieving more sustainable production. This will ensure that future generations have the resources necessary to innovate and create systems, products, and solutions that can contribute to meeting future demands in a responsible and sustainable method.

Reduced resource consumption, lower operating expenses, and enhanced efficiency in the use of available resources all contribute to a more financially and ecologically stable business. The purpose of this paper emphasizes increasing awareness of lean tool practices and implementation in the construction sector.

19.4 Theoretical and Managerial Implications

- In a lean construction environment, the objective is to select the best concept. For the study to be conducted, top-level management support must be provided.
- A cross-functional team comprised of experts from various departments is required. The concept model requires a hierarchical organization. Then, an appropriate MCDM technique must be implemented to solve the model.
- Developing pairwise comparison matrices necessitates the use of manager's knowledge when collecting the necessary data. The computations for determining the pertinent parameters must then be performed.

19.5 Recommedations Suggested

Companies are under increasing scrutiny to operate more efficiently, so toolmakers and project managers must incorporate novel approaches to help their clients reach this goal. This paper's authors propose the following action:

- Lean tools should be implemented in the building industry if waste is to be reduced. Maintaining dynamic and sustainable ecosystems requires learning

how to effectively incorporate Lean initiatives for making improvements to reduce cost and time waste and to increase the benefits of workers and customers.
- The construction industry should understand and incentivize the use of Lean Tools for increased production efficiency and safety.

19.6 Conclusion

Concept selection in lean construction is a multi-criteria MCDM problem. AHP is established as a reliable method for solving MCDM problems. Especially, the AHP method was utilized to select the most effective concept. The practical propensity of employing the AHP method when selecting the optimal lean tool for implementation is shown. The study identified ten significant lean tools used in the construction industry. This research also provided an AHP model for determining the ideal lean tool based on time, cost, quality, and application ease.

- The significant lean tools that can be applied in the construction industry are 5S, PDCA (plan-do-check-act), FMEA (failure mode and effects analysis), JIT (just-in-time), LPS (last planner system), total productive maintenance, Kanban, 5 whys, VSM (value stream mapping), and root cause analysis.
- Ease of application is highly important in selecting a lean tool with 0.390 weightage.
- Root cause analysis is the optimal lean tool that can be applied in construction projects with a 0.176 global priority score or weightage.

When applied in construction projects, the identified optimal lean tool will enhance the project characteristics and optimize the workflow of the construction project. The application of AHP has enabled decision-makers to select the most effective lean concept for implementation, thereby contributing to the success of the business. In the future, a greater number of concept selection studies may be conducted for a variety of manufacturing organizations. Future case studies involving analytical network process (ANP), fuzzy AHP, and fuzzy ANP could also be conducted and their efficacy in concept selection could be compared.

Theoretical and practical recommendations have been proposed based on the findings. Important recommendations include the proposal to expand the assistance offered by construction organizations and the proposal to form a strategic regulatory committee to monitor the issue. This research's recommendations apply to future endeavors as well.

Conflict of Interest The authors declare that they have no conflict of interest.

Financial Interests The authors declare they have no financial interests.

Data Availability The datasets used and/or analyses during the current study available from the corresponding author on reasonable request.

References

Ablanedo-Rosas, J. H., Alidaee, B., Moreno, J. C., & Urbina, J. (2010). Quality improvement supported by the 5S, an empirical case study of Mexican organisations. *International Journal of Production Research, 48*(23), 7063–7087. https://doi.org/10.1080/00207540903382865

al Khalil, M. I. (2002). Selecting the appropriate project delivery method using AHP. *International Journal of Project Management, 20*(6), 469–474. https://doi.org/10.1016/S0263-7863(01)00032-1

Almannai, B., Greenough, R., & Kay, J. (2008). A decision support tool based on QFD and FMEA for the selection of manufacturing automation technologies. *Robotics and Computer-Integrated Manufacturing, 24*(4), 501–507. https://doi.org/10.1016/j.rcim.2007.07.002

Alvand, A., Mirhosseini, S. M., Ehsanifar, M., Zeighami, E., & Mohammadi, A. (2021). Identification and assessment of risk in construction projects using the integrated FMEA-SWARA-WASPAS model under fuzzy environment: a case study of a construction project in Iran. https://doi.org/10.1080/15623599.2021.1877875

Aravindh, M. D., Sriram, N. S., Nakkeeran, G., Jayakeerti, M., Velan, C., & Krishnaraj, L. (2023). Synergistic effect of alliance contract and lean methodology on project performance measures in the construction industry: SEM analysis. *Technological Forecasting and Social Change, 192*, 122545. https://doi.org/10.1016/j.techfore.2023.122545

Aslam, M., Gao, Z., & Smith, G. (2020). Framework for selection of lean construction tools based on lean objectives and functionalities. *International Journal of Construction Management, 22*, 1–12. https://doi.org/10.1080/15623599.2020.1729933

Awad, J., & Jung, C. (2022). Extracting the planning elements for sustainable urban regeneration in Dubai with AHP (Analytic Hierarchy Process). *Sustainable Cities and Society, 76*, 103496. https://doi.org/10.1016/j.scs.2021.103496

Aziz, R. F., & Hafez, S. M. (2013). Applying lean thinking in construction and performance improvement. *Alexandria Engineering Journal, 52*(4), 679–695. https://doi.org/10.1016/J.AEJ.2013.04.008

Belvedere, V., Cuttaia, F., Rossi, M., & Stringhetti, L. (2019). Mapping wastes in complex projects for Lean Product Development. *International Journal of Project Management, 37*(3), 410–424. https://doi.org/10.1016/J.IJPROMAN.2019.01.008

Benmoussa, K., Laaziri, M., Khoulji, S., Kerkeb, M. L., & el Yamami, A. (2019). AHP-based Approach for Evaluating Ergonomic Criteria. In *Procedia Manufacturing* (pp. 856–863). Elsevier B.V. https://doi.org/10.1016/j.promfg.2019.02.294

Bertolini, M., Braglia, M., & Carmignani, G. (2006). Application of the AHP methodology in making a proposal for a public work contract. *International Journal of Project Management, 24*(5), 422–430. https://doi.org/10.1016/J.IJPROMAN.2006.01.005

Chi, C. F., Sigmund, D., & Astardi, M. O. (2020). Classification scheme for root cause and Failure Modes and Effects Analysis (FMEA) of passenger vehicle recalls. *Reliability Engineering and System Safety, 200*, 106929. https://doi.org/10.1016/j.ress.2020.106929

Delisle, D. R., & Freiberg, V. (2014). Everything is 5S: A simple yet powerful lean improvement approach applied in a preadmission testing center. *Quality Management Journal, 21*(4), 10–22. https://doi.org/10.1080/10686967.2014.11918406

Garza-Reyes, J. A., Torres Romero, J., Govindan, K., Cherrafi, A., & Ramanathan, U. (2018). A PDCA-based approach to Environmental Value Stream Mapping (E-VSM). *Journal of Cleaner Production, 180*, 335–348. https://doi.org/10.1016/j.jclepro.2018.01.121

Gupta, S., Dangayach, G. S., Singh, A. K., & Rao, P. N. (2015). Analytic Hierarchy Process (AHP) model for evaluating sustainable manufacturing practices in Indian Electrical Panel Industries. *Procedia - Social and Behavioral Sciences, 189*, 208–216. https://doi.org/10.1016/j.sbspro.2015.03.216

Hamidah, M., et al. (2022). Development of a protocol for Malaysian Important Plant Areas criterion weights using Multi-criteria Decision Making - Analytical Hierarchy Process (MCDM-AHP). *Global Ecology and Conservation, 34*, e02033. https://doi.org/10.1016/j.gecco.2022.e02033

Hashemkhani Zolfani, S., Pourhossein, M., Yazdani, M., & Zavadskas, E. K. (2018). Evaluating construction projects of hotels based on environmental sustainability with MCDM framework. *Alexandria Engineering Journal, 57*(1), 357–365. https://doi.org/10.1016/j.aej.2016.11.002

Improta, G., Perrone, A., Russo, M. A., & Triassi, M. (2019). Health technology assessment (HTA) of optoelectronic biosensors for oncology by analytic hierarchy process (AHP) and Likert scale. *BMC Medical Research Methodology, 19*(1), 140. https://doi.org/10.1186/s12874-019-0775-z

Jaca, C., Viles, E., Paipa-Galeano, L., Santos, J., & Mateo, R. (2014). Learning 5S principles from Japanese best practitioners: Case studies of five manufacturing companies. *International Journal of Production Research, 52*(15), 4574–4586. https://doi.org/10.1080/00207543.2013.878481

Kanamori, S., Sow, S., Castro, M. C., Matsuno, R., Tsuru, A., & Jimba, M. (2015). Implementation of 5S management method for lean healthcare at a health center in Senegal: A qualitative study of staff perception. *Global Health Action, 8*(1), 27256. https://doi.org/10.3402/gha.v8.27256

Kanamori, S., Castro, M. C., Sow, S., Matsuno, R., Cissokho, A., & Jimba, M. (2016). Impact of the Japanese 5S management method on patients' and caretakers' satisfaction: A quasi-experimental study in Senegal. *Global Health Action, 9*(1), 32852. https://doi.org/10.3402/GHA.V9.32852

Keshavarz Ghorabaee, M., Amiri, M., Zavadskas, E. K., & Antucheviciene, J. (2018). A new hybrid fuzzy MCDM approach for evaluation of construction equipment with sustainability considerations. *Archives of Civil and Mechanical Engineering, 18*(1), 32–49. https://doi.org/10.1016/j.acme.2017.04.011

Kumru, M., & Kumru, P. Y. (2013). Fuzzy FMEA application to improve purchasing process in a public hospital. *Applied Soft Computing Journal, 13*(1), 721–733. https://doi.org/10.1016/j.asoc.2012.08.007

Liu, X., et al. (2014). Application of PDCA method in quality management of running medical records at the Department of Orthopedics. *Chinese Medical Record English Edition, 2*(3), 112–115. https://doi.org/10.3109/23256176.2014.910906

Low, S. P., & Wu, M. (2014). Just-in-time management in the ready mixed concrete industry of Chongqing, China, *5*(1), 21–37. https://doi.org/10.1080/15623599.2005.10773065

Lux, A., Mawo De Bikond, J., Etienne, A., & Quillerou-Grivot, E. (2016). FMEA and consideration of real work situations for safer design of production systems. *International Journal of Occupational Safety and Ergonomics, 22*(4), 557–564. https://doi.org/10.1080/10803548.2016.1180856

Makwana, A. D., & Patange, G. S. (2019). Strategic implementation of 5S and its effect on productivity of plastic machinery manufacturing company. *Australian Journal of Mechanical Engineering, 20*, 111. https://doi.org/10.1080/14484846.2019.1676112

Mandavgade, N. K., Kalbande, V. N., Bilawane, R. R., Kanojiya, M. T., & Padole, C. U. (2021). AHP for ranking effect of qualitative factors in uncertainty measurement of material testing. In *Materials Today: Proceedings* (pp. 7921–7925). Elsevier Ltd. https://doi.org/10.1016/j.matpr.2021.02.636

Meekhof, J., & Bailey, A. B. (2017). Failure modes and effects analysis (FMEA) for cataloging: An application and evaluation. *Cataloging and Classification Quarterly, 55*(7–8), 493–505. https://doi.org/10.1080/01639374.2017.1356782

Meira, D., Lopes, I., & Pires, C. (2020). Selection of computerized maintenance management systems to meet organizations' needs using AHP. In *Procedia Manufacturing* (pp. 1573–1580). Elsevier B.V. https://doi.org/10.1016/j.promfg.2020.10.219

Mesa, H. A., Molenaar, K. R., & Alarcón, L. F. (2019). Comparative analysis between integrated project delivery and lean project delivery. *International Journal of Project Management, 37*(3), 395–409. https://doi.org/10.1016/J.IJPROMAN.2019.01.012

Nsafon, B. E. K., Butu, H. M., Owolabi, A. B., Roh, J. W., Suh, D., & Huh, J. S. (2020). Integrating multi-criteria analysis with PDCA cycle for sustainable energy planning in Africa: Application to hybrid mini-grid system in Cameroon. *Sustainable Energy Technologies and Assessments, 37*, 100628. https://doi.org/10.1016/j.seta.2020.100628

Ogrodnik, K., & Kolendo, Ł. (2021). Application of gis technology and AHP to determine the areas with fully developed, compact functional and spatial structure: A case study of Bialystok, Poland. *Land Use Policy, 109*, 105616. https://doi.org/10.1016/j.landusepol.2021.105616

Oral, E. L., Mistikoglu, G., & Erdis, E. (2003). JIT in developing countries-a case study of the Turkish prefabrication sector. *Building and Environment, 38*(6), 853–860. https://doi.org/10.1016/S0360-1323(03)00021-0

Prashar, A. (2017). Adopting PDCA (Plan-Do-Check-Act) cycle for energy optimization in energy-intensive SMEs. *Journal of Cleaner Production, 145*, 277–293. https://doi.org/10.1016/j.jclepro.2017.01.068

Pun, S. K., Liu, C., Langston, C., & Treloar, G. (2014). Electronic waste exchange for just-in-time building demolition, *7*(2), 65–77. https://doi.org/10.1080/15623599.2007.10773103

Ramere, M. D., & Laseinde, O. T. (2021). Optimization of condition-based maintenance strategy prediction for aging automotive industrial equipment using FMEA. In *Procedia Computer Science* (pp. 229–238). Elsevier B.V. https://doi.org/10.1016/j.procs.2021.01.160

Ramesh, V., & Kodali, R. (2012). A decision framework for maximising lean manufacturing performance. *International Journal of Production Research, 50*(8), 2234–2251. https://doi.org/10.1080/00207543.2011.564665

Ranji, A., Parashkoohi, M. G., Zamani, D. M., & Ghahderijani, M. (2022). Evaluation of agronomic, technical, economic, and environmental issues by analytic hierarchy process for rice weeding machine. *Energy Reports, 8*, 774–783. https://doi.org/10.1016/j.egyr.2021.12.028

Reda, H., & Dvivedi, A. (2022). Decision-making on the selection of lean tools using fuzzy QFD and FMEA approach in the manufacturing industry. *Expert Systems with Applications, 192*, 116416. https://doi.org/10.1016/j.eswa.2021.116416

Santos, H., Pereira, M. T., Silva, F. J. G., & Ferreira, L. P. (2018). A novel rework costing methodology applied to a bus manufacturing company. In *Procedia Manufacturing* (Vol. 17, pp. 631–639). Elsevier B.V. https://doi.org/10.1016/j.promfg.2018.10.109

Singh, S., & Kumar, K. (2021). A study of lean construction and visual management tools through cluster analysis. *Ain Shams Engineering Journal, 12*(1), 1153–1162. https://doi.org/10.1016/J.ASEJ.2020.04.019

Song, M. H., & Fischer, M. (2020). Daily plan-do-check-act (PDCA) cycles with level of development (LOD) 400 objects for foremen. *Advanced Engineering Informatics, 44*, 101091. https://doi.org/10.1016/j.aei.2020.101091

Song, B., & Kang, S. (2016). A method of assigning weights using a ranking and nonhierarchy comparison. *Adv. Decis. Sci., 2016*, 1. https://doi.org/10.1155/2016/8963214

Sutrisno, A., Gunawan, I., & Tangkuman, S. (2015). Modified Failure Mode and Effect Analysis (FMEA) model for accessing the risk of maintenance waste. *Procedia Manufacturing, 4*, 23–29. https://doi.org/10.1016/j.promfg.2015.11.010

Tezel, A., Koskela, L., & Aziz, Z. (2018). Current condition and future directions for lean construction in highways projects: A small and medium-sized enterprises (SMEs) perspective. *International Journal of Project Management, 36*(2), 267–286. https://doi.org/10.1016/J.IJPROMAN.2017.10.004

Vinodh, S., Shivraman, K. R., & Viswesh, S. (2011). AHP-based lean concept selection in a manufacturing organization. *Journal of Manufacturing Technology Management, 23*(1), 124–136. https://doi.org/10.1108/17410381211196320

Yilmaz, A., Dora, M., Hezarkhani, B., & Kumar, M. (2022). Lean and industry 4.0: Mapping determinants and barriers from a social, environmental, and operational perspective. *Technological Forecasting and Social Change, 175*, 121320. https://doi.org/10.1016/J.TECHFORE.2021.121320

Zayed, T., Amer, M., & Pan, J. (2008). Assessing risk and uncertainty inherent in Chinese highway projects using AHP. *International Journal of Project Management, 26*(4), 408–419. https://doi.org/10.1016/J.IJPROMAN.2007.05.012

The manufacturer's authorised representative in the EU is Springer Nature Customer Service Centre GmbH, Europaplatz 3, 69115 Heidelberg, Germany. If you have any concerns regarding our products, please contact ProductSafety@springernature.com

Printed and bound by CPI Group (UK) Ltd, Croydon, CR0 4YY
30/09/2026
02225593-0008